MARCHING WITH SPARTANS:
THE LIFE AND WORKS OF ALDEN PARTRIDGE

VOLUME I

MARCHING WITH SPARTANS: THE LIFE AND WORKS OF ALDEN PARTRIDGE

By Franklin C. Annis, EdD

VOLUME I:
CONTROVERSIAL HISTORY & EDUCATIONAL THEORIES OF CAPTAIN ALDEN PARTRIDGE

Foreword by Col. Thomas J. Gordon, USMC (Ret.)

Printed for Franklin C. Annis
Arlington, VA
MMXXIV

Library of Congress Cataloging-in-Publication Data
Names: Annis, Franklin C., 1980–, author
Title: Controversial History & Educational Theories of Captain Alden
Partridge / Franklin C. Annis
Description: Arlington, VA: Printed for Franklin C. Annis, [2024] | Includes
bibliographical references and index.
Identifiers: LCCN 2024922320 (print) | ISBN 978-1-964681-00-9 (Special
Edition Hardcover) | ISBN 978-1-964681-01-6 (Hardcover) | ISBN 978-1-
964681-02-3 (Softcover) | ISBN 978-1-964681-03-0 (eBook)
Subjects: Military Arts and Sciences | Military Readiness | History, Military |
Military Education | Military Training | Education, Higher | Education,
History
LC record available at: https://lccn.lov.gov/2024922320

The author does not control third-party internet websites. There is no
guarantee that any website cited in this work will remain functional, accurate,
or appropriate.

In eternal gratitude, this work is dedicated to my father, Henry Craig Annis.

CAPT ALDEN PARTRIDGE.

Figure 1—Engraving of Captain Alden Partridge by H. W. Smith[1]

---

[1] The Miriam and Ira D. Wallach Division of Art, Prints and Photographs: Print Collection, The New York Public Library, "Capt. Alden Partridge," *The New York Public Library Digital Collections.*

CONTENTS

Figure 2—Painting of Captain Alden Partridge[2]

[2] Original found within Alden Partridge Records, Box 1: Images, non-photographic; Norwich University Archives and Special Collections, Kreitzberg Library, Northfield, VT. Artist Unknown. One of possibly four known images displaying Partridge in civilian attire.

Table of Figures

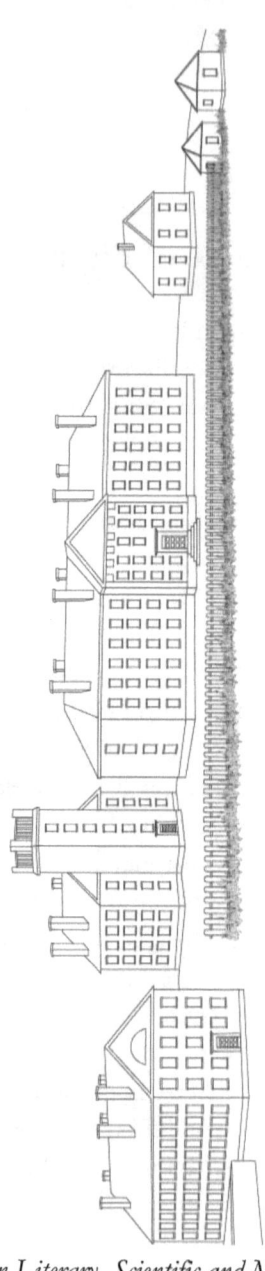

Figure 3—Plan for American Literary, Scientific and Military Academy, Middletown, Connecticut, 1827[3]

[3] This line drawing is a recreation of the original plan for the A.L.S.&M. Academy campus found within Alden Partridge, *Journal of a Tour, of a Detachment of Cadets, from the A.L.S.&M. Academy, Middletown, to the City of Washington, in December, 1826* (Middletown, CT: W. D. Starr, 1827). Failing to gain a state charter, this campus would be sold. The main building is still in use by Wesleyan University.

A Call to Action

I would like to extend a thank you to all those who purchased this work. A portion of the profits will be donated to the Norwich University Kreitzberg Library and Sullivan Museum and History Center. These institutions contain artifacts and documents that are not likely to be found at any other location. Your purchase will help preserve and expand access to these materials for generations to come.

If you enjoy this research and wish other scholars to further investigate the topics discussed within this work, I invite you to provide additional support to the Norwich University Museum and Library to safeguard critical source materials and artifacts. This volume has only scratched the surface of the knowledge held within the museum and library. Historians, active service members, and the public alike will greatly benefit by maintaining and making these materials more readily available.

There is important work ahead at the Kreitzberg Library as they strive to preserve historical treasures. With the recent addition of an archival scanner (with a cost exceeding $50,000), the library can now capture images of the original *Citizen Soldier* newspaper and other significant documents. The *Citizen Soldier* captured the active discussion, in the early years of the American Republic, on how to best form and discipline the militia to provide for the common defense in a manner consistent with the U.S. Constitution. This newspaper, and other materials within the archive, display how National Guard leaders were attempting to improve and professionalize their units prior to the American Civil War. Norwich University served as the intellectual hub of the militia in the early days of the Republic and thus collected volumes of materials on improving national defense. Yet much of the materials these dedicated military leaders produced remains hidden and forgotten in the upper floor of the Kreitzberg Library.

Despite the hefty investment in the said scanner, there are still significant personnel costs needed to scan, transcribe, and digitize these materials for online access. Additionally, the Sullivan Museum faces the ongoing challenge of maintaining and showcasing artifacts dating back over two centuries, ensuring they are preserved for future generations. Your support can make a real difference in safeguarding these pieces of history.

Friends of the Kreitzberg Library Membership
https://alumni.norwich.edu/KreitzbergLibraryMembership
Sullivan Museum Associate Membership
https://alumni.norwich.edu/give/membership-benefits-SullivanMuseum

Please consider becoming a member of the Friends of the Kreitzberg Library and Sullivan Museum Associates. Memberships and donations are tax deductible (Norwich University is a 501(c)3 organization). More information on donating is available on the Norwich University Office of Development at https://alumni.norwich.edu/Giving, development@norwich.edu, or (802) 485-2300. I thank you for any additional support you might provide in preserving the history of our great Republic.

Franklin C. Annis, EdD

Figure 4—Kreitzberg Library[4]

---

[4] Line drawing based on the photo by John H. Knox, Norwich University Class of 1963.

Author's Note

The design of this book owes much to my desire to provide two important services to the reader: the placement of notes on the bottom of pages and the inclusion of verbatim copies of relevant documents (many that have laid dormant in archives for hundreds of years). A traditional publisher would likely find these materials to interfere with the profitability of the work. Being in complete control of formatting through self-publishing allows me to focus this work on the maximization of potential learning of the reader.

I also had concerns with the peer-review process related to modern publishing, as this text crosses multiple academic fields. Traditional publishers would have a tough challenge in finding qualified scholars to review this work, possibly causing months or years of delays. That is not to say that this book was not reviewed. Four trusted scholars from a range of academic disciplines, from philosophy to military history, acted as alpha readers for this work and provided valuable insights on improvements (more in the acknowledgements).

The great authors of the past had no need for our contemporary book publishing process. One could speculate how the works of Bacon, Locke, Defoe, Partridge, and nameless others may have been corrupted and diluted by our contemporary peer-review process. Given that the great authors of the past had little need for publishing services, often paying for the printers themselves, I have no hesitation in assuming the same path.

In the end, I accept total responsibility for this imperfect work. Let it be judged by its merits alone for what it accomplishes or fails to do so. I wish not to receive artificial credibility through the name of a publishing house or a restrictive peer-review process. These books are written to honor the legacy of a remarkable soldier and educator, and therefore, they will be written in a manner as bold and courageous as Captain Alden Partridge was in life.

The views presented within this work are those of the author and do not necessarily represent the views of the U.S. Department of Defense or its components.

CAPTAIN PARTRIDGE.

Figure 5—Engraving of Captain Alden Partridge by the Western Engraving Company[5]

[5] Original found within Alden Partridge Records, Box 1: Images, non-photographic, Norwich University Archives and Special Collections, Kreitzberg Library, Northfield, VT. A similar engraving by J. B. Bartlett appears within John Holbrook, *Military Tactics: Adapted to the Different Corps in the United States, According to the Latest Improvements* (Middletown, CT: E. & H. Clark, 1826).

Foreword

There are few, if any, educators and patriots who had such a profound impact on the American system of education and the military as Captain Alden Partridge yet, very few know his name. Dr. Franklin Annis' work provides the reader with an insightful exploration that illuminates and reshapes our understanding of this pivotal figure in history. Partridge is the father of experiential learning, physical education, and today's Reserve Officer Training Corps (ROTC). His program to develop the mind, body and spirit, first introduced at the United States Military Academy (West Point), was decades ahead of mainstream educators. Though West Point failed, initially, to adopt his approach, his influence laid the foundation for how we prepare and educate the citizen soldiers who defend the Republic today.

There are few scholars willing to take on the ambitious challenge of rescuing Captain Partridge's legacy. Dr. Franklin Annis' in-depth research brings clarity to Captain Partridge's theories while providing practical insight into how reviving these concepts today can address the shortcomings of our current system. As an academic with advanced degrees in education and history, Dr. Annis is uniquely qualified for this task. This first volume establishes Franklin as the authority on the life and works of Captain Alden Partridge while providing the reader with opportunities to apply Partridge's practical wisdom. For the historian, this book is replete with original sourced materials, many of which have gone unstudied in various archives for centuries.

I was first introduced to then Major Franklin Annis in 2019 in my capacity as the Director of the Marine Corps' Command and Staff College. Our newly established Brute Krulak Center for Innovation & Future Warfare at the Marine Corps University took interest in his research on adult learning theory (andragogy). It was during this initial meeting that I discovered Franklin and I shared an enthusiasm for Stoic philosophy and a desire to apply its tenants to better prepare leaders for the modern battlefield. In 2020, I had the pleasure of presenting at an international conference arranged by Dr. Annis for the Army National Guard Chief Surgeon's Office. The conference, later published on the Modern Stoicism YouTube Channel, explored the use of Stoic philosophy to improve the psychological resilience of servicemen. In 2021, as I prepared to assume my current duties as the Commandant at The

Citadel, Dr. Annis enlightened me to the influence of Captain Alden Partridge. He greatly expanded my knowledge of the origins of the land-grant colleges and Partridge's vision in founding The American Literary, Scientific, and Military Academy, and his hand in creating the network of Senior Military Colleges and today's Reserve Officer Training Corps (ROTC) program.

Dr. Annis' anthology of Captain Partridge's theories will be of particular interest to USMA graduates and the alumni of our Senior Military Colleges. After his inglorious departure from West Point, Captain Partridge went on to found multiple private military academies, including Norwich University, the Nation's oldest private military college. His efforts greatly aided in the establishment of VMI and The Citadel. It was not, however, until the National Defense Act of 1916 that Partridge's legacy was codified into law when Congress formally instituted the ROTC program and expanded military training on the Nation's college campuses.

Ivan Pavlov is credited with the adage, "If you want a new idea, read an old book." This remarkable book offers valuable insight, 200 years in the making, for both civilian and military communities on how to improve our contemporary education system. Partridge's emphasis on fitness and Stoic philosophy may be the antidote to the Nation's mental health and obesity crisis. By integrating military arts and sciences, Partridge not only prepared citizens for the common defense of the nation, but he also instilled in them an understanding of the hardships and costs of military expeditions. These lessons offer a strong deterrent to needless military conflicts by ensuring citizens fully comprehend the implications of their votes. Through a polymathic curriculum and experiential learning, Partridge cultivated versatile citizens who contributed both practically and intellectually to their communities. Partridge reminds us that there is no real dichotomy between a classical and practical education. It is possible to educate citizens in higher academic fields while also providing instruction on the practical skills required of citizens. Though the land-grant colleges were not established until the Morrill Act of 1862, Partridge's influence was reflected in the federal government's requirement that these schools include military tactics as part of their curriculum. While Alden Partridge is recognized as the "Father of ROTC," only a portion of his educational concepts were adopted in creating these programs. It would be beneficial for the Senior Military Colleges and the different ROTC programs (including JROTC) to reconsider Partridge's theories and strive to reinvigorate more of his groundbreaking ideas and methods.

Volume I of this series delves into the myriad benefits of Captain Partridge's educational philosophy and uncovers the political complexities that defined the early years of the United States Military Academy. Franklin sheds light on the profound impact these historical events had on Partridge's legacy, challenging the prevailing anti-Partridge narrative perpetuated by his adversaries. Franklin digs deep into dozens of archives to offer the most detailed examination of Partridge's legacy to date. Given the national implications of Captain Partridge's works, this task has been long overdue by military and educational historians.

Most remarkably, Franklin's examination of the underlying philosophical influences within Partridge's curriculum, namely Stoic ethics, can undoubtedly have a powerful impact on the profession of arms. Tracing the evolution of the Stoic tradition from the original adaption of Spartan military practices through Neostoicism and the Enlightenment, Franklin reveals the importance of this philosophic tradition and how it shaped the establishment of the American Republic. Franklin then ties modern psychological research on the benefits of Stoic philosophy for improving resiliency. This work will challenge many of the practices used by the military for the initial and sustained education of service members. Franklin highlights the importance of explicitly stating the Stoic intent behind hardships used in initial military training, like long conditioning hikes, to enhance psychological resiliency. Here, his program goes beyond physical conditioning by including spiritual strength and emotional wellness. By linking military toughening practices to the philosophical roots of our professional ethos, military leaders can help U.S. Service Members grasp the interconnectedness of their profession. This discussion has tremendous potential to influence the educational practices used at the Service Academies and across our ROTC programs nationwide.

Lieutenant Colonel Annis' passion for education and his expertise in history and philosophy shines throughout this work. I look forward to reading future volumes in this series. The American Republic will survive only as long as her citizens are taught their freedoms and have the willingness and ability to defend them. Let this book be the impetus for improving education and strengthening our great Nation.

As Partridge reminds us, "Education is more powerful than the lever of Archimedes, either to sustain or to crush the political institutions of every country—just in proportion as it is, in principle and in practice, in accordance with, or adverse to, those institutions."

Colonel Thomas J. Gordon, USMC (Ret.)

Figure 6—Original Design of ALSMA Cadet Uniform⁶

---

⁶ Drawing of understated Hussar uniforms worn by Partridge's cadets based on a painting attributed to painter Julian Parisen, Object #2015.55.1, Sullivan Museum and History Center Collection. The portrait is one of a pair painted at the Middleton, Connecticut campus. The cadet in the image is believed to be of the Nathan Starr family. Nathan Starr was an influential contract arms manufacturer during the era, with the musket in the painting likely a locally manufactured Starr contract musket. The original painting presents an intriguing mystery as a pentimento of a small boy holding an apron with possible Masonic symbols exists on the right side of the canvas. The cadet also appears to be wearing morning locket at the base of his coatee. Future historians will be challenged to reveal the identity and more details surrounding this phantom child.

Preface

In 1965, a partial biography of Alden Partridge rolled off the press, written by Colonel Lester Webb (PhD in education). Entitled *Captain Alden Partridge and the United States Military Academy, 1806–1833,* this volume was supposed to have been the first half of a two-part study of the life and works of a remarkable American. This footnote appears in the conclusion chapter.

> The author will follow the volume with a second embracing the period from 1818 to 1854 giving the history of Captain Partridge's unique work in the field of education. It will develop the theme that "Captain Partridge was the greatest educator America has ever produced, but few there are who know of his existence."[7]

Unfortunately, the author died before he could complete his second manuscript.

For over a decade, I have studied Army education and leader development theories. Captain Alden Partridge became a hero of mine ever since I discovered his educational theories while writing my doctoral dissertation in 2015. I must admit he was a rather controversial hero to have for reasons that will become apparent in this book. Partridge forged a highly refined educational theory that was generations ahead of his time. There are many aspects of his educational theories that were only shown true by other researchers in the late 20th century. Many of his approaches to education are vastly superior to our present-day techniques. To my knowledge, his educational design is the only one that could fully prepare American citizens for the totality of their civic duties. Partridge's *American System of Education* is the only educational model capable of maintaining the republican system of government intended by the American Founders.

My education and experience in curriculum design and military history places me in a unique position to reveal the suppressed history of this exceptional educational theorist. My aim is to share the ideas and concepts of Partridge with a larger audience in hopes that we may change the American education system to truly support the republicanism intended by our Founding

---

[7] Lester A. Webb, *Captain Alden Partridge and the United States Military Academy, 1806–1833* (Northport, AL: American Southern, 1965), 205.

Fathers. I hope to fulfill the promise of Colonel Webb that he was unable to accomplish in life. It is long past time for American citizens to remember the numerous contributions of Captain Alden Partridge to this great nation. Returning to his educational theories and instructional practices would only serve to build complete citizens for a nobler Republic.

Acknowledgements

This research has been more than a decade in the making. As a result, it is impossible for me to thank the dozens of individuals that have helped find clues and illuminate the forgotten history of Captain Alden Partridge. Below, I would like to thank those who were most influential in lighting the path that resulted in this volume, and I pray that those not mentioned here may forgive my oversight and know that they are still held in appreciation.

I would first like to thank Professor Emeritus Edward Risden of St. Norbert College for mentioning Virgil during an interview concerning the use of archetypal stories. It was in drawing attention to this Stoic-influenced poem that my interest in military philosophy and the educational theories of Captain Partridge collided. Finding Virgil within Partridge's curriculum started me on a journey of tracing philosophical ideas throughout military history.

I would like to extend my gratitude to Professor Jennifer Baker of the College of Charleston. She is a powerful advocate for Stoic philosophy. She has supported my scholarship on multiple occasions and was kind enough to review my theories on military Stoicism contained within this work.

To Dr. Bruce Gudmundsson and Major Donald Vandergriff (Retired), I extend my thanks for both your inspiration and your thoughtful reviews of this work. Gentlemen, my work stands on the foundation that you have built. Thank you for being such powerful examples of officers that truly steward the profession of arms.

My thanks to Dr. Chase Spears for your help in reviewing the first draft of this volume. I pray you remain a powerful advocate for a U.S. Military crafted in the vision of the American Founders.

To Dr. Megan McElheran and Ryan Collyer of Wayfound Mental Health Group, thank you for being such powerful advocates for the return of Stoicism within the uniformed services. It was an honor to march with you, in the tradition of Captain Partridge, through the desert sands of New Mexico during the 2024 Bataan Memorial Death March. I will look forward to marching with you again.

Dr. Andrew Baker, I thank you for the rucks and deep conversation on military education. As recorded in Proverbs 27:17, "Iron sharpeneth iron;

so a man sharpeneth the countenance of his friend." Thank you for keeping me razor sharp, my friend.

For the deep conversations on life and philosophy, I extend my gratitude to Colonel Michael Szczepanski. I thank you for all the mentorship and your warfighter focus. Enjoy this work on your "small engineering school on the Hudson."

I would also like to thank Mr. John T. Hart, former director of the Sullivan Museum and History Center at Norwich University, for his help in editing research concerning the Norwich Cadet Creed.

My deep appreciation extends to Collette Leonard and Amber L. Wingerson of the Sullivan Museum and History Center at Norwich University for their help in providing information on the artifacts and artwork within the museum's collection.

I also extend my gratitude to Gail E. Wiese and Nick Connizzo of the Norwich University Archives and Special Collections for helping to answer my frequent questions and requests for materials.

To the archival staff at the Rauner Special Collections Library at Dartmouth College, the Earl Gregg Swem Library at the College of William & Mary, the David M. Rubenstein Rare Book and Manuscript Library at Duke University, the George Mason University Libraries Special Collections Research Center, the New Hampshire State Library, and the United States Military Academy Library Archives and Special Collections, I extend my thanks.

For his great attention to detail, I would like to thank Dr. Daniel Tortora for his help in editing this work.

I would like to thank all those who followed the journey of my research on LinkedIn and provided encouragement and insights into where additional sources might be found. Among this community, a special thanks to Joyce A. Rivers, Randal H. Miller, Jay King, Dr. David Ulbrich, Colonel Chad Stevens (Retired), Isaac VanMeter, and Michael Morano for their frequent support.

Finally, and most importantly, I would like to extend thanks to my amazing wife, Karolina Annis, for her unwavering support throughout my research.

Character List

Below is a character list of individuals named in relation to Alden Partridge's court-martial. The audience may find this list useful in keeping track of individuals while reading chapters III and IV.

Armstrong, John, Junior (November 25, 1758–April 1, 1843) Seventh United States secretary of war (January 13, 1813–September 1814). Veteran of the American Revolutionary War (major, Continental Army) and War of 1812 (brigadier general, U.S. Army). Armstrong established some of the original regulations for the United States Military Academy (USMA) at West Point.

Berard, Claudius (November 21, 1786–May 6, 1848) French immigrant skilled in language instruction. Served as a professor of French at the USMA for 33 years. Berard allied himself with the cabal against Partridge. During the court of inquiry of 1817, Berard was found guilty of issuing certificates of completion to cadets that were dismissed from the Academy.

Bernard, Simon (April 28, 1779–November 5, 1839) Educated at École Polytechnique, Bernard would serve as the aide-de-camp to Napoleon Bonaparte. Emigrated to the United States and was commissioned as brigadier general of engineers on November 16, 1816. In 1830, he would return to France and be commissioned as lieutenant general by Louis Philippe I. Bernard's service in the U.S. Army created some turmoil in the U.S. Army engineering corps, as his commission signaled doubts about the officers being trained at the USMA.

Brown, Jacob (May 9, 1775–February 24, 1828) A brigadier general in the New York State Militia during the onset of the War of 1812, Brown's battlefield successes would see him commissioned as a brigadier general in the U.S. Army with a promotion to major general the next year. From June 15, 1815, to February 24, 1828, Brown would be the senior officer of the U.S. Army. Brown was a member of the 1815 USMA board of visitors and witnessed the proposal made by the academic staff to convert the USMA into a national university.

Buck, Daniel Azro Ashley (April 19, 1789–December 24, 1841) Graduated from the USMA with the Class of 1808 and resigned his commission in 1811. During the War of 1812, he would raise and command a volunteer company of rangers in the 31st Infantry. After the war, he would hold various political offices in Vermont and would serve in the 18th and 20th U.S. Congress (1823–25 and 1827–29). Buck would strongly support the efforts of Partridge in establishing an alternative to the USMA and was an original trustee for Norwich University.

Burr, Aaron (February 6, 1756–September 14, 1836) Best remembered for his duel with Alexander Hamilton on July 11, 1804. Burr was an American Founding Father with service as a lieutenant colonel in the Continental Army. From 1801 to 1805, Burr would serve as vice president to Thomas Jefferson. In 1807, Burr was arrested for an alleged plot to create an independent country within Spanish Texas and lands within the Louisiana Purchase but was repeatedly acquitted of this charge.

Calhoun, John C. (March 18, 1782–March 31, 1850) Best remembered for his defense of American slavery while the seventh vice president of the United States. While serving as the tenth United States secretary of war, his vision for the future of the U.S. military conflicted with that of Alden Partridge. Calhoun believed the American militia system was unreliable and desired to build a "professional" army based on the Napoleonic model. Calhoun would ultimately deny Partridge access to copies of his court-martial transcripts. This action prevented Partridge from redeeming his public image.

Clinton, DeWitt (March 2, 1769–February 11, 1828) Nephew of George Clinton (U.S. vice president and New York governor). DeWitt Clinton served in many notable political positions, including U.S. senator, mayor of New York City, and New York governor. Clinton was a member of the 1815 USMA board of visitors. He heard the proposal made by the academic staff to convert the USMA into a national university.

Crawford, William (February 24, 1772–September 15, 1834) Ninth United States secretary of war (August 1815–October 1816). After receiving multiple letters reporting misconduct at the USMA, Crawford tasked Captain Samuel Perkins, the West Point quartermaster, to investigate the situation. By 1817, Crawford requested Brigadier General Swift to replace Partridge as superintendent.

Crozet, Claude (December 31, 1789–January 29, 1864) French engineering captain who emigrated to the U.S. in 1816 after the defeat of

Napoleon. Partridge advocated for Crozet's appointment to a professorship at the USMA. Crozet would align himself with the faculty cabal to remove Partridge, attempting to convert West Point into an institution modeled after the École Polytechnique. He left the USMA in 1823 and is notable for several engineering projects, including his work with the Virginia Board of Public Works and the Virginia and Kentucky Railroad. Crozet was one of the founders of the Virginia Military Institute (VMI).

Cullum, George (February 25, 1809–February 28, 1892) 16th superintendent of the USMA. Cullum served in the American Civil War and climbed to the rank of brevet major general in the Corps of Engineers. He was a cadet at West Point in 1830 when Partridge published *The Military Academy at West Point, Unmasked: or, Corruption and Military Despotism Exposed*. Living through the political turmoil surrounding West Point during this era likely biased Cullum's account of Partridge's achievement at the USMA in Cullum's *Biographical Register of the Officers and Graduates of the U.S. Military Academy, from 1802 to 1867*.

Davies, Charles (January 22, 1798–September 17, 1876) USMA graduate, Class of 1815. Veteran of the War of 1812. Accepted the position of assistant professor of mathematics at the USMA in 1816 with a promotion to professor of mathematics in May 1823. He would instruct at West Point until 1837. Best known for writing a series of mathematical textbooks. Lieutenant Davies would witness Partridge's return to West Point on August 29, 1817, and testify about the events of this date at Partridge's court-martial.

Duane, William John (May 9, 1780–September 27, 1865) 11th United States secretary of the treasury and served multiple terms in the Pennsylvania General Assembly. Duane, an influential lawyer of his day, believed the outcome of Partridge's court-martial to be a miscarriage of justice.

Douglass, David Bates (March 21, 1790–October 21, 1849) Graduated from Yale College in 1813 and received a commission into the Army Corps of Engineers the same year. Assigned professor of natural philosophy at the USMA in 1815. In this role, he would come into conflict with Captain Partridge as Douglass aligned his actions with the faculty cabal. In the court of inquiry of 1816, Brevet Captain Douglass was found to have sown discord upon the cadets and to have engaged in near-mutinous activities disrupting the general discipline of the academy. However, with the removal of Partridge, Douglass would continue to serve at West Point through 1831.

Ellicott, Andrew (January 24, 1754–August 28, 1820) American surveyor responsible for surveying the boundaries of the District of Columbia (completing Pierre Charles L'Enfant's plan for the capital city) and territories west of the Appalachian Mountains. Veteran of the Revolutionary War with service as a major in the Maryland state militia. Appointed professor of mathematics at the USMA in 1813. Ellicott was a member of the faculty cabal against Partridge (and later Thayer) that intended to convert West Point into a civilian national academy.

Empie, Adam (September 7, 1785–November 6, 1860) Episcopal priest who became the first Army chaplain for the USMA. Appointed by Brigadier General Swift, Empie reportedly worked well with Captain Partridge. Chaplain (and brigadier general in the U.S. Army Reserve) Herman A. Norton asserts Empie resigned his position at the USMA in response to Partridge's court-martial. Empie is best known for becoming the 12th president of the College of William & Mary.

Eustis, William (June 10, 1753–February 6, 1825) Veteran of the Revolutionary War. As a surgeon, he treated injured soldiers during the Battle of Bunker Hill. Member of the U.S. House of Representatives from Massachusetts from 1801–5 and 1820–23. Eustis was the sixth United States secretary of war (1809–12). Eustis affected the USMA by transferring all active officers, except for Captain Partridge, away from the academy during the War of 1812. He directed the cadets to be commissioned as soon as they had sufficient knowledge and competency to perform field service to support the war effort. The limited appointments to the USMA approved by Eustis placed the academy at risk of dissolution.

Jefferson, Thomas (April 13, 1743–July 4, 1826) Third president of the United States. Primary author of the Declaration of Independence. First secretary of state (under George Washington). Served as vice president under John Adams. Jefferson would establish the USMA in 1802. President Jefferson would have a profound impact on Partridge in scientific and educational endeavors. Partridge and Jefferson would correspond by letter about estimated elevations of mountains.

Madison, James (March 16, 1751–June 28, 1836) Fourth president of the United States. Madison is most remembered for his role in the creation of the Constitution of the United States and the Bill of Rights. He held a commission as a colonel in the Orange County militia during the American Revolutionary War but never had field service. President Madison led the country through the War of 1812. Madison's desire for a professionalized military in the French model would cause conflicts with Partridge's goals for the USMA and its expansion.

Mansfield, Jared (May 23, 1759–February 3, 1830) Mansfield was invited to be a professor of mathematics (and captain of engineers) at the USMA by Thomas Jefferson in 1801 after Mansfield's *Essays Mathematical and Physical* was brought to the president's attention. Jefferson would later appoint Mansfield surveyor general of the United States in 1803 (a position he held until 1812) with the task of surveying the Northwest Territory. Mansfield returned to West Point in 1814 to provide instruction in mathematics until his retirement in 1828. He would join—and possibly lead—the faculty cabal against Partridge and support efforts to remove the military aspects of the USMA.

McRee, William (December 1787–May 15, 1833) Graduated from the USMA in 1805 and was commissioned into the Corps of Engineers. He was a veteran of the War of 1812 with gallant and meritorious service, seeing him receive brevets to lieutenant colonel after the Battle of Niagara and colonel after the defense of Fort Erie. From 1815 to 1817, McRee would travel with Sylvanus Thayer to France and Belgium to survey fortifications and collect materials to improve military education at West Point. In protest of the commissioning of the French engineer Simon Bernard to brigadier general, McRee resigned his commission and ended his military service on March 31, 1819.

Mitchill, Samuel Latham (August 20, 1764–September 7, 1831) Notable American physician, politician, and scientist. Mitchill graduated from the University of Edinburgh Medical School before moving on to study law. He served as both U.S. senator and representative (in six Congresses, from 1801–13) for the state of New York. Mitchill's experience as an educator would include professorships at Columbia College and the New York College of Physicians and Surgeons. In 1818, he served as the surgeon general of the New York State Militia. Mitchill was a founding member and vice president of Rutgers Medical School. As a member of the 1815 West Point Board of Visitors, he was among an audience that heard a proposal made to convert the USMA into a national university presented by the West Point faculty.

Monroe, James (April 18, 1758–July 4, 1831) Fifth president of the United States. Veteran of the American Revolution. Monroe would interact with Partridge while president and secretary of war. Monroe's desire to import a French style of military professionalism would cause conflicts with Partridge's visions for the USMA and an American model of military competency. President Monroe remitted the punishment recommended by Partridge's court-martial and ordered him to return to duty.

Perkins, Samuel (Dates Unknown) USMA quartermaster. In 1815, he was assigned by Secretary of War William H. Crawford to investigate Partridge's actions at West Point.

Snowden, Jno (Dates Unknown) Veteran of the American Revolution. Storekeeper at the USMA.

Swift, Joseph Gardner (December 31, 1783–July 22, 1865) First cadet to graduate from the USMA (October 12, 1802). Third superintendent of West Point from 1812 to 1814. Chief of Engineers of the United States Army from 1812 to 1818. Veteran of the War of 1812. Retired at the rank of brevet brigadier general. Brigadier General Swift's lack of leadership in resolving the internal conflicts at the USMA would limit the progress of Captain Partridge at West Point.

Totten, Joseph Gilbert (August 23, 1788–April 22, 1864) Tenth graduate of the USMA, Class of 1805. Veteran of the War of 1812. Nephew of Jared Mansfield. Died at the Battle of Antietam as a brevet major general. Then Brevet Lieutenant Colonel Totten would be a member of Partridge's court-martial and was not removed despite being junior in rank to Partridge in the Corps of Engineers and a close relative to Joseph Mansfield (one of Partridge's primary accusers).

Tompkin, Daniel D. (June 21, 1774–June 11, 1825) Sixth vice president of the United States. In 1816, while the fourth governor of New York, he would receive a letter from Partridge appealing to New York to lead the way in establishing state military academies to train the militia.

Thayer, Sylvanus (June 9, 1785–September 7, 1872) Remembered as the "Father of the Military Academy." Graduated as valedictorian of Dartmouth College in 1807. Graduated from the USMA in 1808. Veteran of the War of 1812. Superintendent of the USMA from 1817 to 1833. Thayer's refusal to grant Partridge the customary choice of housing at West Point while he awaited a court of inquiry caused Partridge to relieve Thayer of command on August 30, 1817. This event ultimately triggered Partridge's court-martial.

Williams, Jonathan (May 20, 1750–May 16, 1815) In February 1801, appointed major in the Corps of Artillerists and Engineers. Promoted to lieutenant colonel and assigned as the first superintendent of the USMA (1801–3, 1805–12). Promoted to colonel in 1808. Williams resigned his Army commission in 1812 (because of a dispute over a command assignment) and became a brigadier general in the New York State Militia.

Wilkinson, James (March 24, 1757–December 28, 1825) Served in the Continental Army in the American Revolution but was pressured to resign his commission twice. Reached the rank of major general in the U.S. Army and twice was the senior officer of the U.S. Army. Allegedly conspired with Aaron Burr to form an independent nation in Spanish Texas and lands within the Louisiana Purchase. High-paid spy for Spain.

Winder, R. H., Esquire (Dates Unknown) Appointed to the Judge Advocate General's Corps on July 9, 1814. Served as the Army judge advocate in Partridge's court-martial.

Wadsworth, Decius (January 2, 1768–November 8, 1821) Acting superintendent of the USMA (1803–5). Later named the first U.S. Army commissary general of ordnance (chief of ordnance).

Wolcott, Oliver, Junior. (January 11, 1769–June 1, 1833) American politician who served in several key offices, including U.S. secretary of the treasury and judge of the U.S. Second Circuit Court. In 1815, while governor of Connecticut and a member of the USMA board of visitors, Wolcott heard the proposal made by the academic staff to convert the USMA into a national university.

Figure 7—Norwich University South Barracks[8]

---

[8] Sketch of the South Barracks of Norwich University (Norwich, Vermont) based on a photograph by Silas P. Burnham, ca. 1860–66, held within the Norwich University Archives and Special Collections. The barracks were built to Alden Partridge's specifications. A fire destroyed the barracks in 1866 and prompted the university to relocate to Northfield, Vermont, where it continues to this day. (Artist Gala S., @galaartcome fiverr.com)

Chapter I: Introduction

Volume I of *Marching with Spartans: The Educational Theories of Captain Alden Partridge* is the beginning of a series of volumes about a truly remarkable American soldier and scholar whose ideas have been obscured within the historical record of the United States. Captain Alden Partridge was an extraordinary man and America's greatest educational philosopher.[9] His contributions left a lasting mark on the Republic, as he single-handedly sparked a national movement to establish private military academies suitable for educating citizens of the Republic. Colonel Lester Webb, a notable Partridge scholar, recalled in his dissertation:

> [Partridge] believed American democracy was the form of government a free people ought to choose. He believed all citizens should be educated in such a manner as would enable them to perform efficiently and effectively all offices in peace and war. He believed that if the nation was to grow, develop, and survive, it had to produce citizens educated in the sciences for the development of natural resources through internal improvements; educated in literature and history for appreciation of their liberties; and trained in military science so that they as citizen soldiery could defend the nation against all enemies, internal and external. He believed the youth of the land ought to come to maturity under a system which combined the features of his philosophy with duties as citizens. He believed his "American System of Education" was the answer to the educational needs of the young Republic. He established private institutions throughout the country for the promotion of his beliefs and for the honor and glory of his beloved America.[10]

For all his contributions to the Republic, Partridge's achievements have been hidden and obscured, often purposely, by U.S. Army and educational historians. His ideas threatened the survival of the United States Military

---

[9] Webb, *Captain Alden Partridge.*

[10] Lester A. Webb, "The Origin of Military Schools in the United States Founded in the Nineteenth Century" (PhD diss., University of North Carolina at Chapel Hill, 1958), 216.

Academy (USMA) at West Point and the careers of high-ranking Army officers. As a result, U.S. Army historians conspired to transfer Partridge's achievements at West Point onto his replacement (Brevet Major Sylvanus Thayer), as Partridge was painted as a mutinous and incompetent drillmaster. While scholars, like Colonel Webb, have attempted to correct the historical record, the falsified image of Partridge reappears as often as it is refuted. As a result, one of America's greatest thinkers has had his memory marred and his ideas stifled. The ultimate consequence of these actions was that the United States failed to adopt an educational philosophy capable of fully sustaining the nation's system of government.

Setting the Stage

Alden Partridge was born into a nation that was a bubbling cauldron of ideas. Born roughly a decade after the start of the American Revolutionary War, Alden Partridge and his generation faced the challenge of implementing the vision of the American Founders. The American people faced the complex task of redesigning formerly monarchal or monastic institutions to meet the needs and visions of America's constitutional republic. Present-day readers often fail to recognize the extreme complexity of the era. Even when there was a consensus around the philosophical structure of the newly formed Republic, there were often conflicts in how ideas would practically be accomplished. Like today, powerful individuals were not above shaping the government or institutions for their own benefit.

The ongoing Industrial Revolution was also shifting the educational and economic needs of the country. Scientific approaches were proving beneficial throughout American industries. This created an increased demand for educational systems to produce graduates knowledgeable in practical science. While present-day Americans may believe the concepts of higher education have been established for centuries, these concepts are relatively new inventions. Most of our current educational concepts developed after Partridge's lifetime, and many were implemented during the 20th century.

It is unfortunate the period between the American Revolution and the American Civil War is not better covered within public and higher education. The general population is often entirely ignorant of key historical events that shifted the direction of the American Revolution to include events such as Shays' Rebellion, the Whiskey Rebellion, Burr's Conspiracy, and the War of 1812. These events, combined with the continual governmental lack of attention towards the American militia, led to the establishment of a large standing army and the present-day military industrial complex that would horrify the American Founders. Many Americans now believe that our current military institutions cannot be altered and there is no other functional alternative.

However, Partridge presented a much different vision for education. Partridge's *American System of Education* would have supported the constitutional vision for the use of militia as the primary ground defense force for the nation. Partridge's approach to education would have largely geared America to be a more peaceful nation, with individual citizens having a superior understanding of their rights and duties to their fellow citizens.

Philosophical and historical context matters in understanding the foundational freedoms offered within the American Republic. For example, present-day politicians often debate the right to bear arms through semantic arguments over the meaning of the few words found within the Second Amendment while ignoring the hundreds of years of philosophy concerning the use of militias and the numerous materials written by the American Founders. The lack of inclusion of these philosophical and historical materials robs citizens of their ability to fully execute their rights. Today, many citizens do not fully understand or use their rights due to an ignorance of the underlying philosophies enshrined within the American Constitution. Captain Alden Partridge understood America's freedoms were threatened as much by ignorance or miseducation as they were by foreign invaders. As Partridge asserts,

> [I am] satisfied that none of deeper import to the permanency of our republican institutions can engage the attention of the representatives of a great and enlightened people. Education is more powerful than the lever of Archimedes, either to sustain or to crush the political institutions of every country—just in proportion as it is, in principle and in practice, in accordance with, or adverse to, those institutions.[11]

The primary purpose of this work is to reintroduce the reader to the foundational ideas of the Nation and Partridge's unique theories on building an educational system suitable for sustaining the American Republic. As a result, this work will cover significant ground in terms of both time and philosophy. The average reader may find they lack sufficient knowledge to fully conceptualize the discussions that are presented within this work. For this reason, a few recommended books are provided at the end of this chapter to

---

[11] Memorial of Alden Partridge, praying Congress to adopt measures with a view to the establishment of a general system of education for the benefit of the youth of this nation, January 21, 1841, U.S. Congressional Serial Set, H.R. Doc. No. 69, 26th Congress, 2nd Session (1841), 7, Law Library of Congress.

assist with gaining knowledge of the historical and philosophical framing of the early Republic.

Partridge's Curriculum

For those who doubt the scope of the academic materials that would have been presented within Partridge's college curriculum, a quick glance at Partridge's 1826 catalogue will quickly reveal its complexity:

ENGLISH LANGUAGE.

Murray's Grammar, for parsing, Murray's Reader.

LATIN.

Adam's Latin Grammar, Liber Primus, Virgil, Cicero's Select Orations, Cicero de Oratore, de Amicitia et de Senectute, Sallust, Caesar's Commentaries, Horace, Livy, five first books, Tacitus, five books,

GREEK.

Buttman's Greek Grammar, Neilson's Greek Exercises, Delectus, Collectanea Græca Minora, Collectanea Græca Majora, Xenophon's Anabasis, Homer's Iliad, six books.

FRENCH.

Ferry's first Elements, Perrin's Vocabulary, Wanostrocht's Grammar, Boyer's Dictionary, large edition, 2 vols. Telemaque.

SPANISH.

Joss's Grammar, Telemaco, Robinson Crusoe, Newman's Dictionary.

RHETORIC, HISTORY AND GEOGRAPHY.

Walker's Rhetorical Grammar, Blair's Rhetoric abridged, Morse's Universal Geography, large abridgment with Atlas, latest edition; Tytler's Elements of History, Adam's Roman Antiquities.

MATHEMATICS AND NATURAL PHILOSOPHY.

Hutton's Mathematics, Gibson's Surveying, Crozet's Descriptive Geometry, Enfield's Natural Philosophy.

LOGIC, MORAL PHILOSOPHY AND METAPHYSICS.

Hedge's Logic, Paley's Evidences of Christianity, Paley's Natural Theology, Paley's Moral Philosophy, Butler's Analogy, Locke's Essays on the Understanding, Stewart on the mind.

LAWS AND POLITICS.

Constitution of the United States, and of the several States, edition of 1820, by Gales & Seaton; Vattel's Law of Nations, Federalist, Burlamaqui on Natural and Political Law.

For those who contemplate obtaining merely a knowledge of Arithmetic and the Elements of Geography, Worcester's Geography and Walsh's Arithmetic, editions of 1820.

The Testament is superseded in the course of Greek studies by Xenophon's Anabasis, with the belief that the style of the latter is more pure and classical. However, it can be read by those preparing for College.

NOTE… A knowledge of the foregoing authors will be considered as comprising a complete course of education at the Academy, in the several branches of science and literature therein contained. Fortification and the other scientific parts of the military art are taught by familiar explanatory lectures; the practical part is taught in the field.[12]

While some of these texts are outdated, namely Paley's *Natural Theology*, written to refute Jean-Baptiste Lamarck's (pre-Charles Darwin) theories of evolution, the vast majority of these texts remain valid to this day for educational purposes. How many, if any, Americans have encountered an undergraduate curriculum as broad as the one proposed by Partridge? I believe it would be hard to find even one in 100,000 Americans who could pass a comprehensive test on Partridge's curriculum.

Speaking for myself, a man with nearly two decades of military service, having climbed into the field grade ranks, and completed a doctorate in education and a master's degree in military history, I felt ignorant when I first read through the books within Partridge's curriculum and crawled through the archive of Norwich University. It may be natural for us to fall into the *argumentum ad novitatem* fallacy, believing our modern approaches to education are superior to the old simply because we now use them. But that certainly isn't the case.

It would be a grave error to believe that Partridge's ideas are outdated and unable to meet the challenges of the 21st century simply because they are old. If nothing else, I hope this imperfect work might inspire the reader to further investigate the pivotal ideas and history around the foundation of the American Republic. Certainly, time spent investigating the rich history and deep thinkers of this period is time well spent.

Book Series Structure

I originally intended to publish a single book organized into three major parts (Partridge's educational theories, modern military implications, and

---

[12] Alden Partridge, Catalogue of the Officers and Cadets: Together with the Prospectus and Internal Regulations of the American Literary, Scientific and Military Academy at Middletown, Connecticut, 1826, pp. 21–22, Norwich University Archives and Special Collections.

methods to reestablish the *American System of Education*). However, it was quickly determined that the length of such a work with its related appendixes would easily exceed 800 pages. As a result, it was determined that the best course of action was to publish a series of volumes to make this project more manageable for the author while increasing the speed of the reintroduction of Partridge's major theories to the general public.

In this first volume, I will discuss the educational theories and history of Captain Alden Partridge. This includes a detailed discussion of Partridge's major works, including his *Lecture on Education*. Additionally, the conflicts in the historical record will be explored. In future volume(s), I intend to discuss the possible influences on Partridge's educational theories and demonstrate the uniqueness of some of Partridge's ideas. I intend to continue the discussion on Partridge's educational theories, comparing him to the contemporary Prussian and French models of military education. I will also explore Partridge's potential impact on the American transcendentalist movement, specifically on the philosophers Ralph Waldo Emerson and Thomas Wentworth Higginson. Additionally, the modern military implications of Partridge's theories will be examined. This will include what should be the fate of the United States Military Academy and examine whether Norwich University could return to a greater use of Partridge's theories. A discussion on how military education can be used as an effective tool to reinforce national defense instead of national service programs will be presented. The final part will discuss the universal implications of Partridge's theories. This will include discussions on how Partridge's theories could greatly reduce the cost of higher education while creating a truly student-centered curriculum designed for the full needs of American citizens.

Partridge's Achievements

Before investing the time to read an entire book on some obscure soldier and scholar from the early 19th century, one might ask, "Why is Captain Alden Partridge even worth remembering?" It is a valid question. The answer will reveal the reasons for remembering this American genius. The following section will discuss some of the major achievements of Captain Partridge, followed by a listing of his major publications and a timeline of his life. These materials will help the audience understand the importance of Alden Partridge and provide an orientation to the time period that will be addressed by this book.

First Lieutenant

Due to his superior academic performance at the United States Military Academy, Partridge was commissioned as a first lieutenant of engineers on October 29, 1806. He was one of only two cadets to ever be commissioned by the Academy at this advanced rank. Cadets are normally commissioned as second lieutenants.

Long Gray Line

Captain Partridge left a lasting impact on the United States Military Academy at West Point, NY. He served as the first professor of mathematics. From 1814 to 1817, he would serve as the superintendent. Partridge composed the first rules and regulations for the academy.[13] He developed an instructional system that could absorb new cadets as they arrived throughout the year, as there was not yet an established starting term and date. Most visibly, Partridge adopted the first gray uniform for the cadet corps.[14] This uniform design still influences the uniforms worn at the USMA today.

Applied Ballistics

In 1810[15] and 1814,[16] Partridge conducted several experiments in ballistics of both tactical and scientific value. His first set of tests were tactically focused, which involved the rates of fire of cannon and musketry. Partridge was trying to determine how many enemy volleys advancing infantry would likely receive when covering known distances. His second set of experiments attempted to determine the velocity of cannon balls through the use of a ballistic pendulum. He correctly deduced the need for pendulums of appropriate construction to not allow cannon balls to penetrate through them in order to gain accurate measurements of velocity.

Partridge proposed an alternate way of determining velocities by firing cannons level to the ground at a known height. He recognized that gravity would act on a dropped cannonball at the same rate as a cannonball fired horizontally from a level cannon. As a result, the distance to where a cannonball strikes the ground from the muzzle of a level cannon could be divided by the time a static cannonball takes to fall to the ground (when dropped from the same height as the cannon's muzzle) to accurately provide the velocity of the fired projectile. After conducting this experiment from different heights (with taller cannons producing longer ranges), Partridge proved it was possible to create an equation to calculate the impact velocities

---

[13] Alden Partridge, Regulations for Governing the United States Military Academy in New York, Approximately 1810–14, Norwich University Archives and Special Collections.

[14] Ted Spiegel, "The Cadet Uniform Factory," West Point Association of Graduates.

[15] Alden Partridge, Experiments on Artillery and Infantry Firing at West Point, 8 November 1810, Norwich University Archives and Special Collections.

[16] Alden Partridge, Experiments on Artillery and Infantry Firing at West Point, November 1814, Norwich University Archives and Special Collections.

of projectiles fired to various distances. Partridge's research into ballistic coefficients predates the major military research on this subject by half a century.

Comprehensive Educational Philosophy Fully Compatible with the Republic

Captain Alden Partridge wrote several treatises laying out an educational philosophy for America. Partridge understood that European models of education originated from monastic practices and were further refined to support aristocracies. Using European educational models would be inappropriate for the American Republic. Partridge perceived educational practices as a contributing factor to the religious wars (Catholic vs. Protestant) on the European continent. Captain Partridge was a deeply devoted man and embraced the Christian Faith. However, keeping with the spirit of the Constitution, he did not want to see singular religious denominations seize control of education in an attempt to establish an official state religion. Captain Partridge's approach to education blended classical and practical education to instruct citizens in all areas of their civic duties. Unlike competing American and European models of education, Partridge's approach freed education from the control of the aristocracy[17] and influences of the church,[18] while retaining the valuable knowledge and history of past generations. Partridge's model for diffuse military knowledge among U.S. citizens offers significant advantages in national defense and avoids the need for a standing military that was so greatly feared by the American Founding Fathers. As revealed in the following pages, many of Partridge's educational concepts were generations ahead of his time and are superior to our present-day approaches in multiple areas (further explained in later chapters).

First American Educator to Recognize the Importance of Physical Education

While Captain Partridge cannot claim the title of the first physical education teacher in the U.S., it was under his direction that the first one was hired. In 1816, master of the sword and dance instructor Pierre Thomas was

---

[17] The term *aristocracy* here is used to refer to a small privileged ruling class (not denoting the original Greek concept of "rule by the best").

[18] The term *church* here is used to refer to religious bodies that could assert political power/control over the populous. In this context, Partridge is not calling for the removal of theology.

hired at the USMA.[19] Partridge frequently addressed the need for physical fitness in his lectures. His drive to concurrently train the body and the mind within education was decades ahead of notable American mainstream educators such as Horace Mann.

Plan to Expand the Militia

In 1821, Captain Partridge presented a *Lecture on National Defense* that included instructions on how to expand and train a force of two million militiamen. This would have been a force double the size of the U.S. Army that fought in the War of 1812, and the plan would have corrected the training deficiencies that have led to the United States' present-day reliance on a standing army. Additionally, Captain Partridge laid out a plan for building strategic fortifications, being mindful of the cost of maintaining permanent defenses. His plan balanced the need for defense against the risks incurred through excessive defense spending. His preference was for cheaper earthworks that were more durable to artillery fire over the fashionable and costly French-style stone and masonry fortifications. The concepts contained within this lecture are as valid today as the day they were written and would offer a check against current reckless military spending.

Early Experiments in Special Education

Captain Partridge was one of the first educators to provide what we now term *special education*. Partridge instructed at least one student with significant learning disabilities. While Partridge could not accomplish all his educational goals with Thomas Cameron, Partridge taught the boy how to write phonetically and vastly improved his physical health. This instruction occurred in the 1820s. Publicly funded special education institutions did not start to appear in the U.S. until the 1840s.[20]

Experiential Learning

Partridge believed in frequent expeditions to allow his students to practically apply the knowledge they were learning in the classroom. As Larkin reports,

> These field trips provided students with valuable experiences in botany, mineralogy, surveying, engineering, military science, political economy, and history. Student participants observed factories, navy

---

[19] Whitfield B. East, *A Historical Review and Analysis of Army Physical Readiness Training and Assessment* (Fort Leavenworth, KS: Combat Studies Institute Press, US Army Combined Arms Center, 2013).

[20] Penny L. Richards and George H. Singer, "To Draw Out the Effort of His Mind," *Journal of Special Education* 31, no. 4 (1998): 443–66.

yards, arsenals, waterworks, canals, railroads, and bridges. They also
visited legislative assemblies, museums, and historic sites.[21]

As a result, Captain Partridge became a pioneer in experimental learning
(commonly known as the practice of taking "field trips"). Unlike other
educational institutions of the era, this focus on experiential learning provided
real-world experiences for his students and provided opportunities for them to
practically apply the theoretical knowledge they were acquiring in the
classroom.

Land-Grant Colleges

Captain Partridge laid the groundwork for the eventual establishment
of land-grant colleges across the United States. In 1841, he presented a
memorandum to the House of Representatives proposing a plan to grant land
and fiscal resources for states to establish new universities.[22] While there was
little support for such an effort at the time proposed, this action helped
Senator Justin Morrill of Vermont to pass similar legislation in 1862, with
further details found in chapter VIII (see appendix A).[23] The American Civil
War made the need for men with military knowledge readily apparent. This
critical need helped Congress overcome its previous hesitancy to pass such
legislation. While Senator Morrill never credited his action to Captain Partridge,
the parallels between Morrill's and Partridge's plans are noticeable.

Private Military Academies

Partridge founded multiple private military academies, including what
would become Norwich University, the oldest private military academy in the
United States. While Partridge had previously instructed on topics important to
naval warfare in his academies, his establishment of the Virginia Literary,
Scientific and Military Academy at Portsmouth, Virginia, in 1839 can be viewed
as the first attempt to establish a naval academy for the United States (opened
six years before the United States Naval Academy in Annapolis, Maryland).
Portsmouth offered direct access to the Atlantic Ocean for improved use of
vessels for naval warfare instruction. Partridge's efforts in establishing his own
private military academies greatly aided in the establishment of other famous
public military institutions, such as the Virginia Military Institute (VMI) and the
Citadel.[24]

---

[21] Daphne Larkin, "Norwich Museum Presents Historian on Partridge's Rugged,
Outdoor Lessons," VTDigger, February 15, 2018.

[22] Memorial of Alden Partridge, praying Congress to adopt measures with a view to the
establishment of a general system of education.

[23] "Morrill Act (1862)," Milestone Documents, National Archives.

[24] Gordon R. Sullivan, "Remarks by General (Ret.) Gordon R. Sullivan," West Point
Association of Graduates, October 1, 2003.

Hiking Legacy

Captain Partridge was a renowned pedestrian capable of walking over 60 miles in a day. He would take quarterly expeditions hiking through mountain ranges in order to determine the elevation of mountain peaks. Stories of Partridge pedestrian journeys "inspired schoolteacher James P. Taylor to create the Long Trail, which runs the length of Vermont. The Long Trail, in turn, helped inspire Benton MacKaye's vision for the Appalachian Trail, which now runs from Georgia to Maine."[25] Every year, millions of Americans walk on trails inspired by Alden Partridge with little or no knowledge of the great man who led to their creation.

Recognized with Honorary Degrees

Dartmouth College awarded Partridge an honorary AM degree in 1812, with the University of Vermont granting him the same honor in 1821 (along with offering Partridge the position of university president). Present-day academics would recognize the artium magister (or magister artium) as a "master's degree" being intermediate in academic rank. However, this was not the case in early 19th-century America. The doctor of philosophy (philosophiae doctor) (PhD) had yet to appear in the United States. Yale University was the first to award a PhD degree in 1861 in the United States, years after the death of Captain Alden Partridge. The PhD and *magister scientiarum* (MS) degrees, that we are familiar with today, were introduced in the United States after Partridge's lifetime. The award of AM degrees to Partridge, in recognition of Partridge's advancement of the sciences, should be considered the equivalent of a present-day PhD. Dartmouth College and the University of Vermont were not merely honoring Partridge as a "master" of the sciences; they were honoring him at the level of a present-day "professor/doctor."

Reserve Officer Training Corps (ROTC)

The National Defense Act of 1916 further codified the legacy of Captain Partridge into law by expanding military training on college campuses. The military training Partridge started at Norwich University would be extended to colleges and universities across the United States. This act formally established the Reserve Officers' Training Corps (ROTC). To date, ROTC has trained over 600,000 men and women and represents the largest commissioning source of Army officers.[26]

---

[25] Norwich Bicentennial, *200 Things about Norwich* (Northfield, VT: Norwich University, 2019).

[26] Sean A. Gainey, "Army ROTC History," United States Army Cadet Command.

Education of Women

Partridge is a forgotten pioneer in women's higher education, integrating women into college classrooms eight years before Oberlin College. In 1829, Captain Alden Partridge opened the Norwich Female Seminary.[27] While it is unfortunate that much of the history of the Norwich Female Seminary is lost to time, the intent of this institution was to provide women with "All branches of a useful and ornamental education,"[28] with "young ladies [having] the privilege of attending lectures delivered at the University and the use of a well assorted library."[29] According to Norwich University president Charles Plumley, "The (Norwich) Partridge Female Seminary was the first coeducational institution of collegiate grade established in the United States of America."[30]

While Charles Plumley lamented about the loss of the names of the students and other faculty of the Female Seminary in his *Ms. Traditions of Norwich*, at least one name may have been recovered. Norwich University graduate Joyce A. Rivers purchased a copy of François Peyre-Ferry's *The Art of Epistolary Composition* in a charity auction in 2023. On the flyleaf was written the name Hetty B. Hart. Hetty Buckingham Hart (April 19, 1800–August 7, 1876) of Hartford, Connecticut, was the daughter of Major Richard William Hart (1768–1837) and Elisabeth Ball (1772–1843). She was the granddaughter of Major General William Hart (a cavalry officer in the Revolutionary War) and Ester Buckingham. Assuming Ms. Hetty attended or instructed at the Female Seminary when the American Literary, Scientific and Military Academy (ALSMA) was 17 miles from her home in 1826–29, this book might be the oldest surviving artifact of the first coeducational college in the United States.[31]

In Summary

In this section, thirteen major achievements of Captain Partridge were provided to spark the audience's interest in this remarkable educator. It would

---

[27] Alden Partridge, Norwich, VT, Sept. 17, 1829., The Norwich Female Seminary Is This Day Opened to Receive Young Ladies under the Care of Miss Mary B. Ware, as Principal …., Norwich University Archives and Special Collections.

[28] Alden Partridge, Norwich, Prospectus of the Norwich University and Norwich Seminary for Young Ladies, at Norwich, State of Vermont, 1835, p. 16, Norwich University Archives and Special Collections.

[29] Alden Partridge, "Notice," *Vermont Republican & Courier*, May 29, 1835.

[30] Charles A. Plumley, Ms. Traditions of Norwich, chapter 4, p. 16, Norwich University, n.d. Norwich University Archives and Special Collections.

[31] The Connecticut Historical Society maintains some historical records on Ms. Hetty that would be worthwhile for some future historian to explore to see if further information about the Female Seminary could be discovered.

be easy to continue to add dozens of additional points that would make Partridge worthy of remembrance. However, for the sake of brevity and to move on to a more detailed discussion, many of these additional achievements will be further discussed later in this work.

Major Publications

Experiments on Artillery and Infantry Firing at the Military Academy[32,33]
Guidelines for Establishing Multiple Military Academies (see appendix B)[34]
Investigation of Sir Isaac Newton's Binomial Theorem[35]
Lecture on Education (see appendix C)[36]
Lecture on National Defense (see appendix D)[37]
Letter to the Public (see appendix E)[38]
Method of Determining the Initial Velocity of Projectiles[39]
Meteorological Tables[40]
National Education System (see appendix F)[41]

---

[32] Alden Partridge, Experiments on Artillery and Infantry Firing at West Point, 8 November 1810.

[33] Partridge, Experiments on Artillery and Infantry Firing at West Point, November 1814.

[34] Alden Partridge, Guidelines for Establishing Multiple Military Academies, Approximately 1815–1817, Norwich University Archives and Special Collections.

[35] Alden Partridge, An Investigation of Sir Isaac Newton's Binomial Theorem by Alden Partridge, Approximately 1810–1815, Norwich University Archives and Special Collections.

[36] Alden Partridge, "Captain Partridge's Lecture on Education," in *The Art of Epistolary Composition, or Models of Letters, Billets, Bills of Exchange… to Which Are Added, a Collection of Fables… for Pupils Learning the French Language; a Series of Letters between a Cadet and His Father, Describing the System Pursued at the American, Literary, Scientific and Military Academy at Middletown, Conn. … and a Discourse on Education, by Capt. Alden Partridge, Superintendent of the Academy…*, ed. François Peyre-Ferry (Middletown, CT: E. & H. Clark, 1826), 263–80.

[37] Alden Partridge, Lecture on National Defense, 18 July 1821, Norwich University Archives and Special Collections.

[38] Alden Partridge, Letter to the Public, 1820, Norwich University Archives and Special Collections.

[39] Alden Partridge, Letter to Jonathan Williams, 19 January 1812, Norwich University Archives and Special Collections.

[40] Alden Partridge, Enclosure: Summary of Meteorological Observations by Alden Partridge, [ca. 9 December 1815], National Archives.

[41] Partridge, Memorial of Alden Partridge, praying Congress to adopt measures with a view to the establishment of a general system of education, 1–8.

Observations Relative to the Calculation of the Altitude of Mountains, etc., by the Use of the Barometer[42]

The Military Academy at West Point, Unmasked: or, Corruption and Military Despotism Exposed (see appendix G)[43]

Timeline of Major Life Events

February 12, 1785 Alden Partridge was born to Samuel and Elizabeth (Wright) Partridge.[44]

1802–5 Attended Dartmouth College, NH.

December 14, 1805 Became a cadet at the United States Military Academy at West Point, NY.

October 30, 1806 Commissioned as a first lieutenant of engineers upon graduation from the USMA.

1806–13 Assistant professor of mathematics at the USMA.

1812 Awarded honorary master's degree (*artium magister*) from Dartmouth College.

April 1813–September 1813 Professor of mathematics, head of the department, USMA.

1813–16 Professor of civil and military engineering, USMA.

1814–15 Acting superintendent of the USMA.

1815–17 Superintendent of the USMA.

1815 Introduced the iconic gray uniforms at the USMA.[45]

1816–17 Commander of the USMA.

1818 Resigned from the Army.

1819 Founded the "American Literary, Scientific and Military Academy," at Norwich, VT.

1821 Awarded honorary master's degree from the University of Vermont. Declined offer to become the university's president.

1822–23 Vermont surveyor general.

---

[42] Partridge, Enclosure: Summary of Meteorological Observations.

[43] Alden Partridge, *The Military Academy at West Point Unmasked: or, Corruption and Military Despotism Exposed* (Washington, D.C.: Sold at the Bookstore of J. Elliot, 1830).

[44] Some sources list Partridge as simply being born in February 1785. Norwich University has traditionally celebrated February 12 as "Founder's Day" in honor of Partridge's birth. January 12, 1785, is the date provided by Norwich historian William Ellis, *Norwich University, 1819–1911: Her History, Her Graduates, Her Roll of Honor*, vol. I (Montpelier, VT: Capital City Press, 1911), 99.

[45] Frederick P. Todd and Frederick T. Chapman, *Cadet Gray: A Pictorial History of Life at West Point as Seen through Its Uniforms* (New York, NY: Sterling Publishing Co., 1955), 100.

1823 Adopted George Colvocoresses.

1829 Opened the Norwich Female Seminary.

1833, 1834, 1837, 1839 Elected to the Vermont Legislature.

1837 Married Ann Elizabeth Swasey.

1838 Birth of first son, George M. C. Partridge.

1839 Birth of second son, Henry V. Partridge.

January 17, 1854 Died at age 68 in Norwich, VT. He rests at Fairview Cemetery in Norwich, VT.[46]

Significance of the Study

The significance of this study is found in the reintroduction of a model of education suitable for a republican[47] system of government. Partridge's *American System of Education* is free from the restrictive educational models of the past that were negatively influenced by churches and aristocracies. Using Partridge's educational model would have significant impacts on national security. It would increase the Nation's capability in waging wars while also providing a strong bulwark in directing the democratic process to avoid unneeded military engagements. This educational model would address many of the concerns raised by the French diplomat and political philosopher Alexis de Tocqueville in his *Democracy in America* and equip the populace to act through enlightened self-interest and voluntary associations to reduce the reliance on the general government and safeguard against the stifling despotism of the bureaucratic state.

Returning to Partridge's educational model could greatly reduce the cost of higher education while vastly shortening the adolescence of students. The return on investment from abandoning the current model of higher education and returning to Partridge's model would be measured in the trillions of dollars, as students would have years of productivity returned to them while avoiding many of the causes of modern student debt. These are but a few of the significant impacts America might enjoy if we were to only remember the wisdom of one of our first educational philosophers.

Although the pages of this work are primarily intended to reveal the genius of Captain Alden Partridge, they will also reveal how he drew upon the great thinkers of the Western canon to build his *American System of Education*. Many great thinkers will be mentioned within the pages of this book. From the

---

[46] Due to Partridge's focus on fitness, it is estimated that he lived longer than 85 percent of the American men that shared this birth year.

[47] Republican here is being used to describe the design of the United States government and is not referencing the political party.

Greek tradition, great thinkers such as Socrates, Plato, Xenophon, Homer, and the Lacedæmonians (Spartans) will set the stage for the foundations of democracy. The Roman tradition refined the republican concept of government and martial tradition in military leaders, such as Julius Caesar and Marcus Porcius Cato Uticensis (Cato the Younger), and influential thinkers, poets, and historians to include Virgil (Publius Vergilius Maro), Marcus Tullius Cicero, Quintus Horatius Flaccus (Horace), Titus Livius (Livy), and Publius Cornelius Tacitus. European philosophers such as Justus Lipsius, Count Raimondo Montecuccoli, Michel de Montaigne, and John Locke, will set the stage for the birth of the American Republic.

In this story, the reader will find numerous American Founding Fathers and U.S. presidents, including George Washington, Thomas Jefferson, James Monroe, and Andrew Jackson. The folk hero Davy Crockett will even be found advancing Partridge's concerns about West Point in the halls of Congress. If readers are to complete Partridge's curriculum today, they would be shocked to find the breath and width of the ideas in the Western canon that we have simply abandoned teaching citizens in our current educational model.

Summary

The memory and contributions of the greatest American educational philosopher have been obscured by a false history created by Army historians attempting to protect the United States Military Academy at West Point.[48] As a result of pejorative accounts of Partridge within the historical record, the significance of Captain Alden Partridge's contributions is now unknown to the majority of Americans. It is far past time to restore the memory of Captain Partridge and reintroduce his educational theories to the betterment of the American citizen. In these pages, I hope to complete the research first started by Colonel Lester Webb in displaying the brilliance of Captain Partridge's educational theories. This book series will not only attempt to set straight the historical record and recall the historical impact of Partridge's educational model but will also discuss the many positive military and societal effects of reintroducing his *American System of Education*.

Suggested Supporting Scholarship

The following is a recommended reading list, if needed, to bolster the reader's background knowledge in the various historical, political, or philosophical topics that are discussed within this book:

---

[48] Webb, *Captain Alden Partridge.*

Political Debates Surrounding the Founding of the Republic

McClanahan, Brion T. *The Founding Fathers' Guide to the Constitution.* Washington, D.C.: Regnery History, 2013.

Bailyn, Bernard, ed. *The Debate on the Constitution: Federalist and Antifederalist Speeches, Articles, and Letters during the Struggle over Ratification.* Vol. I. New York, NY: Literary Classics of the United States, 1993.

------. *The Debate on the Constitution: Federalist and Antifederalist Speeches, Articles, and Letters during the Struggle over Ratification.* Vol. II. New York, NY: Literary Classics of the United States, 1993.

History of American Education

Gatto, John Taylor. *Weapons of Mass Instruction: A Schoolteacher's Journey through the Dark World of Compulsory Schooling.* Gabriola Island, BC: New Society Publishers, 2017.

Thelin, John R. *A History of American Higher Education.* Baltimore, MD: Johns Hopkins University Press, 2019.

American Democracy in the 1830s (Contemporary to Partridge's Lifetime)

Cook, William R. *Tocqueville and the American Experiment.* Chantilly, VA: Teaching Co., 2004.

Tocqueville, Alexis de. *Democracy in America.* Translated by Henry Reeve. 2 vols. New York, NY: J. & H. G. Langley, 1840.

The Influence of Stoic Philosophy on the American Founders

Goodman, Rob, and Jimmy Soni. *Rome's Last Citizen: The Life and Legacy of Cato, Mortal Enemy of Caesar.* New York, NY: Saint Martin's Griffin, 2014.

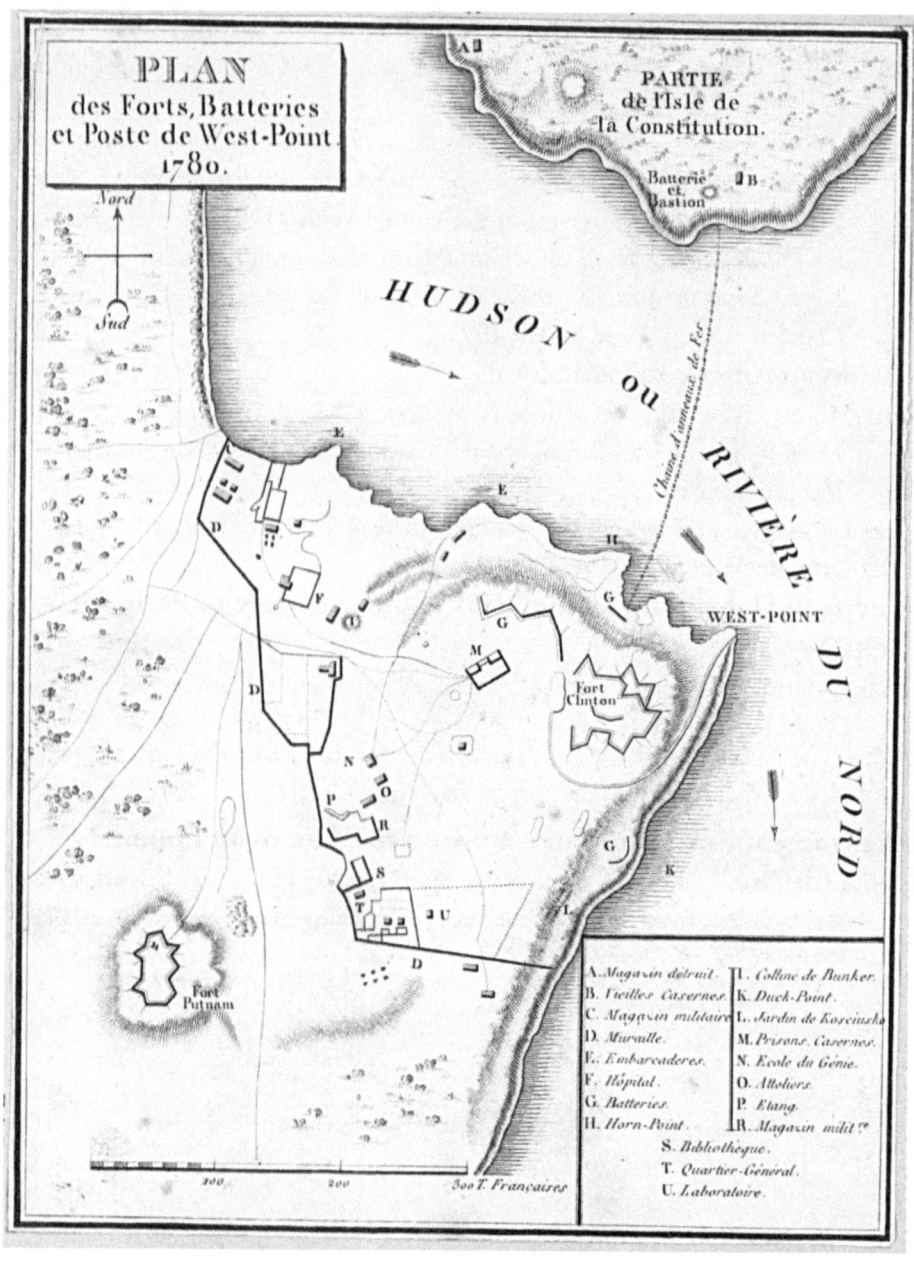

Figure 8—Map of West Point, New York, 1780[49]

---

[49] Lionel Pincus and Princess Firyal Map Division, New York Public Library, "Plan des forts, batteries et poste de West-Point, 1780," *The New York Public Library Digital Collections.* As noted by Webb, the presence of an *École du génie* implies the existence of a military engineering school that predates the establishment of the USMA.

Chapter II: Partridge at West Point

It is unfortunate for the memory of Captain Alden Partridge that he found himself embattled in the raging disputes over the design of the U.S. Army and the purpose of the United States Military Academy (USMA). Seeking to establish a uniquely American vision for education, he found himself caught between a hostile academic staff trying to convert the USMA into a strictly civilian academy and American presidents who sought to import the French model of military education and professionalism. This chapter will provide a brief history of Partridge's roughly twelve years of service at the USMA from 1805 until his relief as superintendent in July 1817. The numerous conflicts within the Academy during its early days will become readily apparent, along with failures in leadership to provide the superintendent position with sufficient authority to establish a clear vision for the institution. My approach in the next three chapters may vary from traditional histories as I present some events out of chronological order as I use a thematic approach. Like Plutarch's *Lives*, I am seeking to weave a larger narrative of events and display the conflicts at the Academy.[50] Other historians have provided more linear

---

[50] "It being my purpose to write the lives of Alexander the king, and of Cæsar, by whom Pompey was destroyed, the multitude of their great actions affords so large a field that I were to blame if I should not by way of apology forewarn my reader that I have chosen rather to epitomize the most celebrated parts of their story, than to insist at large on every particular circumstance of it. It must be borne in mind that my design is not to write histories, but lives. And the most glorious exploits do not always furnish us with the clearest discoveries of virtue or vice in men; sometimes a matter of less moment, an expression or a jest, informs us better of their characters and inclinations, than the most famous sieges, the greatest armaments, or the bloodiest battles whatsoever. Therefore as portrait-painters are more exact in the lines and features of the face, in which the character is seen, than in the other parts of the body, so I must be allowed to give my more particular attention to the marks and indications of the souls of men, and while I endeavor by these to portray their lives, may be free to leave more weighty matters and great battles to be treated of by others." Plutarch, *Plutarch's Lives*, vol. IV, ed. Arthur Hugh Clough, trans. John Dryden (Boston, MA: Little, Brown & Co., 1859), 159.

accounts; however, new research is likely needed to fully tell the history of West Point. That project I leave for other scholars.

Establishing a Military Academy

In the early years of the American Republic, there was a philosophical and political struggle to design an educational model sufficient to "provide for the common defense"[51] of the young Republic that was congruent with the Nation's founding principles. This conflict arose from the need to train and educate military leaders while avoiding a standing army that the Founders feared would threaten the young Republic's liberty. To further complicate this problem, the U.S. Constitution does not provide Congress with an enumerated power to create or maintain a military academy. While serving as vice president, Thomas Jefferson would note, "It was proposed to recommend the establishment of a military academy. I objected that none of the specified powers given by the constn. to Congress would authorize this."[52] Lacking private institutions that could provide instruction in military engineering, government leaders were pressed into action to fill the void in this critical field for national defense.

While President Jefferson did eventually establish the United States Military Academy (USMA) at West Point on March 16, 1802, the true purpose and vision for this institution remained largely in question. This left the USMA in the words of Major Jonathan Williams, the first superintendent of West Point, "A foundling barely existing among the mountains and nurtured at a distance and out of sight, and almost unknown to its legitimate parents."[53] The conflicts in purpose, vision, size, and scope of the academy would deeply shape Partridge's experiences at the USMA.

In retrospect, it is interesting to note Jefferson's original objection to the establishment of a military academy was in the lack of a constitutional authority to do so. It is safe to assume that the intent of the USMA was specific to training land forces, as there was no provision to provide training nor commissions for the U.S. Navy. The constitutional objection concerns Article I, Section 8, Clause 12, which provides Congress with the authority to "raise and support Armies" but was not originally intended to maintain armies outside of times of military conflict. Therefore, Congress was not to maintain a military academy in times of peace. Article I, Section 8, Clause 16, gave

---

[51] U.S. Constitution, art. 1., sec. 8, cl. 1.

[52] Thomas Jefferson, Notes of Cabinet Meeting on the President's Address to Congress, 23 November 1793, National Archives.

[53] Jonathan Williams, as quoted by Elizabeth Dey Jenkinson Waugh, *West Point: The Story of the United States Military Academy Which Rising from the Revolutionary Forces Has Taught American Soldiers the Art of Victory* (New York, NY: McMillian Co., 1944), 46.

Congress the authority to prescribe the discipline of the militia, but the states were responsible for providing this training. Again, we see no allowance for a federal military academy. Apparently, the training of Naval officers of the era was of less concern, as Clause 13 "to provide and maintain a Navy" would have justified the creation of a naval academy.

A naval school (United States Naval Academy/Annapolis Academy) would not be founded until 1845, 20 years after Captain Alden Partridge moved his American Literary, Scientific and Military Academy to Middletown, Connecticut, to offer instruction in naval science.[54] An interesting constitutional solution that Jefferson could have used would have been to establish a naval academy that could also serve a limited role in educating militia officers. Academy-educated militia officers could then be called to active service in the U.S. Army in times of military conflicts.

In his State of the Union address of December 2, 1806, Jefferson would propose a constitutional amendment that would have provided the required authority for Congress to build a "national establishment for education" along with the authority to engage in infrastructure projects to include "roads, rivers, canals, and such objects of public improvement."[55] However, Congress never acted on the suggestion. Therefore, in accordance with the 10th Amendment, the authority over education is "reserved to the States respectively, or to the people."[56]

The opinion that the states owned the domain of education remained well into the 1850s, with this being cited as a major cause of the rejection of Justin Morrill's attempt for a land-grant act in 1857 (further discussed in a later chapter). This lack of authority to maintain West Point has never been resolved. To this day, no enumerated authority has been granted to Congress to create or maintain the United States Military Academy.

An Overview of Partridge's Time at the Academy

Having first attended Dartmouth College from 1802 to 1805, Partridge entered cadet status at the USMA on December 14, 1805, and graduated in the

---

[54] As of 2023, the U.S. Naval Academy Museum omits the existence of Partridge's ALMSA at Middleton, CT, reporting on a display that "Some temporary schools, on board receiving ships and at the Philadelphia Naval Asylum, did prepare some midshipmen to take the examination for lieutenant, but the increasing application of steam power and other new technologies to the Navy, called for improved training of young officer." This oversight is rather remarkable given that Gideon Welles, the U.S. secretary of the navy during the American Civil War, was educated at Partridge's Middletown academy.

[55] Thomas Jefferson, "December 2, 1806: Sixth Annual Message," Miller Center.

[56] U.S. Constitution, amend. X.

following October. He was commissioned first lieutenant in the Corps of Engineers on October 30, 1806.[57] On November 4, 1806, he began duties as the assistant professor of mathematics at the USMA. Less than four years later, on July 23, 1810, Partridge would receive a promotion to the rank of captain.[58] In April 1812, he would assume the duty as professor of mathematics before transitioning to professor of engineering in April 1813. In the superintendent's absence, Partridge fulfilled the duties of this position. Partridge eventually would be assigned to the superintendency of the USMA when it was established as an independent position, separated from the chief of engineers. As recalled by Barnard,

> In 1808, Prof. Partridge was called to act in place of the Superintendent in the absence of Col. Williams, and continued to do so, with brief intervals, until January, 1815, when he was appointed to the office which he filled till March, 1816. [...] In 1808, Capt. Partridge was ordered by Col. Williams to take charge of the internal direction and control of the Military Academy as Superintendent, which post he held until November 25th, 1816.[59]

Army historian George Cullum would downplay Partridge's contributions by reporting he was "in command during the absence of the Superintendent, Jan. 3, 1815, to Nov. 25, 1816 and from Jan. 13, 1817, to July 28, 1817."[60] However, the assignment of the USMA superintendent position separate from the chief engineer was done at President Monroe's direction. Partridge was the first to hold this independent position. Partridge's influence at the USMA would effectively end on July 28, 1817, upon the unexpected relief by then Brevet Major Sylvanus Thayer.

Conflicts of Purpose and Vision for the USMA

There was no clear vision of the purpose of the USMA when it was established by Thomas Jefferson. Colonel Jonathan Williams, Colonel Decius Wadsworth, and Brevet Brigadier General Joseph Gardner Swift—the previous superintendents—failed to provide clear leadership in this area. In the words of Pratt concerning Partridge and his environment; "With the political stage set and full of controversy, it is not difficult to understand how stories of poor

---

[57] George W. Cullum, *Biographical Register of the Officers and Graduates of the U.S. Military Academy, from 1802 to 1867*, rev. ed., vol. I (New York, NY: J. Miller, 1879), 98.

[58] Cullum, *Biographical Register*, I:98.

[59] Henry Barnard, Military Schools and Courses of Instruction in the Science and Art of War, rev. ed. (New York, NY: E. Steiger, 1872), 834-35.

[60] Cullum, I:99.

discipline and administrative abilities overshadowed the contributions of this military leader."[61]

Partridge forged a clear intent for the USMA, but unfortunately it clashed with the vision of a purely civilian "national university" (possibly in the model of École Polytechnique of Paris),[62] that the academic staff wished to establish.[63] This opposing vision for the academy would have the civilian faculty in charge of all student activity with military instruction only provided after graduation (as seen in the French model at Metz and Fontainebleau) with limited graduates entering military service if they either had, "extraordinary endowments either natural or acquired, which are calculated to render such individuals useful to the public" or "exercises in arms, which though mechanical & indicative bodily agility, rather than mental capacity, may be considered as useful."[64]

Hostile staff, including professor Andrew Ellicott, professor Jared Mansfield, and Brevet Captain David B. Douglass, wishing to redefine the purpose of the academy, would work tirelessly and underhandedly against Partridge's goals and vision. Partridge lacked the authority in his position and support of his superiors to enable a clear path forward for the Academy. Many of the problems faced by Partridge at the Academy were only corrected once Brevet Major Sylvanus Thayer took command.

According to Lester Webb, a cabal formed against Partridge, led by Professor Jared Mansfield with the support of Andrew Ellicott, Captain David B. Douglass, and others.[65] There is some indication the cabal was known

---

[61] Richard A. Pratt, "Essence of Military Education: Contributions of Captain Alden Partridge to the United States Military Academy, 1806–1817," 1997, 2, Eisenhower Leader Development Program Papers, United States Military Academy Library.
[62] The École Polytechnique has gone through several reforms. For example, the reforms of 1816 removed many of the military features of this academy. With the academy focused on scientific instruction from 1817 to 1822, students wore a "Bourgeois uniform" (business attire) complete with a top cap. But even in this period, most of the students that graduated from this institution would matriculate into military service. Ambroise Fourcy, *Histoire de l'École Polytechnique* (Paris: L'Auteur, 1828). Further research is required concerning the West Point faculties' references towards the École Polytechnique to determine if they knew of the 1816 reform or if they were referring to the academy when it retained more of its military features. A comparison between the French model for military education (including the École Polytechnique) and the *American System of Education* is planned for Volume II of this series.
[63] Webb, *Captain Alden Partridge*, 60.
[64] Jared Mansfield et al. to Board of Visitors, November 30, 1815, National Archives, as quoted by Webb, 65.
[65] Webb.

outside of the boundaries at West Point with the editor of the *New York Columbian* writing, on September 2, 1817, "We hope that cabal be discontinued at Washington, and the military principle and honor be cherished. This school is too valuable to be sacrificed to intrigue."[66] Historians hostile towards Partridge, such as Waugh, even confirm the existence of a cabal, reporting that "Soon the other members of the faculty formed a cabal against him."[67]

Because of conflicts in the vision and purpose of the Academy, Strobridge reports, "[Partridge] both disliked and distrusted his faculty, which reciprocated in kind."[68] With the faculty not restricted to communication within the chain of command, letters were frequently dispatched directly to the president, the secretary of war, and other influential government officials, often falsely reporting or exaggerating deficiencies and misconduct at the Academy.

Hoping to correct the situation at the USMA and clarify the authority and position of the superintendent, Partridge called on James Monroe, then secretary of war, in January 1815, seeking the approval of an eight-point regulation for the Academy. Partridge's efforts met with success. He returned to the Academy with a signed regulation calling for:

> 1st. A permanent superintendent shall be appointed to the Military Academy who under the direction of the Secretary of War shall have the exclusive control of the Institution and of those connected with it, and will be held responsible for the conduct and Progress of it.
> 2nd. The Commandant of the corps of Engineers is to be the inspector of the Institution. He will accordingly visit it as often as the Secretary of War shall direct, examine into the management and the progress of it, and report to the Department of War accordingly. He will also propose, such alterations, and improvements in the general organization of it as he, and the superintendant may deem necessary.[69]

Monroe not only approved the requested regulations, he also formally appointed Partridge to the superintendent position.

The tool that Partridge needed to correct the situation at West Point was short-lived. Ellicott and Mansfield protested the new regulation to Monroe, causing him to suspend the regulations a mere 19 days later, until he could consult with Brigadier General Swift.

---

[66] *New York Columbian*, September 2, 1817, as quoted by Webb, 123.

[67] Waugh, *West Point*, 53.

[68] Truman R. Strobridge, "West Point, Thayer & Partridge," *Military Review*, October 1989, 80.

[69] Alden Partridge, Regulations for the Military Academy at West Point, 3 January 1815, Norwich University Archives and Special Collections.

On February 2, 1815, Swift would write to Ellicott expressing strong disapproval for Partridge's actions, stating, "I did wish to make things agreeable to Capt. Partridge, I now care not for his opinion or convenience."[70] Historians vary on their opinion of Swift's knowledge and consent for Partridge to advance these regulations for approval. Some historians assert these regulations were a rogue action by the young captain. Others, like Webb, assert Swift was aware of and previously approved the regulation before they raised conflicts with the academic staff.

In either case, it is interesting to note that Partridge only comes to knowledge of Swift's disapproval through a letter from Major Albert of the adjutant general's office.[71] I find this very telling of the quality of Swift's leadership at the USMA. Not only did Swift absent himself from the rightful duty assignment of the chief of engineers, he did so knowing of the continual conflicts between the faculty and staff of the institution. When his acting superintendent, Captain Alden Partridge, attempted to advance regulation to establish unity of command, Swift disallowed this needed correction. In doing so, Swift didn't even take the time to communicate with his direct subordinate who was acting in his stead. Furthermore, Swift would take no meaningful action to resolve the ongoing conflict.

On February 6, 1815, Partridge wrote an apology to Swift, expressing the details of the regulations and Partridge's intent in advancing them. Partridge stated,

> I acknowledge, Sir, to have submitted to the Secretary of War for his approbation all the Articles of those Regulations except the 8th, and in doing this I can assure you, Sir, I have no other object in view than the establishment of a permanent code of Rules for the Academy, which I was convinced would be highly beneficial to it, and which I had not the least idea would be displeasing to you. I presume, Sir, you will recollect the frequent discussions I had with you of the first two Articles. I clearly understood them to meet your approbation and I do not recollect that you at any time urged an objection against them. I remember, Sir, you told me it would be very agreeable to you to stand in the place of the Secretary of War so far as relates to the Academy, and wished my opinion upon the Subject. [...] In short, Sir, you so explicitly declared to me your idea of the necessity of a permanent

---

[70] Joseph G. Swift to Andrew Ellicott, February 2, 1815, National Archives, as quoted by Webb, *Captain Alden Partridge*, 57.

[71] Swift to Ellicott, February 2, 1815, as quoted by Webb, 57.

Superintendent, who should have the immediate control of the Institution that I had not a doubt upon the Subject.[72]

The letter explained that the two additional rules concerned the quartermaster supplying the required materials for the institution and the reception and safe holding of money for young cadets. (This last point was a function that Swift had directed Partridge to complete in previous correspondence.) The remaining articles were all practices that General John Armstrong established when he held the position of secretary of war.

Partridge was so confident that the change in regulation would have a positive impact on the Academy, he was willing to wager two years of his pay, emoluments, and reputation on its success. Partridge would write,

> If, Sir, the Secretary of War or yourself wish any further test of my sincerity or confidence in the Regulations I have proposed, I am willing to undertake the discharge of my duties as Superintendent for one or two years, (but would prefer two) upon a Suspension of Pay and Emoluments, (it being understood that Regulations shall be strictly observed by all concerned during the time), and if the Institution within that period is not evidently benefitted by them, and does not make more rapid progress towards being what it ought to be, than it has ever yet done, my Pay, Emoluments, and Reputation, shall be the forfeiture.[73]

I know few Army officers that would willingly propose or accept such a bargain. However, Partridge was denied the opportunity to demonstrate the wisdom of his plan. History does not record a response from Swift and it is unknown if he ever communicated his specific objection to Partridge.[74]

It could be assumed that the objection to the new regulations surrounded the efforts to unify command under the superintendency. While Partridge retained his new position as superintendent, the efforts to resolve the conflicts in leadership at the USMA were lost. Both Swift and Monroe knew of the ongoing turmoil at the Academy but appeared to do little to resolve the heart of the conflict.

By the spring of 1815, serious efforts outside the Academy were underway to replace Captain Partridge as superintendent of the USMA. Many historians claim that conflict surrounding the advancement of new regulations caused Partridge to lose Swift's favor. Swift desired to make changes to the

---

[72] Alden Partridge to General Swift, February 6, 1815, National Archives, as quoted by Webb, 57.

[73] Partridge to Swift, February 6, 1815, as quoted by Webb, 57.

[74] Partridge to Swift, February 6, 1815, as quoted by Webb, 57.

USMA, possibly seeing Partridge as a barrier to his vision. Swift played a critical role in ensuring funding and support for Brevet Major Sylvanus Thayer and Lieutenant Colonel William McRee to travel to France in the summer of 1815. This well-funded excursion intended to inspect fortifications; gain greater knowledge of the functioning of French military academies; and purchase books, maps, and other materials for the USMA library.[75] With the poor performance of the American military in the War of 1812, the American presidents James Madison and James Monroe intended to import the Napoleonic model of professionalism to strengthen the national defense.[76] With such a significant investment in time and resources, it would be hard to believe that Swift didn't intend for McRee and Thayer to return to the USMA to share the knowledge they learned upon this voyage. Webb suggested Swift intended McRee to assume the position of superintendent and it was only upon the direction of President Monroe that Thayer was assigned to this position.[77] This left Partridge in the awkward position of trying to maintain an academy with a hostile staff and multiple parties searching for a valid excuse to replace Partridge despite his loyal service to the Academy.

The Plot Enacted

The "strange academic faculty" would take many actions from 1815 to 1817 to disrupt Partridge's plans and destabilize the academy. While Webb provides a detailed record covering all the hostile actions of the staff, for the purpose of this research it suffices to provide a few major examples of these events. In October 1815, Partridge took leave and Captain Douglass assumed command and temporary superintendency of the Academy. Captain Douglass, under the believed influence of Ellicott, immediately issued a "Circular" inviting a discussion on changes to the curriculum and management of the USMA.[78]

One key point of discussion was the greater involvement of the academic staff in management of internal affairs. Webb would suggest the primary purpose of this document and meeting was to sow discord among the cadets, causing disarray at the institution. Upon Partridge's return, any imprudence in Partridge's attempts to restore order could be presented by the

---

[75] Matthew Moten, *The Delafield Commission and the American Military Profession* (College Station, TX: Texas A&M University Press, 2000).

[76] Jonathan M. Romaneski, "Importing Napoleon: Engineering the American Military Nation, 1814–1821" (PhD diss., The Ohio State University, 2017).

[77] Webb, *Captain Alden Partridge.*

[78] David B. Douglass, "Circular," October 30, 1815, as quoted by Webb, *Captain Alden Partridge.*

cabal to the board of visitors as evidence of Partridge's unfitness and the failures of the current structure and design of the Academy.[79]

In December 1815, Mansfield would bring the conflict between the staff and Partridge fully out in the open. Mansfield prepared a memorial[80]— without Partridge's knowledge—that was read to the board of visitors. With the board of visitors composed of powerful citizens and politicians (Connecticut governor Oliver Wolcott and New York mayor DeWitt Clinton), a civilian scientist (Dr. Samuel Mitchill), and General Jacob Brown, the "strange academic staff" was clearly making their play to advance the transformation of the USMA into a civilian national university. Within this memorial, a plan was advanced to provide faculty near total control of the students, with authority to direct "the management of every thing relative to the advancement of their pupils in Science."[81] The memorial did not specifically name Partridge as an obstruction to progress, but it claimed the military system advanced by Partridge as being "obstacles, which have resisted improvement, & prevented the approximation of perfection of which the Academy is susceptible."[82]

With multiple letters of complaint flowing in from the academic staff from the summer of 1815, Secretary of War William H. Crawford would task Captain Samuel Perkins, the West Point quartermaster, to investigate the claims. Perkins, collecting complaints from cadets, faculty, and staff, would forward a report to Crawford at the end of December 1815.[83] The charges listed in the March 1816 court of inquiry would stem from this report.

On March 16, 1816, Partridge found himself the focus of a court of inquiry that met with the purpose of "inquir[ing] into the conduct of Capt A Partridge of the Corps of Engineers as Commanding Officer at West Point and Superintendent of the Military Academy."[84] The court would ultimately investigate Partridge's conduct regarding:

---

[79] Webb.

[80] The term *memorial* refers to a written petition and/or statement of facts. While this term is still used in the legal system, an equivalent present-day term with wider usage would be *memorandum* (often shortened to *memo*).

[81] Mansfield et al. to Board of Visitors, November 30, 1815, as quoted by Webb, *Captain Alden Partridge*, 64–65.

[82] Mansfield et al. to Board of Visitors, as quoted by Webb, 63–64.

[83] Samuel Perkins to William H. Crawford, December 28, 1815, National Archives, as quoted by Webb, 70.

[84] Daniel Parker, General Order Appointing a Court of Inquiry, 19 February 1816, Norwich University Archives and Special Collections.

[A]buse of cadets, including corporal punishment and ill-treatment; the administration of public lands so as to invite fraud; nepotism in conduct of services for the cadets; granting monopolies with respect to stores, tailoring and messing cadets, with a personal pecuniary interest involved in the monopolies; and the sale of public property for his benefit.[85]

The allegations quickly deteriorated when members of the cabal rapidly shifted their opinions on the stand to support Partridge's conduct in disciplining the cadets. The court would ultimately state the charge of "unjust punishment of Cadets [...] are not sustained by a Shadow of evidence..."[86]

While the court disapproved of some methods employed by Partridge, they found him consistent and patient in his approach to cadet discipline. Swift testified he authorized the stores, tailor, and related services on post, thus removing any fault from Captain Partridge. Swift also testified that he did not believe there was any misconduct in the cutting wood/selling of fuel on public lands. It was also discovered that Partridge allowed soldiers to cut and sell additional wood in order to pay them for this fatigue duty. While the court disapproved of this action, there was no evidence Partridge took this action with the intent of defrauding the government. One slight instance of poor accounting was noted within Partridge's actions related to receiving money after he had left the position of assistant military agent.

The court of inquiry discovered multiple incidents of misconduct by Captain David Douglass and other members of the academic staff. The official record states, "it appears Discontents have been *Encouraged*, and even Opposition to Legal Authority *Invited*, to gratify personal animosities."[87] Yet these individuals were allowed to remain at West Point. Partridge was exonerated in the court of inquiry of 1816. Captain Douglass, despite evidence of sowing discord upon the cadets and engaging in near-mutinous activities, was not court-martialed. With his misconduct going unpunished, the cabal persisted in their efforts to remove Partridge.

To add to the political turmoil of the period, in 1816 President Madison commissioned Simon Bernard, a former general of engineers under Napoleon, as a brigadier general (second only in the engineering corps to Swift). Claude Crozet, another French veteran (recommended by Partridge), was appointed professor of engineering at West Point in 1816. These events

---

[85] Webb, *Captain Alden Partridge*, 80.
[86] Court of Inquiry, March 16, as quoted by Webb, 80.
[87] Webb, 80.

caused significant discord within the Corps of Engineers as it marked a lack of trust and confidence in American engineers trained at West Point.

Distress over these appointments would eventually lead to the resignation of Brigadier General Swift in 1818 and Lieutenant Colonel McRee in 1819. With the presidents endorsing Francophilia as a means of military professionalism and the USMA academic staff endorsing a vision of a national university, Partridge had few allies in forging a unique *American System of Education*.[88] Thayer was thus selected to replace Partridge, likely because of his recent studies of the French model of military education.[89]

By 1817, Secretary of War William Crawford requested Brigadier General Swift replace Partridge as superintendent. At the time, Swift defended Partridge, writing, "his manner is by no means attractive, but he is an attentive and industrious officer and unless some charge of misconduct is made against him, I shall not disturb him."[90] It is true that of all the early superintendents, Partridge was the only one who wanted the position.[91] Swift likely hesitated to replace Partridge because it would be difficult to find a replacement who would find the position palatable. Sure enough, Brevet Major Sylvanus Thayer would request reassignment from the superintendent position on May 12, 1818, less than a year after assuming the position.[92]

Upon returning from an inspection of coastal fortifications, President Monroe visited the USMA on June 14, 1817. As Romaneski reports, before departing West Point, Monroe received from "Mansfield and his co-conspirators a formal petition against Partridge's leadership."[93] The academic staff would assert that Partridge was turning the Academy into a school for military drills. The report enraged Monroe. President Monroe soon demanded that Brigadier General Swift act against Partridge and called for a court-

---

[88] Romaneski, "Importing Napoleon."

[89] Romaneski.

[90] Swift, as quoted by Jacqueline S. Painter, ed., *The Trial of Captain Alden Partridge, Corps of Engineers: Proceedings of a General Court-Martial Convened at West Point in the State of New York, on Monday, 20th October 1817, Major General Winfield Scott, President* (Northfield, VT: Friends of the Norwich University Library, 1987), v.

[91] John K. Robertson, "Some Preliminary Insights into the Formative Years at the United States Military Academy," presented to the Eisenhower Fellows, West Point, NY, September 12, 1991, 6.

[92] Webb, *Captain Alden Partridge*, 164.

[93] Romaneski, "Importing Napoleon," 143.

martial.[94] Swift would petition the president to lower the authority of the court to that of a court of inquiry.

By July 1817, Partridge would release the cadets early from their summer encampment and arrest the Academy's faculty that were still present at the time in an attempt to restore control over the academy. As reported by Webb, the eight months following the court of inquiry did not change the behaviors of the academic staff that continued their cabal and intrigues. On July 24, 1817, Partridge arrested and confined Mansfield, Ellicott, Douglass, and Claudius Berard and submitted charges to the chief of engineers.[95]

The letter Partridge wrote to Swift after the arrests indicated Partridge believed he had Swift's support in taking these actions.[96] Brevet Major Thayer would arrive at the Academy towards the end of the month, relieving Partridge on July 28, 1817, and releasing the faculty from arrest shortly after. Colonel R. Ernest Dupuy, former West Point public relations officer, would enrich his account of the events with the addition of a significant degree of military drama, recounting Partridge as an "ungainly figure emerg[ing] from the salt-box house [to meet Thayer] [w]ith shambling step."[97] This is a rather odd description of one of America's most capable pedestrians that is not supported by other accounts of Partridge's gait. This and other historical exaggerations were likely added by Dupuy to disparage Partridge's memory within the U.S. Army.

Conclusion

The unclear vision for the USMA created an environment for significant infighting as various factions battled to implement their vision for the future of the institution. Captain Partridge had the misfortune to find himself attempting to retain the primary function of the USMA with a hostile staff wishing to convert it into a civilian university in the model of the École Polytechnique of Paris. While Partridge made inroads in the development of an American model for military professionalism, ultimately his vision clashed with the executive branch's desire to import the Napoleonic model. Partridge attempted to establish unity of command at the institution. Ultimately, he was blocked in these efforts, allowing for continued conflicts between Partridge

---

[94] In review of the historical events, Ambrose would assert, "By now the whole affair was taking on all the elements of a badly written farce." Stephen E. Ambrose, *Duty, Honor, Country: A History of West Point* (Baltimore, MD: Johns Hopkins University Press, 1999), 58.

[95] Webb, *Captain Alden Partridge.*

[96] Ambrose, *Duty, Honor, Country.*

[97] R. Ernest Dupuy, *The Story of West Point* (Washington, D.C.: The Infantry Journal, 1943), 91.

and the academic staff. While Swift had the authority to reject Partridge's methods of resolving the conflicts in the unity of command, failing to provide an alternate solution ultimately condemned Partridge to try his best to lead in a dysfunctional and hostile environment.

Figure 9—West Point Engraving by John Hill[98]

[98] Spencer Collection, The New York Public Library, "West Point," *The New York Public Library Digital Collections.*

Chapter III: Court of Inquiry, October 21, 1817

To highlight the absurd treatment of Captain Alden Partridge, I begin with the results of the court of inquiry of October 21, 1817. This court of inquiry immediately followed Partridge's court-martial and investigated Partridge's charges against the staff. It will become apparent that Partridge had good cause to arrest the faculty and staff for misconduct. This chapter will display that Partridge was indeed used as a scapegoat for the failures of the institution, as others found guilty of disobedience of orders received none of the disciplinary actions applied to Partridge. From the list of findings, it is clear the problems at the Academy were not merely personality conflicts with Captain Partridge, but also included numerous acts of criminal misconduct among the academic staff. As Major General Ernest Harmon would report in an essay for the Newcomen Society in North America, "Modern scholarship is reevaluating the whole situation, and it would seem that President Monroe was more interested in establishing Thayer as successor than in giving Partridge the justice he deserved."[99]

The same body of officers that presided over Partridge's court-martial would the following day compose the court of inquiry to hear Partridge's charges against the staff. The explicit purpose of the court of inquiry was to:

> Enquire into the nature and extent of the complaints and charges against Jared Mansfield, Andrew Ellicott Esqrs., Profs, & Claudius Berard Teacher of French, the Military Academy, & Jno. Snowden Military Storekeeper at West Point & generally to investigate the conduct of officers, cadets & agents of the Military school.[100]

Partridge historian Colonel Lester Webb would assert this approach assisted Partridge, as he had been acquitted of all the allegations made by the faculty and staff during his court-martial. By that time, Captain Partridge demonstrated himself as a man of integrity, while the faculty and staff had shown a propensity to falsehood.

---

[99] Ernest N. Harmon, *Norwich University, Its Founder and His Ideals* (New York, NY: Newcomen Society in North America, 1951), 11.

[100] Court of Inquiry, October 21, 1817, National Archives, as quoted by Webb, *Captain Alden Partridge*, 126.

Webb also noticed a dramatic shift in Brigadier General Swift's attitude towards Partridge. While Swift was rather hostile towards Partridge during the court-martial, during the court of inquiry, he would testify in support of Partridge. For example, Swift would testify that Ellicott and Mansfield, "previous to the month of July 1817, they did not afford to Captain Partridge the Superintendent of the Academy that support and cooperation which in my opinion should have been given."[101] Swift would further acknowledge that the faculty and staff formed a cabal in 1815 with the purpose of removing Partridge from the institution. When questioned about Partridge's hostility to the academic staff, Swift demonstrated support for his attitude, stating,

> It may be remarked here that from communications received from Capt Partridge I had concluded that the Professors were interfering with his authority very much. And I have required him to maintain a strict discipline at the Institution and that if it were necessary to arrest every Professor at it & I would support him in the measure.[102]

Additionally, Swift would also acknowledge his frequent statements of support to Partridge, believing he was the best qualified officer for the office of superintendent. Swift's statements supporting Partridge can be viewed as an admission of Swift's neglect of duty as a superior officer to resolve the ongoing conflicts at the Academy. It was clear from Swift's own testimony that he understood there were personality conflicts at the academy and understood the academic staff were actively colluding to remove an officer that Swift deemed best suited to lead the Academy.

Swift acknowledged that the individual tutorship of cadets under Partridge allowed cadets to advance their course of learning more rapidly than through normal instruction under the other professors. However, because of the offense the faculty were taking from this tutorship, Swift directed Partridge to end the practice. Swift would express his disapproval that Partridge continued to offer personalized instruction. However, Partridge's personalized instruction retained cadets that had personality conflicts with professors and ensured the success of their education when the student-to-teacher ratio was too great to meet the individual learning needs of the cadets. While this action may have offended the faculty, it ultimately benefited the institution.

While the court of inquiry did not investigate all the charges advanced by Partridge, the staff was found guilty of a multitude of charges. One of Partridge's primary accusers, Jared Mansfield, was found guilty of:

---

[101] Webb, 126.
[102] Webb, 128.

(1) disobedience of orders—highly reprehensible. (2) Finds him absent from duty from the date of his appointment in October, 1812, until spring, 1814, & Summer & part of Autumn, 1815, but had permission, except Autumn & claims to have had a furlough from the Secy of War which he lost. (3) Concerned in compiling a course of studies 1 Nov 1815, which course seems to have been deficient in the article of military instruction, but exonerate him from an intention to destroy the mil. Features of the Institution.

Another member of the faculty cabal, Claudius Berard, was found guilty of providing cadet Harvey Brown a certificate of proficiency after he had been dismissed by Swift and Partridge. This effectively allowed Brown and other cadets inappropriately provided with these documents to benefit through credentials from the Academy in which they did not rightfully graduate. Like Mansfield, Berard was also found guilty of disobedience of orders. The full report of the finding reading:

After full investigation & the maturest deliberation on the evidence, pronounce the following opinion: (1) Did neglect to attend the recitation of his class about the first of August, 1815—contrary to orders: Conduct highly reprehensible. (2) By giving certificates [proficiency] to Cadets [Harvey] Brown, Gough, Lagnel, his conduct well calculated to destroy the effect of the order for [their] dismissal. Highly reprehensible. (3) Conduct with respect to harsh language towards Cadets deserves the severest reprehension, because it was calculated to destroy in the Cadets the sense of pride and honor which is indispensable to the character of a soldier and a gentleman. (4) The charge concerning compilation of a course of studies for the Cadets, which course seems to have been deficient in the article of mil. Instruction, but exonerate him from the other part of the allegation, that by such omissions it was intended to destroy the military features of the Institution. (5) Disobeyed order of Commanding Officer August 31, 1817—Highly reprehensible. (6) As to accusation of having… 29th August, 1817, publicly accused the Cadets of mutiny in relation to the transactions of that and preceding pay, the Court find the fact as here stated & are of opinion, not that Prof Ellicott uttered a falsehood, as he is charged, but that he greatly erred in judgment & opinion in the accusation he made against the Cadets. (7) Guilty of profanity—highly reprehensible.[103]

---

[103] Webb, 128.

This confirms that the suspicions that Partridge was reporting to Monroe in 1816 were correct.[104] As Partridge reported to Crawford,

> Before Capt Perkins, Capt Douglass, & Mr. Ellicott came to the Institution, all the internal affairs of the Academy went on smoothly and harmoniously. Since that time it has been a scene of boiling and dissension. These effects must have had a cause, and thus it appears did not exist previous to their arrival.

Mr. Ellicott, believed leader of the cabal, clearly had a different vision of the Academy and was willing to enlist the support of others to bring his vision to life. Not only did Partridge suffer the conflict caused by Ellicott, but similar conflicts would also continue under Thayer until Ellicott's death.

Likely, the most egregious crimes found were those of Captain Douglass, assistant professor of natural philosophy. His near-mutinous actions in sowing discord among the cadets were highly condemned by the court of inquiry. They found the "conduct on the part of Capt Douglass was well calculated to destroy the effects of that order & for this reason highly reprehensible."[105] But yet, the court of inquiry determined it had no authority to refer Captain Douglass to a court-martial. Failure to address this toxic officer would have detrimental effects, as he would continue to cause problems within the Academy, including physically and verbally assaulting cadets. The lack of action against Captain Douglass truly demonstrated the hypocrisy of the events at West Point. Ultimately, the court of inquiry did not fully investigate the crimes that had occurred, nor did they render equal justice under the law.

Thayer's testimony in support of the academic staff likely preserved their employment. According to Webb,

> Despite the Court's findings the scheming and intriguing Ellicott, Mansfield, Berard, Douglass, and others of their gang, were permitted to remain at West Point. There was no change in the "Strange Staff" of the Academy. Thayer took this bunch of academic misfits, after he testified to their wonderful abilities and possibilities, under his immature wing and suffered under their machinations as had Partridge.[106]

Thayer would soon come to regret his support for the "strange academic staff." It did not take long before Ellicott would continue his letter-writing

---

[104] Alden Partridge to James Monroe, July 26, 1816, National Archives, as quoted by Webb, 93.

[105] Court of Inquiry, October 21, 1817, National Archives, as quoted by Webb, 131.

[106] Webb, 131.

campaign about the faults of the Academy now aimed at the deficiencies of Thayer.

Partridge would face the humiliation of his court-martial finding being published in a major newspaper. Yet, the findings of the court of inquiry were not published. While Partridge faced the "court of public opinion," the inability to supply the public with the testimony within his court-martial and the materials related to the court of inquiry prevented the transparency needed to clear his public reputation. Given that the misconduct of the academic staff is rarely mentioned within historical accounts—and is certainly absent within official histories—the public is left with the assumption that the failures at the Academy were primarily caused by Partridge. Historical accounts like that of Robert P. Wettemann, Jr., even cite the relatives of the hostile academic staff to further scapegoat Partridge: "Thayer's arrival brought new life into the academy, for 1819 graduate Edward Mansfield [son of Jared Mansfield] later recalled, 'order took the place of disorder.'"[107] In the end, historical accounts that do not account for the misconduct of the academic staff for the failures and disharmony in the early days of the USMA are engaging in a lie of omission (if this information is known) or incomplete research (if unknown). I will challenge the reader to evaluate what occurred to the faculty as a result of the inquiry in relation to the treatment of Partridge as we move forward to discuss the court-martial.

---

[107] Robert P. Wettemann, Jr., *Privilege vs. Equality: Civil-Military Relations in the Jacksonian Era, 1815–1845* (Santa Barbara, CA: Praeger Security International, 2009), 54.

Figure 10—Battle of New Orleans by Charles Haven Hunt[108]

[108] The Miriam and Ira D. Wallach Division of Art, Prints and Photographs: Print Collection, The New York Public Library, "The Battle of New Orleans," *The New York Public Library Digital Collections.* The Battle of New Orleans demonstrated that a force primarily composed of American militiamen could carry the day against a larger professional European force.

Chapter IV: The Mutiny That Wasn't

Captain A. Partridge
U.S. Engineers
West Point

Portland, 17th July 1817

Dear Sir:

I have received your letter of the 29th ultimate and have reflected much upon the subject of the complaint and accusations which were made to the president at West Point. Very mature reflection has induced him to believe that in justice to the academy and to all those who are connected with it, a court-martial should be instituted to investigate and finally determine upon the whole subject. In this opinion I concur, because I do not believe that characters can stand upon their deserved ground at the Point, in the public estimation, without such an investigation. Measures will therefore be taken to accomplish this object.

You may rest assured that the president does not entertain any opinion derogatory to your character, and that he believes the course pointed out to be the [best] to be pursued. In relation to the existing regulations, no change will take place until the president returns to Washington. Therefore, as <u>soon</u> as you receive this, let the vacation commence.

By the order of the secretary of war, Major Thayer will be sent to West Point to take command of the post and superintendence of the academy. On his arrival I wish you to show him everything and to give him every advice that may enable him to discharge his duty at that place. With respect to yourself, I wish you to retire upon such duty and to such place as may be agreeable to you until this investigation is over.

Upon this subject write me immediately and direct your letter to the care of Colonel Walbach, Portsmouth, N.H. The president and his private secretary leave there tomorrow for Lake Champlain. Comm. Bainbridge and myself go to the Cadtine from thence return hither, and so on to Portsmouth, etc.

I am and shall remain your
Friend and humble servant,
J.G. Swift
Brigadier General[109]

Like the two-faced Roman god Janus, Captain Partridge's educational theories look towards the past to draw inspiration and wisdom while looking forward to the challenges and needs of tomorrow. However, it is also apparent to anyone who has examined the historical records of Partridge that he has "two faces" in another sense. Depending on the bias of the source, you may

---

[109] Letter from Swift to Partridge, July 17, 1817, as quoted by Painter, *Trial of Captain Alden Partridge*, 30–31.

find Partridge recorded as the height of military soldiery or an incompetent drillmaster. For example, the *New York Military Magazine* in 1841 claimed, "[N]o one is more eloquent as a soldier than Capt. Partridge, nor is there a better soldier in the nation."[110] In 1955, R. Ernest Dupuy would refer to Partridge as a "tin soldier," implying that while he could look the part, Partridge was incapable of performing his military duties.[111]

For over 200 years, various parties have dueled over the reputation of Captain Partridge. Hundreds, if not thousands, of conflicting accounts about Partridge exist within the historical record. The legacy of Alden Partridge has come to signify the very real internal conflicts between the use of the militia versus a standing army as the primary land force of the United States. The memory of Captain Partridge has real impacts on the Nation's understanding of the purpose, or lack thereof, of the United States Military Academy (USMA) at West Point. For many, the opinion of Captain Partridge will hinge on his response to an order from Brigadier General Swift displayed at the opening of this chapter (to be further discussed). It is unfortunate for the memory of Captain Alden Partridge that he found himself embattled within the raging disputes concerning the design of the U.S. Army and the purpose of the USMA.

This chapter will explore the events surrounding Partridge's court-martial. I do not intend to fully cover the history of this period, but to reveal the highlights and biases of previous scholarship. The goal is to present an accurate picture of events and recall some of the negative actions of Alden Partridge and others that have colored the historical record. While Sylvanus Thayer is now heralded as the "Father of the Military Academy," he demonstrated less than honorable actions in attempting to publicly shame Partridge by denying him the choice of housing despite established service traditions. This action triggered Partridge to assume command. Thayer's false report of forceful mutiny at West Point placed Partridge at risk of execution in the pursuing court-martial.

Moral Judgement

Those looking to make clear moral judgements about the characters involved in this chapter of history may find themselves at a loss. A detailed examination of all the players from Partridge to Monroe reveals that each man

---

[110] *New York Military Magazine: Devoted to the Interests of the Militia Throughout the Union* (United States: Labree and Stockton, 1841), 67.

[111] R. Ernest Dupuy, "Mutiny at West Point," *American Heritage* 7, no. 1 (December 1955).

may have taken actions that fell far from the ideal behavior of gentlemen. I would caution the audience not to seek simplistic answers in assigning fault. The reality of the situation is complex. The business of forging the Nation was messy. Present-day audiences should not seek to identify "who was to blame" but recognize the successes and shortfalls of all those involved as they struggled to advance what each thought was in the best interest of the Nation. Ultimately, it was the Nation that lost when these powerful men, including Partridge and Thayer, could not put aside personality conflicts to work together in advancing the interests of the young Republic.

The "Mutiny" at West Point

On August 29, 1817, Captain Alden Partridge returned to West Point via a steamboat. The cadets, seeing Partridge's return, cheered his arrival. Lieutenant Charles Davies would testify that he did "not think cheering emanated from any other desire than [the cadets'] wish to exhibit respect" and did not "see more confusion [or disorder] that was necessary to accomplish that object."[112] Partridge asked Lieutenant Davies to "express to the cadets his thanks for the friendship they had manifested towards him."[113]

Partridge went to Brevet Major Thayer on the 29th and 30th of August and requested lodging in his former quarters. According to military customs of the day, with Partridge being the highest-ranking officer within the Corps of Engineers present and West Point being an engineering post, Partridge would have been entitled to his choice of quarters. Thayer offered Partridge lodging in either quarters in a terrible state of repair (with one acceptable room in a damaged building), or in the new barracks (with the cadets). This would have allowed junior officers on post to receive better accommodations than Partridge.

In an age far more sensitive to honor than today, this would have been seen as gross misconduct. Denying a senior captain lodging for the sake of a brevet captain (Douglass) would have been unacceptable. Thayer's actions were nothing short of an attempt to publicly humiliate Partridge by denying him his due honors. In the words of Partridge, accepting lesser accommodations from Thayer would have been "worse than death—degradation by allowing my rights to be taken from me by a junior officer."[114]

---

112 Painter, *Trial of Captain Alden Partridge*, 30–31.

113 Painter, 11.

114 Alden Partridge to John C. Calhoun, March 4, 1818, as quoted by Webb, *Captain Alden Partridge*, 147.

Given Thayer knew Partridge was facing a court of inquiry at West Point and had legal rights to be present, mis-housing Partridge and attempting to interfere with his rights under the Articles of War (An early version of today's Uniform Code of Military Justice) would have been a violation of Article 83. Thayer's actions were "conduct unbecoming an officer and a gentleman."[115] Thayer should have been reprimanded and possibly dismissed from the Army.

Because of Thayer's repeated misconduct, Partridge, after making requests for appropriate lodging two days in a row, relieved Thayer and took command of the post with reluctance on August 30, 1817. Thayer would send word to the secretary of war,

> I have the honor to inform you that Captain A. Partridge of the Corps of Engineers has returned to this Post & has, this day, forcibly assumed the Command & Superintendence of the Military Academy. I shall therefore proceed to New York & wait your orders.[116]

Thayer falsely reported Partridge acting "forcibly," implying a mutiny. Within 48 hours, Brigadier General Swift's aide-de-camp would arrive at West Point to arrest Captain Partridge. A court-martial was convened the following October.

In Thayer's own testimony at Partridge's court-martial, he admitted he understood the reason that Partridge took command, with reluctance, with Partridge doing so <u>without force</u>. Thayer further admitted he believed Partridge had previously intended to restrict his actions at the academy and intended to sow no discord among the cadets as he awaited the court of inquiry. These courtroom admissions were significant as they demonstrated Thayer was engaging in *willful lying* in the initial report of Partridge's actions.[117] This action should have had Thayer prosecuted under Article 83. Due to Thayer's initial reports, Partridge was charged with "Mutiny, and beginning and exciting mutiny." Under the Articles of War,

> Article 7. Any officer or soldier who shall begin, excite, cause or join in any mutiny or sedition in any troop or company in the service of the United States, or in any party, post, detachment, or guard, shall suffer death, or such other punishment as by a court martial shall be inflicted.[118]

[115] U.S. Congress, *U.S. Statutes at Large*, vol. 2 (1799–1813), *6th through 12th Congress*, ed. Richard Peters (Boston: Charles C. Little and James Brown, 1845), 369, Law Library of Congress.

[116] Sylvanus Thayer to the Secretary of War, August 30, 1817, Military Academy Library as quoted by Webb, *Captain Alden Partridge*, 120.

[117] Webb.

[118] U.S. Congress, *U.S. Statutes at Large*, vol. 2 (1799–1813), 360.

Because of Thayer's initial false report, Partridge faced the possibility of execution as a result of the court-martial. Furthermore, if Thayer perceived Partridge taking charge of the USMA as mutiny, Thayer's actions (or lack thereof) should have carried the same penalty. As Thayer's action fell far short of the requirements under Article 8, that would have required him to do "his utmost endeavor to suppress"[119] the mutiny or face a court-martial with the potential punishment of execution. As Webb highlights, Partridge's accurate accounts of occurrences within the court-martial demonstrated his upright and honest nature, while the differences in statements of corroborating witnesses and Thayer's written and oral testimony demonstrated Thayer's tendency towards malice.[120]

Charges

Below is a list of charges to which Partridge would plead not guilty:

Charge 1: Neglect of duty and unofficerlike conduct.[121] Ten specifications provided.

Charge 2: Conduct unofficerlike and to the prejudice of good order and military discipline.[122] Four specifications provided.

Charge 3: Disobedience of orders.[123] Two specifications provided.

Charge 4: Mutiny, and beginning and exciting mutiny.[124] Four specifications provided.

Verdict

The results of the court-martial were published in the prominent American newspaper, the *National Intelligencer*, on November 29, 1817.[125] Partridge would view the release of incomplete court-martial records as a slight against his honor, as it did not explain the full story of the occurrences at the USMA and the misconduct of Thayer and the staff of West Point.

Captain Partridge was found guilty only of the charge of "disobedience of orders," specifically in assuming command at West Point and issuing

---

[119] U.S. Congress, *U.S. Statutes at Large*, vol. 2 (1799–1813), 360.

[120] Webb, *Captain Alden Partridge*.

[121] Painter, *Trial of Captain Alden Partridge*, 4.

[122] Painter, 6.

[123] Painter, 7.

[124] Painter, 7.

[125] R. H. Winder, "General Order," *National Intelligencer*, November 29, 1817, George Washington University Special Collections Research Center.

orders.[126] Partridge was found guilty of two of the four specifications but was ultimately acquitted of the charge of mutiny. Emphasized by Webb,

> It is important that the Court did not find Captain Partridge guilty of a single charge or specification, which the academic staff had for the past three years carried on cabal and intrigue in order to manufacture false and perjured evidence with which to convict him.[127]

This again returns us to the orders that appear at the beginning of this chapter. The famous historian Stephen E. Ambrose misrepresented Swift's order to Partridge as the following: "Swift talked to Partridge, telling him he could either choose any duty he wanted or take a leave until the court could be convened."[128] Swift's actual order read, "With respect to yourself, I wish you to retire upon such duty and to such place as may be agreeable to you until this investigation is closed."[129] A plain reading of Swift's order would imply the court of inquiry would occur *in absentia*. This would have violated Partridge's rights under Article 91 of the Articles of War. Partridge had the right to face his accusers, with neither Swift nor President Monroe holding the power to deny Partridge this right established by Congress. Article 91 reads as follows:

> In cases where the general or commanding officer may order a court of inquiry to examine into the nature of any transaction, accusation, or imputation against any officer or soldier, the said court shall consist of one or more officers, not exceeding three, and a judge advocate, or other suitable person as a recorder, to reduce the proceedings and evidence to writing, all of whom shall be sworn to the faithful performance of their duty. This court shall have the same power to summon witnesses as a court martial, and to examine them on oath. But they shall not give their opinion on the merits of the case, excepting they shall be thereto specially required. The parties accused shall also be permitted to cross examine and interrogate the witnesses, so as to investigate fully the circumstances in the question.[130]

Therefore, it could be asserted that Swift issued an illegal order. It is truly tragic that Partridge was ultimately punished for disobeying an order that he had no legal obligation to obey. This reinforces Webb's claim that Partridge would

---

[126] Interestingly "his commanding officer" was removed from both these specifications in the final verdict. Painter, *Trial of Captain Alden Partridge*, 100.

[127] Webb, *Captain Alden Partridge*, 140.

[128] Ambrose, *Duty, Honor, Country*, 58.

[129] Painter, *Trial of Captain Alden Partridge*, 31.

[130] U.S. Congress, *U.S. Statutes at Large*, vol. 2 (1799–1813), 370.

likely have been cleared of all charges if represented by a lawyer instead of representing himself.[131]

William Duane, a contemporary legal expert, gave the following opinion of Partridge's court-martial: "There can be no rational doubt to the illegality of the sentence, unless an agent under the law professes an authority to abuse the law and institute his will which would be a gross absurdity." Partridge's "disobedience" was a direct result of the conduct unbecoming an officer by Brevet Major Sylvanus Thayer in not allowing Partridge to return to his previous quarters at West Point per the established traditions of the U.S. Army.

> Interestingly, in his memoir, Swift reports,
> [O]n 28th August [1817], Captain Partridge had called and breakfasted with me in Brooklyn, and requested my authority to extend his leave so as to allow him to visit West Point for study. I declined any such consent, and said to him that such a movement would not only contravene the order of the President of the United States to me, but would also injure and defeat at once any purpose he might contemplate of restoration hereafter. The conversation dropped there, and I had not a thought that Captain Partridge would act in opposition to such a purpose on my part.[132]

Scholars like Romaneski treat this memoir account as fact. Yet, it is questionable if this interaction occurred, as no mention of this incident appears within the court-martial documents. It is hard to believe direct interactions between Partridge and Swift would not have been entered as further evidence of disobedience. Between Brigadier General Swift possibly inventing this anecdote and calling attention to Partridge's behaviors as a form of an ad hominem, one might conclude that Swift was attempting to mask his own misconduct within his memoirs. One might ask how Captain Partridge might be brought in front of a court of inquiry for issues such as damages to artillery pieces when Partridge had notified Brigadier General Swift of the situation without resolve. Should Swift not have been questioned about his inaction in this regard and other known problems at West Point? Is it not possible and likely that Swift had purposely misreported facts later in life to ensure Partridge remained the scapegoat for the Academy's and Swift's own failures?

---

[131] Webb, *Captain Alden Partridge.*

[132] Joseph G. Swift, *Memoirs of Gen. Joseph Gardner Swift: First Graduate of the United States Military Academy, West Point, Chief Engineer U.S.A. from 1812 to 1818, 1800–1865* (Worcester, MA: F. S. Blanchard & Co., 1890), 167.

Sentence

Ultimately, the court sentenced Captain Partridge to be "cashiered."[133] Cashiering was a disgraceful dismissal of military service. This sentence was the most lenient sentence the court-martial could assign for officers found in violation of Article 9 (disobedience to a lawful command of a superior officer). As stated in Article 89, the court did not have the authority to "pardon or mitigate" this level of sentence. Under Article 65 and Article 89, the court suspended the punishment "until the pleasure of the President of the United States could be known."[134,135]

The court-martial provided the following paragraph appealing to the president to reduce the sentence:

> The court, in consideration of the zeal and perseverance which the prisoner seems uniformly to have displayed in the discharge of his professional duties up to the period of August last, beg leave to recommend him to the clemency of the president of the United States in the hope that the punishment above awarded may be remitted.[136]

Reading both the simple sentence and supporting paragraph together, we find the court-martial entered the lowest sentence possible by law while appealing to the president to provide a pardon. This comprehensive reading of the sentence will be critical in further discussions of the official historical records of the USMA.

Court Reported Charges Frivolous and Vexatious

The trial judge advocate, R. H. Winder, Esquire, immediately discovered difficulties in the prosecution of the case. In a letter to Adjutant and Inspector General Daniel Parker, Winder reports,

> I found it necessary on conversations with Professor Mansfield to alter and realter and militate the charges framed from those prepared by him. I had been misled by his statement in the papers furnished me and other information. He was almost utterly unprepared with proofs of his charges. They have broken the effect of the principle charges— have occupied, embarrassed and perplexed me almost to the exclusion of any attention to the principle charges. The evidence turns out wretchedly. The Court is dissatisfied. The whole effect of the charges

---

[133] Painter, *Trial of Captain Alden Partridge*, 100.

[134] Painter, 369–70.

[135] U.S. Congress, *U.S. Statutes at Large*, vol. 2 (1799–1813), 367.

[136] U.S. Congress, *U.S. Statutes at Large*, vol. 2 (1799–1813), 367.

not connect with the recent transactions here is to the advantage of Capt. Partridge.[137]

The errors in the case were so significant that Winder did not present evidence for several specified charges.

On the charge of neglect of duty and unofficerlike conduct, the fifth specification was:

> In that being commanding officer and superintendent as aforesaid, he suffered the cadets to the Military Academy at West Point to trade and be trusted so largely at the store and tailoring establishment at West Point that at one period in the year 1816, the debts of only fifty of them at said store and establishment amounted to about seven thousand five hundred dollars.[138]

The fifth specification given up as "Captain Partridge was absent" during the period.[139] The sixth specification under this charge was:

> In that being superintendent of the Military Academy at West Point and commanding officer at the post during [the] greater part of the time between April 1816 and July 1817, he suffered, whilst superintendent and commanding officer in that period, the discipline, police and government of the academy and the discipline and police of the post to be and remain extremely bad.[140]

The sixth specification was abandoned due to "no evidence" being presented to the judge advocate.[141] The ninth specification under this charge was:

> In that being superintendent and commanding officer as aforesaid, he permitted the cadets to leave the academy at West Point at the commencement of the vacation in July 1817 without their "turning in" or depositing the public property in their possession in a place of safekeeping.[142]

The ninth specification was dropped "because Order Book showed cadets had been directed to turn in equipment."[143] Finally, the second charge was conduct

---

[137] R. H. Winder to Adjutant and Inspector General Daniel Parker, October 24, 1817, National Archives, as quoted by Webb, *Captain Alden Partridge*, 134–35.

[138] Painter, *Trial of Captain Alden Partridge*, 5.

[139] R. H. Winder to Adjutant and Inspector General Daniel Parker, November 13, 1817, National Archives, as quoted by Webb, *Captain Alden Partridge*, 135.

[140] Painter, *Trial of Captain Alden Partridge*, 5.

[141] Winder to Parker, November 13, 1817, as quoted by Webb, *Captain Alden Partridge*, 135.

[142] Painter, *Trial of Captain Alden Partridge*, 6.

[143] Winder to Parker, November 13, 1817, as quoted by Webb, *Captain Alden Partridge*, 135.

unofficerlike and to the prejudice of good order and military discipline. The second specification under this charge was:

> In that being commanding officer and superintendent as aforesaid, he permitted Cadets Walker, Donaldson, Wells, and Hunter to join the Military Academy at West Point upon new appointments on or about 21st of June 1817 in violation of the regulations prescribed for the academy.[144]

This specification "was given up because he could find no cadets to testify."[145] Ultimately, 4 of the 15 specifications (over 25 percent) related to charges associated were abandoned by the judge advocate.

The court ultimately found so many faults within the first two charges (the basis of the initial court of inquiry) that they reported the following with the verdict:

> The court cannot conclude this trial without pronouncing the further opinion that the two charges and most of the specifications under those charges appear to the court both frivolous and vexatious. The court, however, do not attach any censure to the judge advocate, whose name is ex officio attached to the charges and specifications.[146]

This clearly stated the court disapproved of the reasons to call for the initial court of inquiry and wanted to ensure these proceedings did not tarnish the reputation of the judge advocate, R. H. Winder.

As summarized by Cunliffe,

> [Partridge] was short of staff, and two members in particular—Jared Mansfield and Andrew Ellicott—seem to have been lazy and malevolent. It was never made clear what authority [Partridge] possessed in relation to Swift, to the staff and to the cadets. Swift appears to have been a moral coward, allowing the President, the Secretary of War, the West Point faculty and Alden Partridge to believe what he thought they wished to believe. At the court-martial, allegations of brutality and incompetence were all disproved. On the face of it the only thing wrong with Partridge was that some people did not like him, or had decided that they preferred Thayer. He was blamed for everything that seemed unsatisfactory at West Point.[147]

---

[144] Painter, *Trial of Captain Alden Partridge*, 6.

[145] Winder to Parker, November 13, 1817, as quoted by Webb, *Captain Alden Partridge*, 135.

[146] Painter, *Trial of Captain Alden Partridge*, 100.

[147] Marcus Cunliffe, *Soldiers & Civilians: The Martial Spirit in America 1775–1865* (Boston, MA: Little, Brown, and Company, 1968), 153.

It is interesting to ponder what course history may have taken if Thayer would have not engaged in misconduct, sparking Partridge's assumption of command and the resulting court-martial. How would the Army have treated Captain Partridge after being found not guilty at the court of inquiry? The ultimate goal was to relieve Partridge regardless of his interest in remaining and his personal sacrifices made for the academy. It may have been highly likely that Partridge would have been sent back to the Academy within a year because of Thayer's failures and lack of desire to maintain the superintendency of the Academy.

Major Faults of the Court-Martial Identified

The scholarship of Colonel Lester Webb is likely the most detailed on the faults and occurrences of Partridge's court-martial. Webb names numerous faults within the proceedings. Two of these faults will be highlighted below. The participation of Brevet Lieutenant Colonel Joseph G. Totten as a member of the court-martial is a significant fault. Webb also took umbrage to the court's lack of consideration for the "custom of the service" concerning the housing of superior officers.[148]

Brevet Lieutenant Colonel Joseph G. Totten

There are two major reasons Webb suggested that Brevet Lieutenant Colonel Totten should not have been allowed to sit as a member of the court-martial. First, Totten was the cousin of Jared Mansfield, one of Partridge's accusers in this case. This family relationship would have likely biased Totten's opinions of the case.[149] Second, Totten was junior in rank in the Corps of Engineers to Captain Alden Partridge, with Totten being commissioned as a captain of engineers two years after Partridge.[150]

Webb makes the argument that Congress' "An Act Fixing the Military Peace Establishment of the United States," passed on the third of March 1815,[151] should have removed brevet ranks within the Army, thus making Totten a junior captain to Partridge at the time of the court-martial (with Thayer in the same position). However, there was disagreement at the time if this act affected brevet ranks awarded under the authority of "An Act Making Further Provision for the Army of the United States, and for Other Purposes," approved on the sixth of July 1812. Both Totten and Thayer would be brevetted under the provisions of this act, although Webb would question if

---

[148] Webb, *Captain Alden Partridge*, 143.

[149] Webb, 143.

[150] Webb, 143.

[151] U.S. Congress, *U.S. Statutes at Large*, vol. 6: *Private Laws and Resolutions (1789–1845)*, ed. Richard Peters (Boston: Charles C. Little and James Brown, 1846), 224–26, Law Library of Congress.

Thayer met the requirements of "gallant action or meritorious conduct." In section 4 of this act:

> [T]he President is hereby authorized to confer brevet rank on such officers of the army as shall distinguish themselves by gallant actions or meritorious conduct, or who shall have served ten years in any one grade: Provided, that nothing herein contained shall be so construed as to entitle officers so brevetted to any additional pay or emoluments, except when commanding separate posts, districts or detachments, when they shall be entitled to, and receive the same pay and emoluments to which officers of the same grades are now or hereafter may be allowed by law.[152]

Ultimately, this debate over brevet ranks would continue well past the court-martial until they were definitively removed from the Army in March 1821.[153]

Partridge made the following written request to the court on October 23, 1817:

> Lieutenant Colonel Totten was challenged by the prisoner, Captain Alden Partridge of the Corps of Engineers, for the causes (1) because Lieutenant Colonel Totten was his junior captain in the Corps of Engineers; (2) because he was relation and friend of Professor Mansfield, whom he, Captain Partridge, considers inimical to him; that he wished it to be understood that in making these objections, he did not mean to question the honor and integrity of Colonel Totten; that he believed him to possess these qualities in as high a regard as any other officer of the army; and that he made the objections from a sense of duty.[154]

While the court ultimately overruled this objection, Brevet Lieutenant Colonel Totten should have recused himself without need of action from the court.

"Customs of the Service" Housing Superior Officers

In the proceedings of the court-martial, it was acknowledged that Partridge was the superior officer at West Point over Thayer. Thayer's brevet rank did not hold authority on the engineering post of West Point. Partridge had been promoted to captain of engineers years before Thayer (Date of rank: July 23, 1810[155] vs. October 13, 1812[156]). Yet while the court recognized

---

[152] U.S. Congress, *U.S. Statutes at Large*, vol. 2 (1799–1813), 785.

[153] James B. Fry, *The History and Legal Effect of Brevets in the Armies of Great Britain and the United States: From Their Origin in 1692 to the Present Time* (New York, NY: D. Van Nostrand, 1877).

[154] Painter, *Trial of Captain Alden Partridge*, 3.

[155] Cullum, *Biographical Register*, I:98.

[156] Cullum, I:107.

Partridge as the superior officer, it did not recognize the custom of the service of the highest-ranking officer having the privilege of selecting their preferred lodging. If Thayer had not disregarded this custom, Partridge would have never assumed command of the USMA, thus avoiding the charge and conviction of disobedience of orders. Webb suggests that because Partridge's misconduct was only driven by Thayer's misconduct, the court should have "called the matter a draw and found Partridge not guilty of disobedience."[157]

Presidential Clemency

On November 27, 1817, the adjutant and inspector general would record that the "[P]resident [James Monroe] is pleased to remit the punishment; and directs that Captain Alden Partridge be released from his arrest, and report for duty to the senior officer of engineers."[158] The president's only disagreement with the proceedings was the comment concerning the "first two charges as frivolous and vexatious." Monroe believed men of "fair character" brought the charges and thus allowed Partridge the ability to prove them groundless.[159] Upon release, Partridge requested a four-month leave that was approved by Swift.[160]

Some sources, noted by Webb to include West Point historians General George W. Cullum and R. Ernest Dupuy, indicated that Monroe only remitted the sentence to allow Partridge to resign without the disgrace of being cashiered. This claim is asserted without evidence. One possible source for this notion comes from Swift's memoirs, as he records,

> I called on the President in reference to the subject of Captain Partridge, and advised a remission of the sentence of the court, provided Captain Partridge would resign.[161]

However, there are questions as to the reliability of Swift's memoir. Given that Swift records other events that he did not present as evidence at the court-martial, it is questionable this passage was not also a fabrication made decades after the events to shape the public's opinion. As asserted by Webb, "This is pure fabrication of Swift, aided and abetted by the writers of the 'West Point' story indifferent to truth."[162] More recent scholarship by Romaneski would reenforce this myth by stating, "Swift convinced the president to allow Partridge to resign his commission in lieu of cashiering him," referencing page

---

157 Webb, *Captain Alden Partridge*, 143.

158 Painter, *Trial of Captain Alden Partridge*, 103.

159 Painter, 103.

160 Webb, *Captain Alden Partridge*, 145.

161 Swift, *Memoirs*, 168.

162 Webb, *Captain Alden Partridge*, 143.

103 of Painter. However, page 103 of Painter is D. Parker's report that President Monroe remitted punishment and directed Partridge to report to the senior officer of engineers for duty. There is no evidence within this document of any arrangement for a resignation instead of being cashiered.

There is an interesting question of why Partridge would be granted leave if the condition of his clemency was resignation. Why was he not expected to do so immediately? Partridge would prepare charges against Thayer and Swift and actively seek an audience with Calhoun during his leave, with the goal of being restored to his position as superintendent at the USMA. This is hardly the behavior of an individual who had accepted a deal of resigning in exchange for a remitted sentence in order to avoid the shame of being cashiered. Likewise, Webb found no evidence of any such agreement outside of Swift's memoirs.[163]

Partridge's Search for Justice

Captain Partridge would travel to Washington, D.C., to seek an audience with John C. Calhoun, the secretary of war, with the purpose of regaining the superintendency of the USMA. Partridge believed Thayer was only sent to West Point to command during the court of inquiry. With the court over and Partridge acquitted on the charges brought by the academic staff, Partridge believed it would be appropriate for him to assume his previous assignment. Partridge was denied an audience with Calhoun but left his request in writing.

Partridge offered sound reasons for his request. He had served at West Point for 12 years, advancing the interest and gaining detailed knowledge of the institution. Partridge had organized and reorganized the Academy. He had endured a hostile academic staff in which he was acquitted of their false claims in two court cases. Due to being short on academic staff, Partridge worked extra duty for four or five years instructing cadets with no additional compensation. The Army maintained Partridge at the USMA during the War of 1812, which "prevented [him from] sharing in any of the honors of the field." Ultimately, Partridge reported, "after having by my own individual exertion brought it to a perfect state of organization and discipline, to be superseded and another one brought in to receive the fruits of my labors, I must consider as very hard..."[164]

---

[163] Webb, 143.

[164] Alden Partridge to John C. Calhoun, December 9, 1817, National Archives, as quoted by Webb, 146.

Partridge's request for reassignment to the superintendency had merit. The USMA would have likely become a victim of the War of 1812 if it wasn't for the heroic efforts of Captain Partridge. The secretary of war under President Madison, William Eustis, had little interest in supporting the Academy.

During the war, every officer except Captain Partridge was reassigned to missions away from West Point.[165] The cadets were commissioned early to fill vacancies in the wartime army. Eustis would resign in December 1812. With limited appointments granted by Eustis, the Corps of Cadets numbered only six cadets at the end of 1812.[166] Partridge, faced with instructor shortages, would expand the subjects he taught to ensure the success of the Academy. During this war, Partridge had also assumed the role as Superintendent with the resignation of Colonel Jonathan Williams in 1812 and his replacement Brigadier General Joseph Swift away on wartime duties. "Alden Partridge," Morris highlights, "provided the only continuity in Academy governance from 1807 to 1817."[167] Anyone with an unbiased view would certainly recognize Partridge's meritorious service at the USMA in overcoming instructor and resource shortages while enduring a hostile staff. In Partridge's eyes, and many others, his laudable services at the USMA should have been allowed to continue.

After finding no satisfaction in D.C., Partridge returned to his home in Vermont. On December 13, 1817, Partridge would write to Calhoun to report Thayer's "violation of established regulations of the War Department relative to assignment of quarters; conduct unjust and arbitrary; and conduct unbecoming an officer and a gentleman."[168] On March 4, 1818, Partridge would send another letter reporting an additional allegation against Thayer of petty peculation said to have occurred in the summer of 1817.[169] On March 17, 1818, Partridge would write again, this time to report Brigadier General Swift for the charge of "extravagant and needless expenditure of public money in the construction of public buildings at West Point" and

---

[165] "On my arrival I found the officers had all been ordered from this post except myself; of course, I, am left entirely alone, without any assistant which renders my duty somewhat tedious." Alden Partridge, Letter from Alden Partridge to Samuel Partridge, 12 April 1813, p. 2, Norwich University Archives and Special Collections.

[166] Dupuy, "Mutiny at West Point."

[167] John F. Morris, "Crucibles of Virtue and Vice: The Acculturation of Transatlantic Army Officers, 1815–1945" (PhD diss., Columbia University, 2020), 146.

[168] Webb, *Captain Alden Partridge*, 146.

[169] Webb, 146.

unmilitary and ungentlemanlike conduct in urging Capt Alden
Partridge [...] to arrest several of the Instructors of the Military
Academy—pledging himself to support such measure should it be
deemed necessary but after the Instructors had been arrested, he
basely abandoned the cause, changed sides, and insidiously endeavored
to attach blame to Capt Partridge for his conduct.[170]

This last charge involved Swift's release of Brevet Captain Douglass in
violation of the assurances he gave to Captain Partridge.

Partridge was operating under the assumption that Thayer and Swift
would be arrested under Article 77 of the Articles of War and face trial.[171]
Certainly, Partridge had demonstrated himself to be an honorable man of "fair
character." If the president didn't deem the charges brought against Partridge
"frivolous and vexatious" because the accused brought "to trial has afforded
him an opportunity to prove the charges groundless,"[172] then the same
standard should have been applied to Thayer and Swift. Certainly, Calhoun had
direct access to evidence (Thayer's testimony in Partridge's court-martial) to
substantiate the justification for Thayer's trial. However, President Monroe and
Secretary Calhoun would choose to violate the law to protect Thayer, Swift,
and the USMA from further intrigue.

It became clear to Partridge, by Calhoun's lack of action against
Thayer and later Swift, that the U.S. Army was not enforcing a just standard in
the application of the law. On March 20, 1818, Calhoun would write, "After
maturely considering the 77th Article of the rules and articles of war, I am of
the opinion that the rule is not imperative."[173] Partridge would challenge
Calhoun on his authority to override laws established by Congress. Webb
asserts that Calhoun viewed the charges against Thayer and Swift "as an
outgrowth of the clash between him and Thayer in August, 1817."[174]

---

[170] Partridge to Calhoun, March 17, 1817, as quoted by Webb, 147.

[171] Article 77 reads, "Whenever any officer shall be charged with a crime, he shall be
arrested and confined to his barracks, quarters, or tent, and deprived of his sword, by
the commanding officer. And any officer who shall leave his confinement before he
shall be set at liberty by his commanding officer, or by a superior officer, shall be
cashiered." U.S. Congress, *U.S. Statutes at Large*, vol. 2 (1799–1813), 368–69.

[172] Painter, *Trial of Captain Alden Partridge*, 103.

[173] Interestingly, Calhoun also reports, "The reasons for not ordering the court, have
already been verbally explained to you," a fact that Partridge would later refute. John C.
Calhoun to Alden Partridge, March 20, 1817, National Archives, as quoted by Webb,
Captain Alden Partridge, 148.

[174] Webb, 147.

After gaining no satisfaction during his leave, Partridge would tender his resignation on April 1, 1818 (effective April 14, 1818) to the president.[175] In the words of Lester Webb, "Thus ended one of the most dramatic military careers of any graduate of the United States Military Academy."[176] One can only imagine the emotions that Partridge was experiencing as he realized the organization he had dedicated over a decade of his life to displayed no interest in living up to standards of the law, the U.S. Constitution, or ensuring Partridge's public reputation accurately reflected his true actions.

Partridge did not end his search for justice with his resignation. He continued to seek an audience with Calhoun and Monroe. Partridge would write at least seven more letters to Calhoun, calling on him to live up to the law and legal precedent of the era, and arrest and court-martial Thayer. Webb points out that Calhoun was not ignorant of his duty, as he had previously pushed Andrew Jackson to court-martial General Edmund P. Gaines in a similar situation when Colonel W. A. Trimble persisted in advancing charges.[177] In a letter dated October 27, 1818, Calhoun would quibble over interpreting Article 77,

> I see no reason to alter the construction which I had given to the 77th Article of War. It cannot be doubted, that when an officer is charged with a crime, that the arrest, confinement, & etc. are necessary consequences; but the enquiry is still left, what is meant by charging an officer with a crime under the Article? Can a mere preferment of allegations against him be considered a charge & a crime within its meaning? Such a construction has never been given to it; and, in fact, could not be given without the greatest absurdity and most mischievous consequences.[178]

Calhoun attempted to assert that Partridge's claims are nothing more than allegations, yet the transcripts of Partridge's court-martial validated the claim. Calhoun falsely claimed that Partridge's allegations had already been investigated within his own court-martial and therefore "proceeding in relation to them… would be not only useless, but pernicious."[179] However, this was clearly false. Thayer was not the accused in Partridge's court-martial and his

---

[175] Webb, 154.

[176] Webb, 154.

[177] "[I]t has been judged necessary, as well as from regard to his character, as to the reputation of the army, to institute a court martial for the investigation of those charges." William H. Crawford to Andrew Jackson, May 21, 1816, National Archives, as quoted by Webb, 152.

[178] Crawford to Partridge, October 27, 1818, as quoted by Webb, 150.

[179] Crawford to Partridge, October 27, 1818, as quoted by Webb, 151.

actions were not fully examined, nor was any corrective discipline applied to his misconduct.

Partridge wrote to Calhoun again on November 17, 1818, making one last appeal to his honor:

> I trust, Sir, that we have laudable ambition of leaving behind an honorable and lasting reputation… But … If I live, and justice be not done, all the facts relative to the subject which has been under discussion between us will be so rendered as to exist as long as our names be remembered and will then attach honor or disgrace to our memories according to the unbiased decision of future generations.[180]

Partridge's words here were prophetic. While the truth surrounding the events of August 30, 1817, was suppressed well into the 20th century, with the publication of Webb's research in 1965, the image of Partridge was ultimately redeemed while the legacy of Calhoun was further diminished.

Finding no hope in action from Calhoun, Partridge attempted to gain official copies of Thayer's false report of mutiny at West Point and related papers. Publishing these documents would allow Partridge to repair the damage to his personal reputation. Partridge came close to gaining a copy of these documents. However, between the time Partridge was given the opportunity to select the documents he wished to have copied and the replication of these documents by a scrivener, Calhoun intervened. This caused Partridge to make an appeal for these documents directly to President Monroe on November 20, 1820:

> I have the honor to inform your Excellency that on the 30th of March last [1820] I called upon the Secretary of War at his office in Washington, and requested of him a Copy of Major Thayer's official report relative to my assuming the Command at West Point on the 30th of August 1817. I also at the same time informed him what use I intended eventually to make of this report, viz, that I intended to publish it in conjunction with the other facts and circumstances connected with that event; that I had charged Major Thayer with *Willful lying*; that I know he had been guilty of the offence, and further that I knew he had been guilty of having made a false one, when contrasted with facts as subsequently proved—even by his own testimony. The Secretary observed, in reply to my request, that he could not then determine whether he would grant me a Copy, but would let me know in a short time. On the following day I called at the

---

[180] Alden Partridge to John C. Calhoun, November 17, 1818, National Archives, as quoted by Webb, 152.

War-office, when Major Van De Venter immediately presented me with a Bundle of Papers and wished me to select such as I was desirous to obtain Copies of. Observing that they contained Major Thayer's Report, and some other Documents relative to the transactions at West Point at the time before mentioned concerning which I had spoken to the Secretary the day preceding, I accordingly selected Major Thayer's Report and one or two other Communications on the same subject. These were immediately placed by Major Van De Venter into the hands of one of the under Clerks to be Copied, and I was informed by the Major that if I called at the office at any time before three o'clock P.M., the Copies should be furnished me. I would here observe that after I had perused Major Thayer's Report, I observed to Major Van De Venter, that I was perfectly astonished that a man in his [Thayer's] situation should, on the very day when the events referred to occurred, deliberately sit down and state, as facts, what he must have known was absolutely and totally false. About two hours after I left the office, I received a Note from Major Van De Venter from which following is an Extract: 'The Secretary of War [John C. Calhoun] directs me to inform you that he cannot decide whether or not he will furnish you with Copies of the papers you requested until he understands explicitly from you the object for which you require them. If you desire to use them before a Military Committee, they shall be furnished to the Committee on its application for that purpose.' I immediately waited on the Secretary and again stated to him the purpose for which I wished these Copies. But which he finally refused to grant me. I now deem it my duty to apply to your Excellency upon the subject, and respectfully to request that you will order the Copies, then promised, to be sent to me. Of my right to them I believe there can be no doubt. Major Thayer in his report has unequivocally asserted facts, implicating in the highest degree my Conduct as an officer. His statement has subsequently been proven, even by his own testimony, to be totally false. I have regularly, and in due legal form preferred Charges against him, *embracing the Charge of falsehood*, which Charges I then did and still do hold myself ready to place before any tribunal of Justice; but the Secretary of War in utter disregard and contempt of the laws in his Country, has refused to even order any investigation into those Charges, (and I consequently have been debarred the privilege of proving them). And to cap this climax of injustices has finally refused to grant me a Copy of the report in which my Conduct was falsely and maliciously misrepresented, but which was made a ground of public proceeding against me. It remains for your Excellency to decide whether proceedings so evidently unjust and illegal shall be permitted and sanctioned, for the purpose of screening from merited punishment, an individual, who, were his Character and Conduct fully developed to the public, would be pronounced by every impartial person a *disgrace* to the Army of the United States, or whether justice shall finally be done.[181]

Partridge's appeals fell on deaf ears. Neither Calhoun nor Monroe was willing to assist Partridge at the expense of Thayer and the image of the USMA. Richard Pratt, while participating in the USMA Tactical Officer Education Program would "deduce that [Partridge] became the scapegoat for the politics revolving around the existence of the fifteen-year old Academy."[182] Monroe and Calhoun were willing to overlook Thayer's misconduct to advance their goals for the USMA. This essentially shut the door for Partridge to clear his name. Without proof of Thayer's misconduct, Partridge could not pursue the matter further. Partridge would wisely shift his efforts to creating a system of private military academies that would render the USMA obsolete.

The Reality of Partridge's Personal Behavioral Traits

There is a significant focus on Partridge's personal behavior among his distracters. For instance, the following passage is found within Swift's memoirs.

The circumstance that induced the Secretary of War to desire a superceding of Captain Partridge, was not his want of ability, for he was a good teacher of mathematics, and a good infantry and artillery drill officer; it was because his aspect was uncouth, a want of what is called genteel carriage, and awkwardness of manner that gave a repulsive first impression. But Captain Partridge had good qualities as well as good sense. He was said to be a graduate of Dartmouth College, and was there deemed a good scholar; and it cannot be denied that many of the youthful officers of the army in the war of 1812 owed much of their success in the field, to the patient training which they received from "Old Pewter" — Captain Partridge's soubriquet among the cadets.[183]

---

[181] Alden Partridge to James Monroe, November 20, 1820, Partridge Papers, Plumley Collection, as quoted by Webb, 154.

[182] Pratt, "Essence of Military Education," 14.

[183] Swift, *Memoirs*, 170.

While Partridge may indeed have been ungainly in his deportment, Swift's passage here is likely an *ad hominem* attack in an attempt to cover for Swift's own leadership failures. As noted by Webb, Partridge received little mentorship as an officer in his Army career. He was largely left at the USMA to manage the affairs of the academy in the absence of Colonel Williams and, later, Brigadier General Swift. If Partridge was truly a disgrace in terms of conduct and etiquette, wouldn't it be the obligation of Partridge's leaders to take actions to resolve these shortfalls?

If Partridge had negative behavioral traits at the USMA, wouldn't it follow that these behavioral traits would also continue after leaving the USMA? Yet, we find later accounts of his behavior to praise and not admonish. M. E. Goddard would recall Partridge in high esteem,

> Never, perhaps, did a human voice and manner have perfectly combine inconceivable decision with courtesy and kindliness without condescension, [Partridge] met every student as an equal, taking the poor country boy, as well as the older and current cadet in his counsels and treating his pupils, one and all, as men upon their honor—capable of good judgement and manly sentiments.[184]

It is likely true that Partridge may have had some social awkwardness similar to other geniuses. He was naturally introverted, with many recalling his quiet nature.

However, to whatever degree Partridge was lacking in a genteel carriage, it did not seem to stop him from being highly successful in areas normally associated with high-social skills. Partridge maintained relationships with multiple U.S. presidents, governors, congressmen, and other men of high social standing. He was able to build coalitions within communities to establish multiple military academies. Partridge successfully ran on multiple occasions for the Vermont House of Representatives. He was able to host multiple conventions for the militia. Partridge was loved by his cadets at the USMA and his own institutions. With all of Captain Partridge's accomplishments that would stand in stark contrast to the assertion he lacked social skills, an unbiased audience must be brought to the conclusion that Partridge's detractors overemphasized these behaviors as a form of fallacious argument.

Chaplain Adam Empie

In most official historical accounts of the USMA, the conflicts in the early days of West Point center entirely around Captain Partridge. This may

---

[184] M. E. Goddard, "Reminiscences of Captain Alden Partridge," ca. 1884, Dartmouth Library Archives, 3.

lead one to the conclusion that Partridge was the single rogue player who lacked any support from other staff, outside his direct relations. However, this was not the case. Partridge appears to have had the support of Chaplain Adam Empie.

Chaplain[185] Adam Empie was recruited to West Point in 1814 to become the first army chaplain of the institution. Brigadier General Swift hired the Episcopalian minister to address some of the moral failures noted among West Point cadets. It is reported that Swift selected Empie "for this position since his reputation as a devout Episcopalian and capable administrator preceded him."[186] Outside of providing religious services, Empie also served as the professor of ethics, history, and geography with the additional duties as treasurer.[187]

Partridge and Empie worked well together and shared a strong moral focus in their instruction at the Academy. Dean of the Disciples Divinity House at Vanderbilt University (and brigadier general in the U.S. Army Reserve) Herman A. Norton asserts, "The chaplain [Adam Empie] supported Captain Partridge so diligently and consistently that when the Captain was replaced as superintendent by Sylvanus Thayer, Empie felt compelled to resign."[188] This certainly adds to the credibility of Partridge's position and gives credence to the validity of his complaints. Clearly, this situation was more than a personal vendetta. Like Partridge, Adam Empie would have a highly successful career in education after leaving the Army. Empie eventually became the president of the College of William & Mary,[189] from 1827 to 1836.

On an interesting note, Chaplain Empie was one signatory on a petition to James Monroe to establish a five-year curriculum at the USMA (with cadets not graduating faster than four years).[190] While this suggestion

---

[185] Adam Empie would have used the honorific of *Reverend* during his employment at West Point. In this word, I use the honorific of *Chaplain* employed in the present-day Army Chaplain Corps.

[186] Kenneth E. Lawson, "Religion and the U.S. Army Chaplaincy and the Florida Seminole War 1817–1858" (2004), The College of William & Mary Special Collections Research Center.

[187] George S. Pappas, *The Cadet Chapel, United States Military Academy* (West Point, NY: Association of Graduates, United States Military Academy, 1987).

[188] Herman Albert Norton, *Struggling for Recognition: The United States Army Chaplaincy, 1791–1865* (Washington, D.C.: Dept. of the Army, Office of the Chief of Chaplains, 1977), 26.

[189] The College of William & Mary is the second-oldest college in the United States.

[190] James Monroe, Letter from James Monroe to Alden Partridge, 22 January 1815, Norwich University Archives and Special Collections.

would retard Partridge's goals of allowing cadets to progress as rapidly through their course of study as possible, it appears to have been a suggestion to protect the best interests of cadets. In the War of 1812, many cadets were pressed into service (thus ending their time to develop at the Academy). While this may have been a useful device in a government institution to prevent early termination of scholarship, it would never have been needed at one of Partridge's private institutions, as students could have returned to their studies after any event of a militia activation. There are no student advantages for civilian institutions to follow a fixed four-year curriculum. This practice only burdens students with prolonged periods of study with related tuition charges, a debt in time, and funds that may not be needed. This is especially true when most students are not guaranteed a full scholarship (or other time-based benefits) that require protection by a fixed time period of study.

The Honorable Daniel Azro Ashley Buck[191]

Chaplain Empie would not be the only West Pointer who would take action in response to how Partridge was treated. Other members of the West Point community would support Partridge in his endeavors to seek justice and create an alternative to the USMA. West Point graduate David Azro Ashley Buck[192] was one of these individuals. Buck was a former student of Partridge, a veteran of the War of 1812, and one of the few West Point graduates that would later accept a commission in the militia. More precisely, Buck would eventually climb to the rank of colonel in the Vermont militia.[193] Buck served for many years within the Vermont House of Representatives. He would be elected as a Vermont state congressman during the Eighteenth Congress (1823–25) and the Twentieth Congress (1827–29). While in Congress,

> He made a speech in the house in favor of abolishing the office of the major-general of the army, and against the appointment of a Board of Visitors at West Point, and detailing the history of that institution. It

---

[191] My gratitude goes out to Jay King, a descendent of D. A. A. Buck, for contacting me on LinkedIn and directing me to some of the resources that explained the relationship between Congressman Buck and Alden Partridge.
[192] D. A. A. Buck graduated from the USMA in 1808. Given his time of attendance, it is assumed that he would have received instruction in mathematics under Partridge.
[193] While some sources claim D. A. A. Buck was eventually commissioned a general in the Vermont militia, the following source recalls his rank as colonel at time of death. Hamilton Child, *Gazetteer of Orange County, Vt., 1762–1888*, vol. I (Syracuse, NY: Syracuse Journal Company, 1888), 87–88.

would seem as if his military training and experience had not impressed favorably towards the academy of its management.[194] Buck's actions would echo many of the desires of Partridge in improving the militia and restraining the power of Active Army officers. Buck was hailed as "the most popular man that ever lived in Chelsea, and one of the most popular in the state."[195] With Jay King suggesting his popularity came from his views closely aligning with those of his constituents.[196]

Buck would use his influence to support Partridge in building and improving an alternative solution for the education of officers than the USMA. Congressman Buck, in 1824, would assist Partridge in securing cannon ordnance for use in instructing cadets at Partridge's American Literary, Scientific and Military Academy.[197] In the same year, he would send his seven-year-old son of the same name (or similar initials),[198] to attend Partridge's academy. In 1834, D. A. A. Buck was elected as one of the original 11 trustees for Norwich University.[199] Buck's support for Partridge's activities further demonstrated that Partridge was not a lone "bad actor" or singular problem at West Point. Other members of this community were also aware of the problems and took actions to either remove themselves, as in the case of Empie, or acted hoping to correct the situation.

Official Record

There are frequent repeated errors in scholarship on the findings of Partridge's court-martial that have continued well after the facts of the case have been fully made public. For example, the famed historian Stephen E. Ambrose in 1999 falsely records, "Partridge was found not guilty of eight charges which concerned his conduct as Superintendent but guilty of disobedience of orders **and mutiny**."[200] While the court-martial found Partridge guilty of disobeying orders, the court ultimately acquitted Partridge of

---

[194] Merritt Elton Goddard and Henry Villiers Partridge, *A History of Norwich, Vermont* (Hanover, NH: Dartmouth Press, 1905), 176.

[195] Child, *Gazetteer of Orange County*, 87.

[196] Personal correspondence

[197] David Azro Ashley Buck, Letter from D. A. A. Buck to Alden Partridge, 15 January 1824, Norwich University Archives and Special Collections.

[198] List of Cadets and Where They are Billeted, [1824?], Norwich University Archives and Special Collections.

[199] George Gary Bush, *History of Education in Vermont* (Washington, D.C.: Government Printing Office, 1900).

[200] Ambrose, *Duty, Honor, Country*, 60.

the charge of mutiny.[201] These errors could be written off as poor scholarship, but unfortunately they continue to propagate the idea of Partridge as a mutineer—something he was clearly not even by Thayer's own testimony. Partridge's achievements are constantly downplayed, with multiple accounts excluding his formal assignment as superintendent. Examples range from an 1818 history that removed Partridge entirely to the previously mentioned Cullum to the current USMA mathematics department, reporting,

> From September 1813 until December 1816 [Partridge] was the Head
> of the Engineering Department as well as Acting Superintendent in
> the absence of the Superintendent. Unfortunately, his administration
> of this prestigious position was lax and unsatisfactory.[202]

Partridge's attacks on the USMA are further detailed in a later chapter. However, his attempts to end or change the USMA likely fueled the bias among Army historians. Major General George Cullum was a USMA cadet from 1829 to 1833 and would have lived through the political turmoil that resulted from Partridge's 1830 pamphlet against the academy. This likely caused Cullum to heavily bias his history against Partridge while writing three decades later. It is assumed that many historians of later generations may have simply repeated Cullum's bias without recognizing its existence. As long as Partridge's character and achievements at the USMA remain disparaged, his logical arguments against the USMA remain hidden from the general public.

Discussion

This chapter covered considerable territory detailing the events and records of Captain Partridge's court-martial. With all the related factors that play into the memory of Partridge and their complex relationships, trying to establish a logical, unbiased history on this topic is like trying to trace the lines of the Gordian Knot. It is also acknowledged that copious materials on this subject resting in archives have yet to be examined. In lieu of a conclusion, I will end with a discussion of major themes within this chapter.

Use of History

One of the striking aspects of this chapter is the influence of historians on our understanding of the past and their role in "meaning making." Traditionally, we may view the duty of the historian as to explore events in search of patterns of cause and effect. However, as seen within this chapter, historians are not free from bias and do not always operate through virtue. Both implicit and explicit bias affect the conclusions drawn by historians.

---

[201] Painter, *Trial of Captain Alden Partridge.*
[202] David C. Arney et al., "Alden Partridge," Math Biographies of Department Heads.

Falsified histories repeated frequently become accepted as truth, with many present-day historians unknowingly echoing falsehoods. We may question what should be required of historians, especially those employed by the U.S. government, to reduce the bias within their works and ensure the products they present are as truthful as possible. Given that it is impossible for one person to study or know all aspects of history, society depends highly on historians to honestly summarize the past. The great people and events of the past are important to highlight, with lesser events saved for those more interested in specific events or occurrences. Historians should not hide events or people within history simply to avoid tough questions and philosophical arguments.

Much of the early "historical" accounts of the USMA from the 19th century might be best classified as legend. These accounts artificially transfer many achievements of Captain Alden Partridge onto Major Sylvanus Thayer in an attempt to downplay Partridge's importance in history. While this may allow the USMA to abbreviate their historical account, there is a question of the appropriateness of this action. Especially in an institution designed to produce future Army leaders displaying high degrees of moral courage and strength of character. Elevating leaders to legendary status and removing their faults prevents cadets from learning the true difficulties related to human interactions and establishes unrealistic standards of behavior. Even the finest U.S. military leaders have their faults as the result of human nature. Leadership training and education should not attempt to cover these mistakes; rather, they should be discussed honestly in a search for lessons learned to enable the development of better leaders.

Ultimately, Captain Alden Partridge provided more than sufficient positive effects on the American Republic to earn a mention even in summarized histories of the USMA. Present-day historians are challenged to not simply accept institutional historical accounts about Partridge on face value but seek to confirm these accounts through multiple sources to reduce the undue political bias created by the controversies of his day. An honest recounting of the successes and failures would ultimately allow many of Partridge's ideas on education to continue to have positive effects on shaping the American Republic.

Chapter V: Partridge, the Army, the Militia, and Professionalism

Before entering a larger discussion on Partridge's attacks on the United States Military Academy, it is worthwhile to examine the nuances of his opinions towards the active (or "standing") army and the militia. This will be done to further understand his intellectual arguments and determine if Partridge presents any hypocrisy in shifting opinions upon departing the USMA. This chapter will seek to answer three major questions: 1) Did Partridge support the Active Army until he resigned his commission? 2) Did he have concerns about the training of the militia prior to leaving West Point? 3) How congruent was Partridge's *American System of Education* with the Founding Fathers' opinion of the militia and peace establishment? This chapter will conclude with a discussion of professionalism and how this term carries multiple meanings that are often unrecognized or conflated.

Partridge and the Active Army

It is common for distractors to accuse Partridge of being hypocritical in his opinions towards the Active Army, seeming perfectly fine with the idea when he oversaw the USMA, yet he ardently opposed the institution and the Army once he left the service. However, these arguments can quickly be demonstrated as nothing more than strawman fallacies when the contents of Partridge's documents are reviewed. Partridge was never universally opposed to the Active Army. Partridge's concern was specific to the size of this force.

A quick review of the U.S. Army in Partridge's youth reveals that in 1790, the Army only consisted of 1,273 men (including officers).[203] While the Army would certainly surge in size during perceived risk or open military conflicts, it would be trimmed back to little more than a police force in times of relative peace. In 1797, the peace-time Army would be reorganized to 3,359 enlisted and officers.[204] In 1801, with President Thomas Jefferson in office, the

---

[203] John R. Maass, *Defending a New Nation, 1783–1811* (Washington, D.C.: Center of Military History, United States Army, 2013), 13.
[204] Maass, *Defending a New Nation*, 33.

Army reached 4,051 (however the authorized strength was 5,438), with this strength reduced to 3,289 by 1802.[205] Partridge recalled this troop level in 1840: "During the first term of Mr. Jefferson's administration, [the Army] was reduced to three thousand men, commanded by a brigadier-general."[206] Alden Partridge would enter West Point as a cadet under the Jeffersonian philosophy of a small and inexpensive standing army or "peace-establishment" sufficient to maintain security over the seacoast that could be reinforced as required by the militia.

Partridge adopted this Democratic-Republican view on the structure and composition of the military, possibly because of the admiration he had for President Jefferson. While Partridge carried the belief that the United States should be protected by a large militia and a small Active Army force, Partridge himself recommended a larger peace establishment than what Jefferson had established, likely because of the geographical expansion that had occurred since the early days of Jefferson's presidency:

> [Y]our memorialist would respectfully, but urgently, recommend, that the peace establishment be reduced to a number not exceeding six thousand men, (officers included) being fully convinced that all the duties necessarily devolving on such establishment, can be just as well discharged for the public benefit as by the present number.[207]

Furthermore, Partridge clearly asserts functions and need for an active army, "that it constitute, simply, a police-guard, to protect, in time of peace, the necessary fortifications on the seaboard and inland frontier, as well as other public property that requires such protection from lawless depredation."[208] Therefore, it can be strongly asserted that Partridge saw an acceptable role for the Active Army and would even allow for the expansion of such a force in proportion to the growth of the nation. However, this force would never replace the militia in its primacy for national defense.

Similarly, the claim that Partridge was hypocritical for accepting an appointment at the USMA to become an Engineer can also be dismissed. Partridge acknowledged the need for highly trained Engineer officers serving

---

[205] Maass, 39.

[206] Alden Partridge et al., "Memorial of a Committee of the Military Convention at Norwich, Vt. Praying the Revision and Alternation of the Military Defense of the United States, to the Honorable Congress of the United States," *Citizen Soldier*, July 22, 1840.

[207] Partridge et al., "Memorial of a Committee of the Military Convention at Norwich, Vt."

[208] Partridge et al.

on active duty.[209] Therefore, it cannot be claimed he wanted to prevent others from becoming part of the small elite group actively serving as Army engineers. At the time of his appointment at the USMA, West Point was the only academy in the Nation that provided instruction in engineering. However, with the establishment of Partridge's American Literary, Scientific and Military Academy in 1819, this was no longer the case. Furthermore, Partridge provided several plans to expand the success of Norwich University across the states. Therefore, it would not be hypocritical for Partridge to call for the closing of the USMA after a superior educational model more suitable for the Republic had been tested and could be implemented across the nation.

Partridge and the Militia

Another common argument against Partridge was that his concern for the militia only appeared after his relief from the USMA. This asserts that supporting the militia was the best line of attack to seek revenge against the USMA. In this argument, it is asserted that Partridge didn't have an issue with West Point, only producing active-duty officers until after he left the academy. It also suggests that Partridge was hypocritical in spending years enjoying the benefits of active federal service only to destroy these privileges for others after his departure from service. However, there is evidence to demonstrate that while Partridge was still assigned to the USMA, he was trying to propose an expansion to the academy that would have made cadets contracted to commission into active service a minority (40 percent) within the proposed institutions.[210] It is legitimate to note that in Partridge's proposal, the 250 civilian students that would pay for their own education at each academy are not explicitly stated to improve the training and discipline of the militia. However, the Militia Act of 1795 would have applied to these civilian graduates, and thus this plan would have had that effect. Additionally, there are records of Partridge writing his father concerning the preparation of the militia as early as 1807.[211]

An article appeared in the *National Intelligencer* on April 21, 1814, under the pen name Americanus (a pen name frequently used by Partridge) concerning the District of Columbia militia encouraging militia officers to

---

[209] Partridge et al.

[210] Partridge, Guidelines for Establishing Multiple Military Academies.

[211] Alden Partridge. Letter from Alden Partridge to Samuel Partridge, 2 October 1807, Norwich University Archives and Special Collections.

engage in "voluntary preparation [self-study]" in the profession of arms.[212] Further, Partridge wrote Daniel D. Tompkin, governor of New York, in 1816, asking if New York would take the initiative in establishing a training system for the militia that Congress could then adopt:

> The state of New York, preeminent in wealth, in population, and in enterprize, seems destined to take the lead in this, as well as in many other usefull undertakings. The great advantages that would result, not only to the state, but to the country generally, from the establishment, upon proper principles, of such a seminary, are to me self evident.[213]

Evidence clearly shows that Partridge was concerned with the training of the militia throughout his career, and he was actively petitioning for militia concerns with senior members of government more than a year before his relief from West Point and in newspapers dating back to at least 1814. Considering this evidence, the claim that Partridge did not care about the militia until his resignation from the Army can be dismissed as historically false. Interestingly, the USMA mathematics department even reinforces this belief,

> While an instructor at USMA, Partridge became engrossed with a vision of his. Much of his free time was occupied with his development of a plan for military and technical education. It was his premise that "a large standing army was a menace to the country and urged rather that the nation should train a large 'citizen soldiery' in the art of war."[214]

American Founders' Support for the Militia

The American Founding Fathers supported the primary use of a militia system for ground defense for several reasons. While the need for a peace establishment [small active force] was required to maintain key permanent fortifications, the active-duty army was never intended to be so large as to overpower the local militia. Standing armies are a direct threat to the people. Historically, they could easily be corrupted to serve the will of tyrants. Standing armies are expensive, stripping the resources from the nation and often driving nations to war to justify their expense. In the words of George Washington,

> Altho' a large standing Army in time of Peace hath ever been considered dangerous to the liberties of a Country, yet a few Troops,

---

[212] With its similarity of writing style and content, I assert there is a high probability it was composed by Partridge. Americanus, "District Milita," *National Intelligencer*, April 21, 1814, George Washington University Special Collections Research Center.

[213] Alden Partridge, Letter from Alden Partridge to Daniel Tompkins, 8 August 1816, Norwich University Archives and Special Collections.

[214] Arney et al., "Alden Partridge."

under certain circumstances, are not only safe, but indispensably necessary. Fortunately for us our relative situation requires but few. [...] Were it not totally unnecessary and superfluous to adduce arguments to prove what is conceded on all hands the Policy and expediency of resting the protection of the Country on a respectable and well established Militia, we might not only shew the propriety of the measure from our peculiar local situation, but we might have recourse to the Histories of Greece and Rome in their most virtuous and Patriotic ages to demonstrate the Utility of such Establishments. Then passing by the Mercinary Armies, which have at one time or another subverted the liberties of allmost all the Countries they have been raised to defend, we might see, with admiration, the Freedom and Independence of Switzerland supported for Centuries, in the midst of powerful and jealous neighbours, by means of a hardy and well organized Militia.[215]

As displayed in Washington's quote, the desire for a militia-based defense also allowed for the moral development of the Nation. If every able-bodied man was expected to answer the call to arms, the populace would be determined in vote and personal action to avoid the perils of unneeded war. To reinforce this point, Anderson and Mc Chesney suggest that the increase in the number of battles occurring during the American Indian Wars just prior to and following the American Civil War was in at least part due to increased use of standing armies over militia forces.[216] Citizens that were now released from the dangers of the battlefield grew to advocate for the employment of violence through standing military forces to resolve conflicts.

A great percentage of military leaders seek to live at the expense of the general government depending on the constant state of war—or threats thereof—to ensure their livelihoods and chances for advancement. This creates a powerful incentive for military officers to strongly advocate for military solutions to problems that could likely be resolved through more peaceful means. As explained by Kern,

Once established, a government's military, its bureaucrats and leaders, as well as laymen all face a different set of incentives. Those with a job related to the military have an incentive to keep their job. In most cases, they probably also desire to see the scope of their power

---

[215] George Washington, Washington's Sentiments on a Peace Establishment, 1 May 1783, National Archives.

[216] Terry L. Anderson & Fred S. Mc Chesney, "Raid or Trade? An Economic Model of Indian-White Relations," *Journal of Law and Economics* 37, no. 1 (April 1994): 39–74.

expanded and their pay increased. The support for war, then, is the ideal policy for achieving those goals. These incentives may not transform a champion of peace into a war-loving bureaucrat, but they can have effects on the margins. It's much easier to rationalize a war if your job depends on it.[217]

The toxic effect of America's large-scale abandonment of a militia-based ground defense force can be seen throughout the 20th century into the present with the rise of the military-industrial complex, unsustainable military spending/national debt, and constant calls for military involvement in conflicts around the globe.

While the events of the Burr Conspiracy are not widely known by our present-day general public, it would have been a powerful confirmation of the warning of the American Founders of the risk of standing military forces and the ambitions of professional (monomathic) military men. In 1804, Vice President Aaron Burr would plot to create an independent country within Spanish Texas and lands recently acquired through the Louisiana Purchase. Burr enlisted the support of Brigadier General James Wilkinson, the commanding general of the United States Army. Burr would persuade Jefferson to appoint Wilkinson as the first governor of the Louisiana Territory. In 1806, Wilkinson was set to supply the U.S. forces under his command to support Burr's actions.[218]

Fearing the failure of the mission, Wilkinson altered the ciphered letters he had received from Burr to minimize his involvement and turned them over to President Jefferson, leading to Burr's arrest. Wilkinson's doctored evidence and self-serving testimony in Burr's trial helped Burr escape conviction. Wilkinson would later be court-martialed by James Madison in 1811 but was acquitted, with Wilkinson retaining his command, as the "federal administrations in the early United States feared that if they did not command the loyalty of the army, it could threaten to depose them."[219] Thus, the young American Republic found itself in a similar position to the Roman Republic after the Marian reforms, fearing that the loyalties of their soldiers lay more in their commanders than their nation.

Brigadier General Wilkinson would serve as a longtime spy for Spain, attempting to not only succeed Kentucky and Tennessee into the Spanish

---

[217] Andrew Kern, "Why the Founding Fathers Feared a Standing Army," Foundation for Economic Education, February 6, 2024.

[218] Maass, *Defending a New Nation*.

[219] Andro Linklater, *An Artist in Treason: The Extraordinary Double Life of General James Wilkinson* (New York, NY: Walker, 2010), 195.

Empire[220] but also to betray the Lewis and Clark Expedition. In 1854, letters were found in Spanish archives confirming that Wilkinson actively engaged in treason.[221] Wilkinson stands as a powerful example of the extreme dangers to the Republic from active duty officers in command of standing military forces. Partridge's proposals to improve the militia and limit the size and power of the U.S. Army were an attempt to control this risk.

The Semantic Argument of Professionalism

There is a frequent assertion that Partridge stood in the way of professionalizing the Army officer corps. For example, a textbook used to instruct ROTC cadets in military history recounts:

> [John C.] Calhoun early turned his attention to the Military Academy, where [William H.] Crawford's attempts at rehabilitation had been impeded by controversy stirred up by the arbitrary actions of Superintendent Capt. Alden Partridge. After Partridge was removed and Bvt. Maj. Sylvanus Thayer was appointed Superintendent in July 1817, the academy became a vital force in maintaining a corps of professionally trained officers.[222]

This assertion is little more than sophism, a false argument. It is an equivocation based on the industrial monomathic definition of a "professional" being an expert in only one specific task. In historical context, during the War of 1812, militia officers would be frequently replaced (regardless of competency) with active-duty officers. This often led to disastrous results, as seen on August 24, 1814, in the Battle of Bladensburg. The replacement of militia leaders immediately upon activation destroyed trust and confidence within the chain of command, often leading to desertion and poor battlefield performance. This fueled the false assertion that only "professional" standing military forces should be considered as a basis for national defense, thus ensuring employment for individuals that wanted to live at the expense of the state as soldiers and officers. Historian William Weber blames the failure of Bladensburg on Madison's and Monroe's desire for a linear engagement with

---

[220] Robert Brammer, "General James Wilkinson, the Spanish Spy Who Was a Senior Officer in the U.S. Army during Four Presidential Administrations: In Custodia Legis," Library of Congress, April 21, 2020.

[221] Linklater, *An Artist in Treason.*

[222] Ironically, in making a sematic argument against Partridge in regard to military competency, the authors use the incorrect term "trained" over the more appropriate "educated." Richard W. Stewart, ed., *American Military History: The United States Army and the Forging of a Nation*, vol. I (Washington, D.C.: Center for Military History, U.S. Army, 2005), 165.

British forces instead of the use of the militia engaging in the more suitable guerrilla warfare tactics on the British flanks and rear.[223]

Amateur historian Michael Morano notes, "The historical record does not yet show the degree to which Bladensburg influenced Alden Partridge, but hopefully research continues."[224] While Partridge's opinion of the aftermath of Bladensburg is unknown, it appears he was aware of the need to further professionalize the D.C. militia in the *National Intelligencer* article he published a couple months earlier. Lieutenant governor of Connecticut John Cotton Smith would relay the governor's objection to the replacement of militia officers in an article published in the *National Intelligencer*:

> His excellency [Governor Roger Griswold] conceives then that an order to detach a number of companies, sufficient for the command of an officer of the United States, cannot, with propriety, be executed, unless we were prepared to admit, that the privates may be separated from their company officers and transferred into the army of the United States; thus leaving the officers of the militia without any command except in name, and in effect impairing, if not annihilating the militia itself, so sacredly guaranteed by the constitution of the several states.[225]

No sane military leader would willfully break the chain of command of military units just prior to combat. This is especially true in militia units that have longstanding relationships within their groups that have lasted for years, if not entire lifetimes. These actions were not taken to improve combat efficiency but were a deliberate attempt to enforce a system of credentialism to ensure full-time and expanded employment of active-duty Army officers.

These rhetorical devices may not be recognized by present-day readers that live in a society largely created for monomathic careers and professions. Carefully examining the following quote by Frank Morn can drive home this point: "West Point demonstrates two conflicting trends [...] First, there was a desire to professionalize the leadership ranks of the military. Second, at the

---

[223] William Weber, *Neither Victor nor Vanquished: America in the War of 1812* (Washington, D.C.: Potomac Books, 2013).

[224] Private correspondence.

[225] John C. Smith, "Documents Accompanying the Report of the Secretary of War to the Committee of the Senate Respecting the Conflicting Jurisdiction of the General and State Governments over the Militia—Continued. Sharon, Connecticut, July 2, 1814," *National Intelligencer*, May 2, 1815, George Washington University Special Collections Research Center.

same time the public dislike the notion of a professional army."[226] Morn uses the term *professionalize* in his first trend to equate to education or competency. In his second trend, he uses the term *professional* to refer to full-time employment status. The intended meanings between these two uses of the concept of professionalism are significant. So different as to refute the claim that the two trends are conflicting, as it would be possible to have a highly educated/competent officer corps within the militia.

	Monomathic "Expert"	Polymathic "Generalist"
Professionalism Defined by Competency (Ability)		Partridge's *American System of Education*
Professionalism Defined by Full-Time Employment (Job Status)	French Model of Military Professionalism	

Figure 11—Use of the Term Professionalism in Relation to the Specialization/Generalization Spectrum

It would be hard to point to a single American officer during the early 19th century that was more interested in improving the military competency of the Nation than Captain Alden Partridge. The renowned American educator and reformer Henry Barnard asserted that Captain Alden Partridge "did more than any other individual to introduce military instruction and exercise in schools not national or professionally military."[227] Partridge would establish several academies intending to spread military knowledge throughout the populace. He would advance systems to provide military education within the halls of Congress. Partridge would host military conferences and take an active role in training the militia. He frequently provided lectures to the public on military topics and contributed heavily to militia related newspapers.

The legacy of Captain Alden Partridge within private military academies and the ROTC program continues to "professionalize" (improve the ability of) the U.S. military. The only thing that was different about Partridge was he did not endorse the French monomathic approach to military "professionalism" defined through full-time employment. Partridge wished to see an American model of education that produced polymathic individuals fully capable of not only military service but employment in all other necessary industries and institutions of American life. To drive home this point, if

---

[226] Frank Morn, *Academic Politics and the History of Criminal Justice Education* (Westport, CT: Greenwood Press, 1995), 6.

[227] Barnard, 854.

Partridge is thought to be less "professional" than Thayer, where is Thayer's educational philosophy?[228] Where was Thayer's plan to spread military competency throughout the Republic? Having correctly framed Partridge in his relations to the Active Army and militia, along with examining the concept of type professionalism advanced by his *American System of Education*, we can move on to examine Partridge's attacks against the USMA in the following chapter.

---

[228] Webb, *Captain Alden Partridge.*

Chapter VI: Lifelong and Vociferous Enemy of the Military Academy

Partridge is well remembered for his persistent, and often hostile, attacks on the United States Military Academy, being noted as a "lifelong and vociferous enemy of the Military Academy" by R. Ernest Dupuy.[229] That claim should be judged by the truth of Partridge's assertions and the actual evidence of the conditions at the USMA that he reported. It is unfortunate the civil and military leaders criticized Partridge's arguments, largely avoiding serious discussions on these topics, by asserting that Partridge was simply a disgruntled former superintendent, with the truth about events surrounding his removal from West Point not released to the public within his lifetime. In reviewing some of Partridge's negative assertions about the USMA, we may find that, while possibly false at the time provided, they became powerful predictions of what would occur. This chapter will review the major claims made by Partridge against West Point to determine their validity and examine how the USMA weathered these accusations.

In 1830, under the penname Americanus, Partridge published a pamphlet entitled, *The Military Academy at West Point, Unmasked: or, Corruption and Military Despotism Exposed*. This is likely the most frequently addressed publication by Partridge's detractors, yet Partridge's arguments within this pamphlet are never addressed comprehensively. Most sources provide counterarguments to Partridge's claim that the USMA was building a military aristocracy, with most of those educated being sons of wealthy or governmental men. Some sources will discuss what is viewed as Partridge's use of strictly pejorative terms, calling West Point cadets *effeminate*, *pedantic*, and *military dandies*. These counterarguments fail to address the underlying claims and simply address Partridge's tone. Below, seven of the dozens of assertions Partridge made within this pamphlet will be examined to determine if his claims are as easily disregarded as official histories claim.

---

[229] Dupuy, "Mutiny at West Point."

1. Charity School

Partridge labeled the United States Military Academy as a "Charity School." This term is typically associated with 18th-century English elementary schools established to instruct the children of the poor and were supported through private charity or religious orders. While many viewed the use of this term in relation to West Point as pejorative, the USMA was engaging in this model of education, except for it being federally supported. The cadets that attended West Point were not charged tuition and were even provided with an allowance and a guarantee for gainful employment after graduation.

Partridge would make the false claim that the USMA was primarily educating the sons of wealthy and influential men that had no need for scholarships to support their children's education. As asserted by Partridge,

> If, however, the rolls of this institution for the last twenty years be examined, it will be found that many more of the *rich* and *influential* have been educated there, than the *poor*.[230]

As noted by Professor Carlton F. W. Larson, a review of West Point graduates by the 1843 Board of Visitors found that over 83 percent of cadets came from "moderate circumstances" or poorer families.[231] The Board of Visitors reported the cadets "are the sons, in most cases, of the farmers and working men of the country."[232] Ironically, this report may have demonstrated Partridge was correct in asserting the USMA was indeed a charity school.

The USMA was accepting more cadets than there were possible commissions within the U.S. Army. A considerable number of graduates would never serve within the U.S. military. As a result, Representative Charles Wickliffe of Kentucky in 1831 would call upon the "Committee on Military Affairs to inquire into the expediency of dismissing extra officers unnecessarily educated at West Point."[233] Wickliffe would assert that the USMA had

---

[230] Alden Partridge, Memorial of Alden Partridge, relating to the Military Academy at West Point, and praying that young men educated at other military schools may have an equal chance for admission to the army as those young men have who are educated at West Point, January 21, 1841. U.S. Congressional Serial Set, H.R. Doc. 68, 26th Congress, 2nd Session (1841), 3, Law Library of Congress.

[231] Carlton F. W. Larson, "Titles of Nobility, Hereditary Privilege, and the Unconstitutionality of Legacy Preferences in Public School Admissions," *Washington University Law Review* 84, no. 6 (January 2006): 1375–1440 (quote on 1430).

[232] Report of the Board of Visiters [*sic*] of the U.S. Military Academy, *Army & Navy Chronicle*, Feb. 23, 1843, at 201, as quoted by Larson, "Titles of Nobility, Hereditary Privilege, and the Unconstitutionality of Legacy Preferences," 1430.

[233] Wettemann, *Privilege vs. Equality*, 59.

exceeded Jackson's vision of educating officers for the U.S. Army and had become a national university.[234]

While many attempt to defend the excess graduates from the USMA, stating the national need for engineers in the expanding nation, this justification runs afoul of constitutional authorities. It may have been excusable for Thomas Jefferson to establish the USMA, with a small student body population of forty cadets, in 1802 for the immediate needs to provide for the common defense in the absence of equivalent private educational institutions. It may have even been justified for James Madison in 1812, with a looming war with Great Britain, to expand the cadet population at the Academy to 250 cadets due to the lack of private educational institutions suitable for the task.

But is it truly acceptable that the Academy was continued, with a cadet population far exceeding the Army's demand for officers, when Partridge had established and proven an educational model suitable for spreading military knowledge throughout the country without the need of federal funding? On what constitutional grounds was the USMA continued when, by the 1830s, hundreds of citizens were educated at or above the level of USMA graduates and were available for the needs of the common defense? In the end, Partridge's claim here is true that the USMA did function as a charity school. However, his claim concerning the demographic that made up the majority of graduates was false.

2. Military Aristocracy/Hereditary

One of Partridge's primary concerns about the USMA and commissions within the U.S. Army is as old as the Articles of Confederation and the foundational principals that sparked the American Revolution. As asserted by Partridge,

> And whereas, it is a fundamental principle in every free Government that stations of honor, trust, and emolument shall be equally open to all; and whereas, by the regulations of the Military Academy at West Point, (sanctioned by the War Department,) all the offices of honor, trust, and emolument in the military service of the United States are monopolized by the *proteges* of that institution, who are *educated at the public expense*, to the *utter exclusion* of those who are equally well qualified, and who are *educated at their own expense*.[235]

---

[234] Wettemann, 59.

[235] Alden Partridge and Edmund Burke, Memorial of Alden Partridge and Edmund Burke, in behalf of the State Military Convention of Vermont, praying the adoption of a plan proposed by them for the reorganization of the militia of the United States, U.S. Congressional Serial Set. Sen. Doc. No. 197, 25th Congress, 3rd Session (1839), p. 6, Law Library of Congress.

Partridge was concerned that the USMA monopolized on active duty commissions. These restrictions prevented a true meritocracy, in which all male citizens could compete equally for commissions in the U.S. Army. Partridge was particularly concerned with the loss of talent from individuals who sought entrance into military service later in life. As he explains,

> [N]o youth can be admitted into the military academy, who is under fourteen, or above 21 years of age; and by the established regulations, and the usage of the War Department, no person can be commissioned in the military establishment of the U. States, who has not been educated at West Point.[236]

Partridge feared that educated men who had reached adulthood would be permanently excluded from consideration for a commission, regardless of their academic and military competency. Partridge feared few citizens would be in the position to gain recommendations from their congressional representatives and presidential approval to enter West Point. Those who did secure nominations enjoyed free education and resulting profitable long-term employment derived from being an Academy graduate. As noted by Romaneski, "Partridge's belief that West Point had helped create a military aristocracy was not, however, entirely without merit, even if his conclusions were a bit extreme."[237]

Andrew Jackson would ignore the law and grant dozens of commissions to civilians educated outside of West Point, especially in newly formed dragoon units.[238] So while the complaint was valid in terms of the letter of the law, the practice of granting commissions conditional to a West Point education varied between presidents. Captain Partridge would compose a memorial to Congress on this subject, in 1841, requesting men educated at other institutions also be considered for active commissions (see appendix H).

Partridge also expressed concerns that the USMA was not equally benefiting the states. As the United States expanded, Western and Southern states did not have the educational infrastructure nor established educational traditions as New England. Therefore, it is not surprising that most West Point candidates who failed the entrance examinations came from the West and South. As asserted by Partridge,

> But, should any one still doubt on this subject, your memorialist would observe that Congress has already adopted the principles of the proposed plan, in practice, by repeated acts of legislation,

---

[236] Partridge, *Military Academy at West Point Unmasked*, 4.

[237] Romaneski, "Importing Napoleon," 294.

[238] Cunliffe, *Soldiers & Civilians*.

appropriating portions of the public domain for the purposes of education, and even by appropriating millions of dollars of the public revenue, to be drawn directly from the Treasury of the United States, to support the Military Academy at West Point, under the plea that it was necessary to provide for the common defence. Now, all the appropriations hitherto made for purposes of education have been partial, benefiting some States at the expense of the others, and have also aided to sustain the present anti-republican system; while the plan proposed by your memorialist would do equal justice to all the States, and insure the establishment of a system of education in the United States, that would prove the strongest support and safest protection of the liberties of the people.[239]

The disparities in the ability to pass the USMA entrance examinations reinforced the belief that West Point existed for the benefit of New England and the wealthy families of the West and South that could afford superior educational opportunities to their sons compared to what was encountered by the common family.

Is there a hereditary tradition within West Point? One of the strongest claims Partridge made against West Point is that it was becoming restricted to powerful American families. In his own words, "But your memorialist will go further, and aver that the regulations at West Point have not only constituted an aristocracy in the United States, but that this aristocracy has already become, in a great degree, hereditary."[240] Partridge's exact argument and why it violates the U.S. Constitution is often lost on the present-day audience.

There most certainly are strong family traditions within this institution. In 2012 *West Point Magazine* celebrated the legacy graduates in an article that reported,

Together [the descendants of Charles Raymond and William R. King] has become one of the longest and largest legacy families in the Academy's history. Seven continuous generations, stretching from the Class of 1825 to the Class of 2009, have made the same trek to West Point. [...] Six and seven generations are where certain families break from the pack. Three- and four-generation families crowd the pages of the West Point Association of Graduates' Register of Graduates. Even five generations are increasingly common as the years go on. Six and

---

[239] Partridge, Memorial of Alden Partridge, praying Congress to adopt measures with a view to the establishment of a general system of education, 7.
[240] Partridge, Memorial of Alden Partridge, relating to the Military Academy at West Point.

seven generations, however, require a genealogical succession that starts within the first few decades of the Academy's founding. Rene DeRussy, as the 89th graduate of West Point, is the earliest graduate to start one of these elite legacy families. After he graduated in the Class of 1812 (and then served as Superintendent from 1833 to 1838), six generations of his direct descendants—most recently James Lincoln Jr. '90—tossed their hats on graduation day.[241]

A family with a seven-generation tradition of attendance at the USMA isn't sufficient proof to assert the institution is hereditary. While some may question if the descendants of Raymond and King have not developed advantages in connecting with members of Congress to secure appointment recommendations, considerations for appointment do not award any advantage in acceptance to candidates that had parents attend the institution. In short, West Point does not operate under the same legacy system employed at most American colleges and universities. Does this disprove Partridge's claim? The answer is no.

Professor Carlton F. W. Larson may have best summarized the argument in the article, *Titles of Nobility, Hereditary Privilege, and the Unconstitutionality of Legacy Preferences in Public School Admissions*. The origin of this objection lays at the very beginning of the American Republic. The sixth article of the Articles of Confederation reads:

> No State, without the Consent of the united States, in congress assembled, shall send any embassy to, or receive any embassy from, or enter into any conference, agreement, alliance, or treaty, with any King prince or state; nor shall any person holding any office of profit or trust under the united states, or any of them, accept of any present, emolument, office, or title of any kind whatever, from any king, prince, or foreign state; nor shall the united states, in congress assembled, or any of them, grant any title of nobility.[242]

While the language within this article specifically addresses the *titles* of nobility, it was understood at the time as the prohibition against the naming of a superior class or the granting of privileges to any class in the traditions, often though heredity means, of the Old World. This article was later transferred into the U.S. Constitution:

> No Title of Nobility shall be granted by the United States: And no Person holding any Office of Profit or Trust under them, shall, without the Consent of the Congress, accept of any present,

---

[241] Marissa Carl, "West Point's Legacy Families," *West Point Magazine*, Fall 2012, 41–43.
[242] "Articles of Confederation (1777)," National Archives.

Emolument, Office, or Title, of any kind whatever, from any King,
Prince, or foreign State.

The prohibition of nobility, and its practice through heredity, seems to have
been well understood and accepted, as it generated little conversation during
the ratification process. As asserted by Larson, "there are enough statements in
the ratification period to lend strong support to an interpretation of the
Clauses as a substantive prohibition on hereditary privilege."[243]

In 1783, a test of the prohibition against nobility clause within the
United States, relating to the military community, occurred with the founding
of the Society of Cincinnati. This organization was founded by officers of the
Continental Army (many still serving within the U.S. Army),

> To perpetuate therefore, as well the remembrance of this vast event, as
> the mutual friendships which have been formed under the pressure of
> common danger, and in many instances cemented by the blood of the
> parties, the Officers of the American Army do hereby, in the most
> solemn manner, associate, constitute, and combine themselves into
> one Society of Friends—to endure as long as they shall endure, or any
> of their eldest male posterity, and in failure thereof, the collateral
> branches, who may be judged worthy of becoming its supporters and
> members.[244]

While the organization may have had noble goals, the wearing of
symbols normally associated with European nobility and hereditary
membership created public outrage. Quotes opposing this organization from
key American Founders are numerous. For example, Samuel Adams asserted
that society made as "rapid a stride towards an hereditary military nobility as
was ever made in so short a time."[245] As reported by Larson, because of
intended hereditary membership within this society,

> [I]t was denounced as illegitimate and inconsistent with the Articles of
> Confederation by people who disagreed about almost everything else.
> On what other subject did John Adams, John Jay, Thomas Jefferson,
> Samuel Adams, Benjamin Franklin, and, I think we can safely say,
> George Washington, all agree.[246]

---

[243] Larson, "Titles of Nobility, Hereditary Privilege, and the Unconstitutionality of
Legacy Preferences," 1402.
[244] "The Institution of the Society of the Cincinnati," The Society of the Cincinnati.
[245] Letter from Samuel Adams to Elbridge Gerry (April 23, 1784), as quoted by Larson,
"Titles of Nobility, Hereditary Privilege, and the Unconstitutionality of Legacy
Preferences," 1391.
[246] Larson, 1400.

The society only escaped public scrutiny by suggesting the removal of its hereditary requirements for membership. While this helped to ease the pressure, this suggestion was never ratified within the society, and it continues to operate on a hereditary membership to the present day.

Although with good intent, the USMA would also run afoul of the trapping of nobility by awarding appointments to applicants based on the service of their fathers. By 1828, Secretary of War James Barbour would state, "I eagerly seize the opportunity of canceling a debt of gratitude by the appointment of the descendants of those who have been thus distinguished by such services, *civil or military*."[247] Given that West Point appointments are more than simple scholarships but also lead to positions within the government of trust and honor, restricting any quantity of seats within the USMA for this purpose ultimately asserts that not "all men are created equal." While Andrew Jackson replaced Barbour's vision with a more egalitarian perspective, the practice of providing hereditary privileges to military families has only increased since the 19th century.

For example, in 1926, in the wake of World War I, Congress created an additional 40 cadetships at West Point that were restricted to the sons of those who died during the war. The law was extended in 1945 to grant eligibility for these cadetships to the sons of those who died fighting in World War II. In 1954 and 1966, the law was modified to extend the eligibility to children of Korean War veterans that had either died or become 100 percent disabled during the conflict. In 1972, the law was amended to also include the children of those listed as missing in action. Larson summarized the problem and provided an alternate solution:

> [T]he current law raises significant problems under the Nobility Clause, because it continues to use ancestry as a simple proxy for need. A better way of providing for the children of fallen soldiers would be to offer college scholarships to those who could demonstrate that the loss of their parent resulted in financial hardship. Such scholarships could be used at any university in the country, and would thus be a form of compensation for the lost income and opportunities due to the parent's death while in government service. They would not limit anybody else's opportunities [to gain commissions through the USMA], and thus would not be part of a zero-sum game.

---

[247] James Barbour, Letter from the Secretary of War 3 (Washington, Gales & Seaton 1828), as quoted by Larson, "Titles of Nobility, Hereditary Privilege, and the Unconstitutionality of Legacy Preferences," 1427.

A law from 1945 allowing for the president to extend unlimited West Point appointments to the children of Medal of Honor recipients pushed the institution into a new level of hereditary privilege. Unlike appointments for those who may have lost their parents because of military service (resulting in the children needing some form of welfare to compensate for the lost financial resources of a parent), providing unlimited appointments for the children of living Medal of Honor recipients could not be justified by financial need but are an explicit hereditary privilege. While the War Department objected to the proposed bill in 1942, citing the lack of need for a new admissions category and the inappropriateness of providing this small group preferential treatment, by 1945, the official response to the bill was "no objection."[248] While ultimately the U.S. military is controlled by our civilian government and will answer to the will of Congress and the direction of the president, it is disappointing that the Army does not continuously advocate for the removal of any aspect of the law that may compromise the U.S. Constitution the profession is sworn to defend.

Partridge's proposals to either abolish the USMA as it existed or convert it to a training ground for commissioned officers would have resolved these constitutional conflicts of heredity. Ultimately, the children of killed, wounded, or missing in action service members today could be compensated through scholarships and grants for education to further show the nation's appreciation for this community. These scholarships could certainly perpetuate family traditions for military service if used in institutions that have adopted Partridge's *American System of Education*. Graduates applying for military commissions within this model would do so without compromising the spirit of the Constitution. In the end, Partridge's assertion that the USMA was growing a military aristocracy has only grown increasingly true with time.

3. Fitness

The fitness level of cadets dropped dramatically after the departure of Captain Partridge from the USMA, as confirmed by the Board of Visitors. This led Partridge to claim West Point cadets were reduced to *military dandies*, interested in little else than their personal appearance. Without the experiential learning experiences provided by Partridge's pedestrian adventures, the cadets were reduced to philosophical learning that was devoid of practical application, leading to the claim they had been reduced to *military pedants*. There is little doubt that Partridge's pamphlet played a role in Representative Davy Crockett introducing legislation to end the USMA in the same year. Noting the decline in physical fitness, in 1830, Crockett asserted that USMA cadets were "too

---

[248] H.R. Rep. No. 79-898, at 2 (1945), as quoted by Larson, 1435.

delicate, and could not rough it in the army like men differently raised."[249] As asserted by Partridge,

> The truth is, (and it cannot be much longer concealed from the view of the people, by the reports of *boards of visiters*,) that the whole system of education at West Point is well calculated to form *military pedants* and *military dandies*, but will never form *efficient soldiers*. Much more important to them is their attention to the *cut* of the *coat*, the placing of a *button*, and the *snowy whiteness of gloves and pantallons*, than those *physical* and *moral qualities* which are absolutely necessary to the correct and efficient discharge of the active duties of the field.[250]

As the academic and administration requirements of the USMA increased after Partridge departed, the time investment in physical fitness training and fitness level of the cadets decreased. This led to the Board of Visitors for 1826 making the following remark: "A thorough and careful physical education is of more importance to a military officer than to any other person; but it is not offered at this Academy."[251]

Furthermore, Thayer lacked the innovation and emphasis on physical education that was integrated into Partridge's educational methods. Soon the fitness program at the USMA "denigrated into purely military drills."[252] Appearing to suffer from risk paralysis, in 1818, Thayer banned the "playing at ball" in fear of cadets injuring themselves and damaging government property.[253] This only further degraded the physical fitness of the cadet corps. It was not until 30 years after Partridge's departure from the USMA that the emphasis on physical fitness returned.[254] During this time, Partridge would build Stoic hardiness into the cadets at his American Literary, Scientific and Military Academy through frequent marches and other demanding physical activities.

The loss of physical fitness ultimately led to the claim that USMA cadets were *effeminate*. The request for a $6,000 appropriation to build a large

---

[249] David Crockett, "Resolution to Abolish West Point," February 25, 1830, United States House of Representatives: History, Art & Archives.

[250] Partridge, Memorial of Alden Partridge, relating to the Military Academy at West Point, 3.

[251] Edward S. Holden and W. L. Ostrander, *The Centennial of the United States Military Academy at West Point, New York* (Washington, D.C.: Government Printing Office, 1904), I:896.

[252] Robert Degen, *The Evolution of Physical Education at the United States Military Academy* (West Point, NY: United States Military Academy, 1967), 21.

[253] Degen, *Evolution of Physical Education*, 22.

[254] Degen, 21.

field house for cadets to exercise comfortably during the winter only furthered this claim. As future Army officers, one would expect these cadets to train in all weather conditions. Partridge would allude to the hardships of the Army during the Revolutionary War to drive home the point in proving the requested field house was not supportive of developing the hardiness that military officers require.[255]

While the present-day reader could assume that *military dandies*, *effeminate*, and *military pedants* were nothing more than gross pejorative terms used by a disgruntled former employee, there was considerable truth in these labels. This is especially true when comparing the difference between cadets educated under Thayer and Partridge. As reported by the *New York Military Magazine* in 1841,

> We will put the scholars of Captain Partridge against the West Point graduates, and they will be found in point of energy, effectiveness, and sterling worth, much superior as staunch and able officers- possessing qualities as men – qualities which nature has given them. [...] We protest against the institution at West Point, and declare it to be obnoxious to all our notions of equal rights.[256]

While the lack of physical fitness training appropriate to soldiers has been and remains corrected at the USMA, ultimately Partridge's assertion of the lack of physical fitness was true when made.

4. Violations of Law

Partridge's 1830 pamphlet also highlighted the multiple violations of law occurring at the USMA. The perceived strength of these arguments in the public eye likely varied. Highlighting the use of infantry officers in lieu of engineer officers, as specified by law, as instructors was likely perceived as an issue of little importance outside the Corps of Engineers. However, the gross expansion of the academic staff to 30 positions, when the law only allowed for eight, along with additional pay provided to cadets serving in instructor roles, would likely draw the attention of taxpayers and congressmen for the unauthorized draw on the treasury.

In a personal slight against Thayer, Partridge points out it took seven instructors plus Thayer to fulfill the duties Partridge had previously done by himself.[257] On a related note, Partridge always ensured cadets were provided instruction on all required subjects, even if he had to instruct the materials

---

255 Partridge, *Military Academy at West Point Unmasked.*
256 *New York Military Magazine*, 329–30.
257 Partridge, *Military Academy at West Point Unmasked.*

himself. Thayer, on the other hand, would simply allow cadets to miss instructions on topics in the absence of instructors.[258]

In his 1841 memorial to Congress, Partridge expressed extreme concern about the cost and utility of the board of visitors:

> Your memorialist considers the Board of Visiters that annually assemble at West Point a grievance. This board never has any *existence whatever in law*, but was established by Executive usurpation; yet, to pay the expense of this illegal board, your memorialist believes that more than fifty thousand dollars has been drawn from the public Treasury. Your memorialist earnestly solicits that this appropriation, the making of which is a direct sanction of Executive *usurpation*, should be discontinued.[259]

Partridge here could be claimed as hypocritical, as his plan to extend the military academy through the establishment of two other institutions near large cities called upon the president to establish similar boards of visitors.[260]

Captain Partridge would also hypocritically point out that the West Point regulations (which he largely composed) called for a two-month summer encampment, short of the three-month encampment required by law. Partridge had previously received a reprimand from Brigadier General Swift for ending the encampment early in order to return focus to academic studies.[261] However, there is a just question on the equal application of the law. If Captain Partridge received a reprimand for shortening the encampment, shouldn't Major Thayer have suffered the same discipline for equally violating the law? In the end, Partridge's claim of violations of law can be deemed truthful. However, it must be acknowledged that Partridge himself often disregarded these same laws for the sake of expediency.

5. Cost

In multiple documents, Partridge would assert the cost to maintain the USMA annually as $200,000 (approximately $6,746,000 adjusted for inflation to 2024 dollars). For example, in Partridge's and Burke's proposal to Congress to reorganize the militia:

---

[258] Webb, *Captain Alden Partridge.*

[259] Partridge, Memorial of Alden Partridge, relating to the Military Academy at West Point, 6.

[260] Partridge, Guidelines for Establishing Multiple Military Academies, 63–64.

[261] "For any individual who either alters, or changes the provisions of a law, assumes to himself the Sovereign authority of the nation, which is vested exclusively in the law-making power." General Swift to Alden Partridge, March 30, 1815, as quoted by Webb, *Captain Alden Partridge*, 63–64.

> Resolved, That, in the opinion of the members of this convention, the
> Military Academy at West Point is well calculated to establish a military
> aristocracy in the United States and to constitute the nucleus of a
> permanent standing army, and that, consequently, it ought to be
> abolished, and the amount which it costs the people to support it
> (about two hundred thousand dollars per annum) ought to be
> appropriated to the furnishing of instruction, and improving the
> discipline of the militia of the United States.[262]

Some scholars have suggested that Partridge exaggerated these figures. But
even if these figures were inflated, the fact remains that the government does
not have the same incentives to control costs compared to equivalent civilian
institutions.

Partridge highlighted the cost of government expenditures for building
at West Point, noting an expenditure of $115,354.15 by 1817 growing to over
$300,000 in 1830.[263] He challenged the expenditures at the USMA compared to
the academies he had established. In the words of Partridge,

> Now I would ask, could economy have been consulted, when
> $300,000 is expended in preparing an establishment for the
> accommodation of 250 young men? I have known buildings to have
> been erected, at the expense of private individuals and under private
> direction, which afforded ample accommodations for 250 pupils, with
> the greater part of their instructors, were as permanent and more
> elegant than those at West Point; and yet, with all their
> appurtenances—including the grounds attached to them—cost less
> than $40,000.[264]

The amounts provided by Partridge would assert that similar-sized
student bodies in private military academies could be supported through
infrastructure at one-eighth of the cost. Even if Partridge's report of $300,000
is to be doubted as an overestimate, if the total official expenditure of 1817 was
considered including the east barracks, it would still suggest that Partridge's
academy was 400 percent more cost-effective than West Point.

Ultimately, Partridge's argument concerning costs could be boiled
down into a single question that has no need for specific figures to answer.
Why was the federal government spending money on a military academy that
was no longer needed while the education and discipline of the militia went

---

[262] Partridge and Burke, Memorial of Alden Partridge and Edmund Burke, in behalf of
the State Military Convention of Vermont, 6.
[263] Partridge, *Military Academy at West Point Unmasked.*
[264] Partridge, 9–10.

largely unsupported? Even if one were to argue that the states and not the federal government were responsible for the disciplining of the militia, at least federal support in these areas would be in closer alignment with the spirit of the Constitution.

Neglecting the militia and building a restrictive service academy designed to "professionalize" the Active Army officer corps in a manner to specifically justify a standing military larger than the minimum required peace establishment most certainly is not within the letter or spirit of the U.S. Constitution. This question is only amplified when it is acknowledged that many USMA graduates never served on active duty. How could federal spending be justified at all on cadets when they were accepted in numbers exceeding military need?

If the purpose of the USMA was to produce competent military officers, then that role was already being fulfilled by Partridge's *American System of Education* and the private military academies he had established. There was no need for continued federal spending on the USMA that the general government lacked an enumerated power to establish or maintain. In the end, Partridge's claim here is true. There was no need for the federal government to maintain the USMA when equal educational opportunities could be supplied through civilian institutions, to include those using the *American System of Education*.

6. False History

One of Partridge's strongest claims against the USMA was in presenting a false history of the Academy. As Partridge asserts,

> I know that repeated efforts have been made, to impress the public with a belief that the whole system of organization, discipline, and indeed every thing else connected with the military academy, owes its origin to [Thayer]. I know also, that the commandant of engineers declared in an official report, made several years ago, in substance, that previous to the year 1818, which was shortly after he took charge of the institution, there was little or no organization or system in the seminary—that the cadets were admitted into it, and commissioned from it without any examination. &c. &c.[265] The assertions contained in that report on this subject, I positively deny, and assert that they are untrue in point of fact. Previous to the re-organization of the academy, under the law of the 29th of April 1812, a code of general and internal regulations, and a course of studies suited to the state of the institution

---

[265] *&c* is the archaic typesetting abbreviation for the term *et cetera* (Latin for "and other similar things"), in the present day abbreviated as *etc.*

at that time, had been drawn up by the superintendent, the immediate predecessor of Lieut. Col. Thayer, which were sanctioned by the proper authority. Under this system, the cadets were regularly examined, previous to being recommended for commissions; and it was under this system that some of the most valuable and distinguished officers, that have ever gone from that institution (several of whom were particularly distinguished in the last war) were formed.[266]

Partridge was correct in his assertion, as the organization and curriculum of the USMA was largely established by Partridge prior to Thayer's arrival. Thayer made modifications to the system; however, an unbiased review of the changes made by Thayer could hardly be asserted as a complete overhaul. According to Webb, "Generation after generation, class after class, from the first educational institution founded by the Federal Government, have been tortured into the belief that because of Thayer the Academy still exists."

This false narrative about the early days of the USMA continues its perpetuation to the present day. As of September 2024, the USMA website's "A Brief History of West Point" omits Partridge and makes the following claims about Thayer:

Major Sylvanus Thayer, considered the "Father of the Military Academy," served as Superintendent from 1817–1833. He improved academic standards, instilled military discipline, and emphasized honorable conduct and officership. Aware of our young nation's need for engineers, Thayer made civil and military engineering the foundation of the curriculum and the top few in each class became engineers. The Academy was the first engineering school in the United States. Other officers served in the infantry, cavalry (dragoons), artillery and other branches in the frontier Army.[267]

Like the false history of 1818, we see continual claims that military discipline and an engineering-focused curriculum did not predate Thayer's arrival. The desire for Army historians to scapegoat Partridge at the USMA has artificially and inappropriately raised the grandeur of Sylvanus Thayer. Given Thayer's conduct leading to Partridge's court-martial, one is left to ponder how the USMA squares its history with its own honor code. While Thayer deserves accolades in history, these are best reserved for his actual achievements with the American Academy of Arts and Sciences and his philanthropy associated

---

[266] Partridge, *Military Academy at West Point Unmasked*, 13.

[267] "A Brief History of West Point," United States Military Academy West Point, para 3.

with the Thayer School of Engineering at Dartmouth College. Conclusively, Partridge was truthful in his claim that the USMA was presenting a false history of the institution.

7. Military Training Camp

Partridge proposed an interesting solution for replacing the USMA that would avoid all the complaints he raised within his 1830 pamphlet. Partridge suggested that the USMA campus should be converted into a training ground with officers and enlisted soldiers rotating through training in six-month intervals. As explained by Partridge,

> Let such numbers of officers, non-commissioned officers, and privates as can be spared from each regiment or corps of the army, embracing all the recruits (at least) from the northern department, be sent in rotation to West Point, and constitute a school of practical instruction, let the non-commissioned officers and privates be organized into a brigade, to be officered by the officers on detachment, and by those on new appointments, for the purpose of practical military instruction. Under the present organization of the army, one thousand non-commissioned officers and privates could be detached constantly, including the recruits, for the purpose of instruction, and still leave enough to do all the necessary garrison duty. One third of the officers might also be detached for this purpose, including those on new appointments, the whole would constitute a handsome brigade, and would enable the officers to practice, and become well acquainted with all the evolutions of the line.[268]

The training interval would suggest that Army officers would rotate through this training once every 18 months, spending roughly one-third of their careers constantly engaged in brigade training cycles to truly become masters of their profession. Likewise, enlisted personnel would benefit from immediately experiencing a brigade-level training exercise that would repeat roughly every 12 years of their career.

Under this plan, officer vacancies would be filled from states in proportion to their representation in Congress, reserving 17 percent of the commissions for meritorious promotion of non-commissioned officers. New officers would be immediately put into a training rotation upon commissioning. These men would receive instruction in "practical mathematics, of field engineering of infantry and artillery duty, and also the duties of light troops—the duties of camp and garrison, and a complete course

---

[268] Partridge, *Military Academy at West Point Unmasked*, 12.

of practical military tactics."[269] (This initial officer training would be similar to the modern Army Basic Officer Leadership Course.)

Only members of the military would receive training, thus eliminating the unjustified use of funds at the USMA for cadets that would never enter military service. This shorter period of training (six months versus four years) would supply lieutenants to units faster while focusing on the practical skills needed for line officers. Engineers would be selected from the officers that demonstrated superior academic and scientific merit.

Finally, some of the newly commissioned officers would enter the Army with a preexisting bachelor-level education (both Partridge and Thayer being examples of this occurrence). Partridge would have certainly endorsed a universal exposure to his *American System of Education* prior to commissioning. The continual growth in the number of American colleges and universities removed the need for the U.S. Army to directly supply this level of education. Similarly, today one might ask why the Army continues to run an undergraduate program when there are approximately 4,000 regionally accredited colleges within the United States, with over 1,000 offering the Army ROTC program (through 273 battalions).

The adoption of Partridge's training model would have likely produced the superior level of professionalism and standardization throughout the Army that was desired. Using the USMA facilities in this manner would have had a significant impact on the proficiency of not only the Army officer corps but enlisted soldiers as well. We can only imagine what the Army might have become if they had adopted this model to place their soldiers into a rotation similar to the modern National Training Center and Joint Readiness Training Center rotations. It must be acknowledged that Partridge did indeed present a plan for the USMA that would have resolved constitutional conflicts and enhanced the training and professionalism of the Army.

In the final analysis, it must be acknowledged that Partridge had several valid claims against the USMA. Many of which continue to this day. It is unfortunate that Partridge included many other petty arguments within his 1830 pamphlet (like infantry officers being used as instructors when the law called for engineering officers), allowing USMA supporters to further paint him as a malcontent, disgraced former employee. One may speculate that if he had focused his arguments on the most egregious faults, Partridge would have had more success in resolving the errors he perceived in educating the population for national defense.

---

[269] Partridge, 11.

Survival of the United States Military Academy

Cunliffe asserted, "A closer look reveals that [Partridge's] attacks upon the Academy were more noisy than powerful."[270] The attempts to abolish the USMA failed for numerous reasons. As discussed, the restricted information disclosed to the public concerning the true occurrences surrounding Partridge's court-martial allowed the USMA supporters to cast Partridge's criticism as an attempt for vengeance from a disgruntled former employee. As noted by Cunliffe, "Those in official circles dismissed [Partridge] as a man unhinged by brooding over private wrongs."[271]

While many congressional representatives expressed support for closing the academy at West Point, none stopped making recommendations for appointments. These men did not want to lose a powerful tool for the spoils system, nor did they want to be seen as denying a free education to the less fortunate young men within their states. As explained by Ambrose, "Even congressmen who complained most about West Point usually ended up voting for it, since its elimination would deprive them of an importance source of patronage."[272]

Not to mention the West Point graduates that climbed into positions of political power within the Army and federal government, who strongly advocated in the Academy's defense. Even President Andrew Jackson, the president who came closest to closing the Academy, realized he could reward the officers he served with in the War of 1812 by appointing their sons to West Point. Many also wanted to preserve the prestige of the USMA. With the increased notoriety of European military academies, such as the École Polytechnique, many Americans wanted to maintain their national military academy to compete against their European equivalents. In the end, there were simply too many incentives to maintain the USMA than to end it.

With so many efforts by Partridge to compete against and ultimately end the USMA, there are little incentives for Army historians to create a positive record of Partridge's achievements. From historical accounts from 1818 to the military history books used within the current Army ROTC curriculum, U.S. Army historians created and maintain a false and incomplete history of Partridge's actions and achievements at West Point that has been perpetuated into the 21st century. This false history continues to tarnish Partridge's reputation, prevents full examination of his arguments against the

---

[270] Cunliffe, *Soldiers & Civilians*, 151.
[271] Cunliffe, 155.
[272] Ambrose, *Duty, Honor, Country*, 119.

USMA, and ultimately prevents the nation from recognizing the brilliance of Partridge's educational philosophy.

Arguments Never Answered / Philosophy Versus Legality

Did USMA historians perform *damnatio memoriae* (historical negation) on Captain Alden Partridge? Partridge's success in establishing private military institutions likely prevented him from being written out of history. However, false and hostile historical accounts created by the USMA community prevented Partridge's arguments against West Point from being fully examined. While this may have preserved the Academy, it prevented the needed conversations about how the United States could structure education to better embrace the foundational philosophy of America.

It is fully recognized that there is a gap between philosophical visions for the Nation and the processes that can be practically applied. However, temporary solutions and compromises in philosophy should be routinely reexamined to determine if superior methods, better aligned with our foundational philosophy, can be found. Neither avoiding the conversation nor appeals to tradition will lead the United States to become a more perfect union. In 1804, it may have been justifiable to establish the USMA, without an enumerated power, to address a small and specific unanswered need for national defense.

However, with the expansion of the purpose of the Academy and alternate educational solutions found through the creation of private military academies, we might ask if the USMA's existence can still be justified under the spirit of the Constitution. For over 200 years, alternate educational institutions providing similar curriculum to the USMA have existed within the United States. Alternately, the Army Reserve Officers' Training Corps, currently offered at over 1,000 American colleges and universities, has greater capacity to produce commissioned officers than the federal service academies. It is not an unreasonable question to ask if the USMA, with its associated costs, is even needed.

It is further recognized that as America has drifted away from its foundational philosophy, there has been an increased dependence on the letter of the law and the interpretation of the courts. This has led many Army officers to frequently reference a broad interpretation of Article I, Section 8 of the Constitution (Congress' authority to tax and spend) as justification for the maintenance of the USMA. However, this interpretation would have exceeded the interpretation of Federalism established by the American Founders. I believe that in the wake of the Long War, the U.S. Army Officer Corps could repair the lost trust of the American people by closely examining itself and petitioning Congress to remove anything found not keeping with the original spirit of our Constitution. This would include petitioning for the removal of unconstitutional hereditary advantages now offered to the children of the military community.

Figure 12—Alden Partridge's Gravestone[273]

---

[273] Captain Alden Partridge was laid to rest in the Fairview Cemetery, Norwich University. His individual gravestone simply reads, "father." The appearance of the original family gravestone is unknown to the author. In the will of U.S. Navy Captain George Musalas "Colvos" Colvocoresses, Partridges' adopted son, he directed the erection of the obelisk than now stands as the family marker. (Artist Gala S.)

Chapter VII: Alden Partridge's Educational Philosophy

This chapter will explore the educational philosophy of Captain Alden Partridge. The primary source that will be used to structure this chapter is Partridge's manuscript entitled *Lecture on Education*.[274] This document succinctly merges his major educational theories. In this lecture, Partridge lists six principal errors that he perceived in American education, followed by a detailed proposition to correct these defects. The faults noted by Partridge include incomplete education for a citizen of the Republic, lack of physical education, idle time, excess of money, universal course of study, and a prescribed length of time for completion. Examination of this lecture provides a useful means of understanding Partridge's vision of a higher education system suitable for instructing students on all aspects of American civic duties and responsibilities. This lecture is key to understanding his design for a "system of education [that is] perfectly American."[275]

This examination will continue with a discussion of additional educational theories found in other works by Captain Partridge. When appropriate, further discussions will be provided on Partridge's theories in relation to current educational practices and his larger historical impact on the nation. Possible historical/philosophical inspirations for Partridge's theories will also be briefly explored.

Defects in Education

1. Incomplete Education for a Citizen of the Republic

Partridge's primary objection to the state of education was that it was insufficient in educating students for their complete role as citizens; "It is not sufficiently practical, nor properly adapted to the various duties an American citizen may be called upon to discharge."[276] He dreamt of an educational system that would fully prepare citizens for their role in society. In his pointed

---

[274] Partridge, "Captain Partridge's Lecture on Education."

[275] Alden Partridge, Talking Points by Alden Partridge for a Speech or Publication, Approximately 1820–1829, p. 1, Norwich University Archives and Special Collections.

[276] Partridge, "Captain Partridge's Lecture on Education," 264.

questions, the responsibilities that Partridge believed the American education system should execute can be seen:

> Have they been instructed in the science of government generally, and more especially in the principles of our excellent Constitution, and thereby prepared to sit in the legislative councils of the nation? Has their attention been sufficiently directed to those great and important branches of national industry and sources of national wealth— agriculture, commerce, and manufactures? Have they been taught to examine the policy of other nations, and the effect of that policy on the prosperity of their own country? Are they prepared to discharge the duties of civil or military engineers, or to endure fatigue, or to become the defenders of their country's rights, and the avengers of her wrongs, either in the ranks or at the head of her armies? It appears to me not; and if not, then, agreeably to the standard established, their education is so far defective.[277]

Taking Partridge's objections to contemporary education and stating them in the positive, five major themes concerning the requirement to develop a complete citizen become apparent. Partridge would expect educated American citizens to 1) Be prepared to participate as members of legislative councils through an understanding of history and the field of political science to include knowledge of the U.S. Constitution and how proposed legislation would affect the United States and foreign powers; 2) Understand economics and the importance of national industries to include a working knowledge of agriculture, commerce, and manufacturing; 3) Comprehend international relations with the ability to understand how the policies of foreign nations will affect the U.S.; 4) Be capable of functioning as civil or military engineers; 5) Possess sufficient military knowledge along with physical and mental toughness to serve in the defense of the Nation, to include the capacity to serve in military leadership roles.

Before an examination of Partridge's asserted educational requirements for U.S. citizens begins, Partridge's curriculum for fulfilling these requirements drawn from his 1841 memorial to Congress will be presented.

> 1. Every thing of a sectarian character, in religion as well as politics, to be utterly and entirely excluded therefrom.
> 2. An extensive course of mathematics, theoretical and practical, with their application to civil and military engineering, and geometric operations generally, to the various departments of physical philosophy [physics], and astronomy, navigation, &c.

---

[277] Partridge, 265.

3. A complete course of physical philosophy, embracing mechanics, hydrostatics and hydraulics, pneumatics, optics, chemistry, magnetism, electricity, natural history, &c.

4. Political economy, embracing the three great departments of national industry—agriculture, manufactures, and commerce; with an examination into their mutual relations, and combined effects upon the welfare and prosperity of a nation. The subjects of currency, monopolies, and labor, would necessarily be involved in the preceding.

5. The science of government generally, in its various forms, with an examination of the objects for which it is professedly instituted; and the consequences which have resulted from it, under its several forms, in different ages and in different countries. To the foregoing should be added a thorough acquaintance with the constitution of the United States, and the principles of our republican institutions generally, embracing civil administration, &c.

6. An extensive course of history, ancient and modern, of geography, &c.

7. Moral science, mental philosophy, logic, natural and political law, laws of nations, elocution, and an extensive course of ancient and modern literature; the study of the ancient and modern languages, (not excepting the English language, to which all the students should be required to give a full share of attention) to be left optional with the students, or their parents or guardians.

8. Civil engineering, embracing the construction of roads and canals, aqueducts, viaducts, locks, bridges, topographical drawing, &c.

9. Military science and instruction, embracing permanent and field fortification, artillery, the attack and defence of fortified places, the construction of batteries, sea coast and harbor defence, garrison and camp duty; tactics, including the grand and minor tactics, the schools of the soldier, platoon, company, battalion, and the evolutions of the line, military drawing, &c.

10. Architecture.

11. Each student should be allowed to progress as rapidly as possible in his studies, consistently with the thorough understanding of the same, and not be retarded, to be kept in college with such, as might have less capacity, or be less studious than himself.

Your memorialist is well convinced that, under such a system, at the least twice as much useful knowledge would be acquired by any given number of students, in the same time, as is now acquired under the present restrictive system.

12. A course of physical education, which would preserve, the health of the students, render them vigorous and active, and prevent injury to their constitutions, however intense might be their application to study. Regular military exercises, including fencing, &c., would constitute the best system of physical education. These could be attended to at such hours of the day as would otherwise be spent in idleness or useless amusements, for which they would be a pleasing and useful substitute.[278]

Having presented both Partridge's asserted educational requirements for U.S. citizens and his proposed curriculum for addressing these needs, a detailed examination of Partridge's theories may begin.

Polymathic

The most noticeable aspect of Partridge's educational philosophy is its polymathic nature. Students were not intended to receive training in one specific profession, but to develop skills in multiple industries and in all aspects of the duty and responsibility of citizenship. Partridge's educational model would have students learning a much broader range of subjects than our present-day university system. This approach to education offers many advantages, including introducing students to a great variety of tools and assumptions used in various different academic fields. As a result, Partridge's approach to education lends itself well to drive innovation, as students may apply tools normally associated with one academic discipline in other areas. While the present-day system is built more for the purpose of gaining specialized employment, Partridge's model would offer a broad range of advantages to students in fulfilling the totality of their social roles as citizens and community members.

In our present-day system, general education requirements of a baccalaureate degree may compose up to 50 percent of a student's academic program with these 60 or so credit hours being completed in subject such as reading, writing, research, mathematics, natural sciences, social studies, critical thinking, literature, history, arts, or humanities. The remaining 60 or so credit hours would be invested into a single academic discipline representing a "major area of focus." In Partridge's model, outside the exception of a prodigy, students would not normally carry a major and would study across a broader range of academic disciplines.

Graduation in Partridge's model would occur when students demonstrated they could exceed a threshold of competency composed of the

---

[278] Partridge, Memorial of Alden Partridge, praying Congress to adopt measures with a view to the establishment of a general system of education, 4–5.

sum of competencies across a range of subjects. In other words, students would be considered learned men when they could demonstrate their ability to solve problems (theoretical and practical) across a spectrum of academic fields. While students would be encouraged to study in all the subject areas suggested within Partridge's curriculum, Partridge recognized that because of variance in individual student temperament and ability, a universal level of performance in each subject could not be expected nor demanded. By allowing students to maximize learning in areas of interest and allowing students to avoid areas of total aversion, Partridge could increase the achievements of his students while safeguarding the critical aspect of internal motivation and a love of learning. More on this subject will be presented further in this chapter.

Partridge's approach to education has many features not seen in present-day education models. These features include the universal focus on science of government, national industry, foreign relations, engineering, agriculture, military leadership, and military knowledge and ability. A brief discussion on these topics will be supplied before moving on to Partridge's second objection to the contemporary education of his era.

Science of Government

The educational model presented by Captain Partridge supports American democratic processes more than any model of education in use today. Not only would students be instructed in the U.S. Constitution and the functioning of the U.S. government; they would also be instructed on how to take an active role in local legislative councils' The level of civic engagement by common citizens in Partridge's era was significantly higher than our present-day engagement. There was strong resistance to the centralization of power, with citizens attempting to resolve problems at the lowest levels of government possible. The French diplomat Alexis de Tocqueville, visiting the United States in the 1830s, noted, "The public duties in the township are extremely numerous and minutely divided."[279] The importance of participating with and even assuming office within township, county, state, and federal government would be a significant part of Partridge's curriculum. Partridge himself served as a powerful role model of these activities, having held the office of Vermont surveyor general and multiple appointments in the Vermont House of

---

[279] Alexis de Tocqueville, *Democracy in America*, trans. Henry Reeve, 4th ed., vol. I (Philadelphia, PA: J. & H. G. Langley, 1841), 64.

Representatives, along with engaging in other unsuccessful political campaigns.[280,281]

National Industry

Partridge provided his students with instruction on the major industries within the United States. Through this process, his students would learn about the relation of industries, how Americans were being employed, current market conditions, and opportunities for future growth. Through experiential learning opportunities, Partridge's students could try their hand at various industries, from agriculture to manufacturing. This provided students with a healthy respect for industry. Additionally, students would learn the importance of and threats to American industries' interactions with foreign nations. Unlike our contemporary graduates, Partridge's students would be able to quickly assess the potential impact of political policies on the American economy.

Foreign Relations

Partridge's students were not merely to have a domestic focus; they also learned foreign policy. Partridge expected his graduates to be able to examine the policies of other nations and predict the impact of those policies on the American Republic. This international awareness would allow citizens to predict conflicts or opportunities with other nations and to absorb or avoid the progress or faults of other nations. Partridge's curriculum would allow citizens to learn how nations might best interact through cooperation and find effective means of conflict resolution.

Engineering

Engineering was one academic discipline the young Republic critically needed. The young United States was in desperate need for roads, bridges, canals, dams, and other works of civil engineering. As reported by Miller, "Partridge most favored engineering, for which he, because of the irregular and limited training of American engineers, saw enormous opportunity."[282] Given that Partridge intended for the vast majority of his students to not take up active military service, this education and training in engineering would position them well for profitable employment in this in-demand field.[283] When

---

[280] Partridge made five unsuccessful attempts in running for the United States House of Representatives from 1834 to 1840, losing to Horace Everett each time.

[281] Alden Partridge announced his candidacy for President of the United States of America in Portsmouth, Virginia, in 1841. Alden Partridge, "The Next President," by Alden Partridge, 21 May 1841, Norwich University Archives and Special Collections.

[282] Jonson Miller, *Engineering Manhood: Race and the Antebellum Virginia Military Institute* (Amherst, MA: Lever Press, 2020), 96.

[283] Miller, *Engineering Manhood*, 96.

combined with a knowledge of architecture, citizens trained under Partridge's model would have been capable of designing and building the infrastructure required for industry.

Today, engineering remains a critical field, with civil engineers in high demand. Just as our Nation needed this important field in the early days of the Republic, today our Nation struggles to keep up with infrastructure projects as our population continues to increase, exceeding the capability and lifespans of previous generations' building projects. Informed citizens knowledgeable in this academic field will be critical to securing appropriate levels of public funds to maintain and grow the American economy.

Agriculture

Agriculture was of critical importance to the early American Republic. Not only was it a major industry, but the scientific advancement in agriculture occurring in the 19th century was also dramatically affecting the productivity and cost-efficiency of this economic sector. Captain Partridge was positioning his cadets well to leave his academy and become highly successful and wealthy as farmers in the young Republic. Captain Partridge's instruction on the scientific knowledge of agriculture would predate the establishment of Michigan State University, credited as the first agricultural college in the United States.[284] His instruction would predate the local New Hampshire College of Agriculture and the Mechanic Arts (NHC) by over four decades.[285]

While modern audiences may think it is archaic to have a general focus on practical agriculture in a postindustrial society, one must remember that food security is deeply tied to national security. In both world wars, average citizens were asked to start gardens to supplement food production. By 1943, 40 percent of U.S. produce was being produced in *victory gardens*.[286] The recent supply chain interruptions caused by the coronavirus (COVID-19) demonstrated the need for practical gardening skills among U.S. citizens to build greater food security. Not to mention all the economic and environmental advantages of citizens locally producing food that would include lower costs, decreased infrastructure requirements, fewer negative environmental impacts, and increased employment opportunities.

Partridge's approach clearly defines educated Americans as ones of practical skill. In this regard, there is a significant degree of humility and

---

[284] Neil E. Stevens, "America's First Agricultural School," *Scientific Monthly* 13, no. 6 (December 1921).

[285] K. R. B. Flint, C. V. Woodbury, and C. N. Barber, eds., *Norwich University: A Record of the Celebration of the One Hundredth Anniversary of Its Founding* (Northfield, VT: Norwich University, 1930).

[286] "Gardening for the Common Good," Smithsonian Libraries.

willingness to find value in a wide range of professions that is not typically associated with intellectuals and aristocracies. In the generations after Partridge, this sentiment would be echoed by the former slave and great American educator, author, and orator, Booker T. Washington. According to Booker T. Washington, "No race can prosper till it learns that there is as much dignity in tilling a field as in writing a poem."[287] Our contemporary higher education presents an almost complete divorce from practical labor, preferring to avoid association with those seen as common journeymen, machinists, manufacturers, and mechanics (the common "jacks" of the trades).

Military Leadership

In our current university/college model, instruction in leadership skills is not typically included within the curriculum outside a select group of majors. While some students may encounter leadership experiences in extracurricular activities, many students may graduate with no formal instruction on leadership. In Partridge's curriculum, students would not only be presented with a host of materials on leadership (to include the works of Xenophon, Cicero, Tacitus, etc.), they would be expected to rotate through leadership roles. Through the course of military duties, Partridge's students learned leadership skills and practiced on the very difficult audience of their own peers. This leadership training would position Partridge's graduates well in starting businesses, entering the military, or running for government offices.

Military Knowledge and Ability

When the concept of universal military education in college is brought up to a modern audience, the pejorative term *militarism* is often used. Militarism is defined as the "desire of a government or people that a country should maintain a strong military capability and be prepared to use it aggressively to defend or promote national interests."[288] However, Partridge never intended the military training of his academy to be the primary focus, and neither he nor his students intended for the United States to become a hostile nation. Captain Partridge addresses this topic in some detail in his 1820 Letter to the Public:

> I mean nothing more than that the military should constitute an appendage to their civil education and thereby qualify them for the correct and efficient discharge of their duties as soldiers, when their country requires their services in that capacity.[289]

---

[287] Booker T. Washington, "Atlanta Compromise" (Speech presented at the Cotton States and International Exposition, September 18, 1895), para. 4.
[288] "Militarism," Oxford Reference.
[289] Partridge, Letter to the Public, 3.

Military training was an "appendage" required for educating young men in their civic obligations to protect and defend their nation if called to war. It also had additional benefits of being useful to build physical fitness and mental resilience, a subject addressed further in this chapter. Partridge makes this emphatically clear: "I beg not to be misunderstood as recommending a system of education for our youth purely military."[290] This lesser emphasis on military education can even be seen in the original title of Norwich University. Partridge elected to list the military emphasis last in his "American Literary, Scientific and Military Academy."

Our current educational model completely lacks any compulsory military training, even though American men ages 18 through 25 are required by law to register in the Selective Service System (draft). With the U.S. Military actively engaging in warfare for most of the 21st century, one might question how Partridge's model displays more of the threat of militarism than what already exists. Certainly, the total neglect of educating the American populace in the military arts and sciences did not prevent the American government from engaging in the recent decades of warfare.

Bishop Samuel Fallows, in a benediction at the U.S. Military Academy in 1910, presented the alternate term of militancy that correlates far better to Partridge's theories than Prussian militarism. According to Fallows,

> 'Militarism is the maintenance of national power by means of standing armies. Militancy is the martial spirit which should animate every breast.' I held that the American soldier is pre-eminently the man of peace. He does not seek war; he does not provoke war. It is his mission to end war in the speediest manner possible. The militant American soldier will forever prevent the militarist's rule, for he is the servant of the whole American people. He is the incarnate American flag.[291]

Unfortunately, the nuanced distinction between militancy and militarism seems to have been lost from common understanding in both the general public and the military community.

If anything, Partridge's educational model supported a militia system that would have discouraged the United States from engaging in prolonged or aggressive warfare. In his 1821 "Lecture on National Defense," Captain Partridge identifies the conversion of a land-owning militia force to a

---

[290] Partridge, 3.

[291] Samuel Fallows, "Report of Bishop Fallows," in *Journal of the Proceedings of the Twenty-First General Council of the Reformed Episcopal Church* (Philadelphia: Reformed Episcopal Publication Society, 1897), 79.

professional army fueled by military conquest as the downfall of Rome. As reported by Partridge,

> The liberties of Rome were safe while every Roman citizen, considered and felt himself a Soldier. But how fatal was the reversal, when by the operations of a system organized by Gaius Marius – the savior and scourge of Rome, and matured by Julius Caesar, with a view doubtless to the accomplishment of his ultimate object, – the final prostration of the liberties of his country – those noble and patriotic legions, which had so often in times of peril and of danger, proved the shield of their country and the terror of its enemies, became transformed into mere mercenary bands, alienated from their country, and identified in views in feelings and in interests with their leaders alone. Under the former system she was enabled to set at defiance that Prince of Generals – the great Hannibal himself – though encamped under her very walls at the head of a veteran and victorious army, while under the latter she fell an almost unresisting victim to the mere advance guard of a Roman army led on by one of her degenerate sons.

The military reforms of Gaius Marius allowed for the modernization of the Roman military and increased recruitment to win the Jugurthine War and counter an invading force of Germanic raiders; it created a standing army in which the recruited *capite censi* (unpropertied lower class) looked towards military campaigns as a way to win wealth and status. The Marian military reforms accelerated the fall of the Roman Republic and the rise of the Roman Empire as her legions became more loyal to military commanders (such as Julius Caesar) than the will of the senate and the people of Rome. As reported by Cui,

> The downfall of the Roman Republic was certain as long as it continued to ignore the necessity for changes, but without the new Marian army, leaders might not be able to centralize power so quickly; therefore, the new Marian army was a central component of Rome's transformation. Marius's leadership was the precursor of the later dictatorships, and the progression from Pompeius to Caesar to Octavius gradually eroded the republican system and replaced it with a military autocracy. The causality was quite subtle, and it took more than eighty years for this profound impact to fully reveal itself;

however, the new Marian army certainly played a crucial role in the eventual demise of the Roman Republic.[292]

Partridge's educational model was specifically constructed to counter the fear of a general loss of the "military spirit" among the American militia that has only accelerated since the War of 1812.

> If those so necessary requisites be not attended to, if the great body of American citizens do not feel that they are something more than merely nominal soldiers, our population will gradually degenerate; our militia, so emphatically styled the bulwark of our liberties and independence, will lose their military spirit, will decline and finally be destroyed; on the ruins will spring up the standing army, detached by feeling and by interest from the great mass of the people, and when this crisis arrives, it will not require the spirit of a prophecy to predict our fate from that of the most celebrated republics of antiquity.[293]

Partridge's desire was to build a large reserve of military knowledge and ability into the populace of America, who would perceive that a large standing army was both dangerous and a waste of resources. If the United States fought wars primarily with a militia force, it would discourage offensive warfare and prolonged conflicts. The language of the U.S. Constitution describes the role of the militia as a primarily defensive force; the Congress shall have the power to "provide for calling forth the Militia to execute the Laws of the Union, suppress Insurrections and repel Invasions."[294] Once the United States would employ her militia force into combat, she would immediately feel the economic and social impact of losing thousands of men that would normally be engaged in critical economic and civic functions. Because of the social effects of military conflict, there would be a general avoidance of warfare or a focus on rapidly resolving military conflicts when they could not be avoided.

Captain Partridge's approach to military education better informs citizens of the importance of their voting rights. In our current day, it is common to find citizens that cannot name a single member of the relatively small professional military maintained by the United States. These voters have little to no knowledge of the trials and hardships of military service. They cast their votes with little or no consideration of how their vote will affect this unknown professional military force. In Partridge's model, students would be

---

[292] Siwen Cui, "The Marian Military Reform and Its Effects on the Roman Republic," *Proceedings of the 2021 International Conference on Public Relations and Social Sciences* (ICPRSS 2021) (2021): 997.

[293] Partridge, Letter to the Public, 2–3.

[294] U.S. Constitution, art. I, sec. 8, cl. 15.

exposed to military training and tactics. They would come to learn of the hardships of military life and would come to a greater understanding of how their vote would affect the lives of the U.S. Military. Having a direct understanding that a vote cast by a citizen may cause the same citizen to be called to war would likely have a significant impact on the ballot box.

The best defense of the military aspects of Partridge's educational model may have come from "Norwich men" of a later era. While many in academia may constantly assert the idealist notion that the world is only a few years away from outlawing warfare and it therefore should not be taught to modern students, Partridge's model realistically deals with human nature and mankind's continual reliance on warfare. In response to a mass mailing of a pacifist pamphlet on the Norwich campus, the following appeared in the *Norwich Guidon* in 1926:

> The charge that the R.O.T.C. breeds a desire for war is hardly more sensible than to say that surgical training breeds in a doctor the desire to witness suffering and agony. The surgeon exists because suffering and agony continue. The soldier exists because the idea of force continues. But the man of medicine is no more anxious to banish pain and sickness than the soldier is anxious to see warfare go upon the scrap heap. The pacifists argue that the R.O.T.C. hampers the progress of serious efforts to outlaw the idea of force. So far as Norwich is concerned it can safely be maintained that no member of the Military Faculty is striving to produce in his students a love of war. [...] A little military like a little medicine, goes a long way—but that little is sometimes essential. [...] The Hague, prior to the great war had apparently eliminated such things as poison gas. Then it was found that between nations, as between individuals, contracts are sometimes futile. The League of Nations is a worthy attempt to make war impossible, but until the League had demonstrated indubitably that it has made war impossible the soldier will occasionally be called upon. Norwich fails to see how the R.O.T.C., in building up a citizen reserve and thereby limiting strictly the size of the professional soldier class, is hamstringing any serious and intelligent effort to make war impossible. The R.O.T.C. may interfere with maudlin attempts, but when has the maudlin been anything but a source of failure and disaster. [...] Those who insist that the change in human nature which will irrevocably outlaw force is already at hand should be asked when they expect humanity to take unto itself the attributes of the archangels.[295]

---

[295] "Editorial," *Norwich Guidon* IV, no. 31 (June 4, 1926): 2.

Educating students on the correct applications of military arts and sciences should not be viewed as equivalent to warmongering. Partridge presented a cogent argument about how this knowledge would make America less vulnerable to unneeded military conflicts and prevent the loss of civil liberties through the appropriate application of military instruction.

Brigadier General Leigh Robinson Gignilliat, superintendent of Culver Military Academy from 1910 to 1936, would echo Partridge's assertion and demonstrate how military training could be used to both motivate young men towards learning civic virtues while ultimately safeguarding against needless military action:

> The military instinct is natural to most boys, and it may be utilized to teach him valuable lessons of loyalty, patriotism and discipline without making him bloodthirsty or warlike. Several years of military training are usually quite sufficient to gratify his curiosity and satisfy his desire for military life. He is then content to enter upon commercial or professional pursuits, a citizen prepared to serve his country either in peace or war. He has gained some small conception of what the horrors of war may be, he has some taste of the arduous side of military life, and he will be logically a greater lover of peace than the boy not so trained and a most stable citizen when the hysteria of war threatens the nation.[296]

For many young boys, the thought of military adventure can be idealized. It is often seen in the early days of war how men totally ignorant of military life will flock to the nation's call. But a man who has experienced the hardships and fatigue of military life quickly has their illusions shattered. As a student grows in their knowledge of the horrific realities of warfare, their desire to charge into combat decreases. Military knowledge becomes the temper that allows men to justly weigh the cost of war against its purposes and intent. This judgement will allow men to avoid unneeded military violence while preparing them well for warfare if the call to duty is justified.

2. Lack of Physical Education

Partridge perceived the lack of "improvement of the physical powers of the students"[297] to be a grievous error in education. Partridge viewed physical fitness as something that went beyond just physical development.

---

[296] Leigh Robinson Gignilliat, *Arms and the Boy: Military Training in Schools and Colleges, Its Value in Peace and Its Importance in War, with Many Practical Suggestions for the Course of Training, and with Brief Descriptions of the Most Successful Systems Now in Operation* (Brooklyn, NY: Bobbs-Merrill Co., 1916), 81–82.

[297] Partridge, "Captain Partridge's Lecture on Education," 265.

Facing physical hardships was a means of developing a "vigorous constitution" and the capability to endure "exposure, hunger and fatigue."[298] While Partridge acknowledged how it wasn't always possible to develop this constitution, in "nine cases out of ten,"[299] he asserted it was just as easy to develop strong men as it was to allow for boys "to grow up puny and debilitated, incapable of either bodily or mental exertion."[300]

Captain Partridge is credited as being the first American education philosopher to address physical fitness.[301] Partridge prided himself on his ability to restore the health and constitutions of students that were just a short time before "feeble and debilitated."[302] While other physical education pioneers of the time, most notably Catharine Beecher, introduced calisthenics into her Hartford Female Seminary in 1827, physical exercise did not become an accepted practice in education until after the late 1830s and Horace Mann's contributions to the establishment of public education.[303,304,305] Harvard University would construct the first American college gymnasium in 1826, with medical professors appealing to students to take part in Jahn gymnastics.[306] Yale University had students petition for and receive gymnasium equipment in 1827, with President Mark Hopkins later commenting, "[O]ur students will no longer, as in former years, leave college with emaciated frames and pallid countenances, through want of proper exercise."[307]

---

[298] Partridge, 266.

[299] Partridge, 267.

[300] Partridge, 266.

[301] Degen, *Evolution of Physical Education*.

[302] Barnard, *American Journal of Education*, vol. XIII (Hartford, CT: Henry Barnard, 1863), 68.

[303] Rebecca Noel, "'No Wonder They Are Sick, and Die of Study': European Fears for the Scholarly Body and Health in New England Schools before Horace Mann," *Paedagogica Historica* 54, no. 1–2 (February 2017): 134–53.

[304] Horace Mann was appointed first secretary of the Massachusetts Board of Education. He is remembered as "The Father of American Education," as his work in establishing tax-supported elementary public education would eventually spread across the nation. Due to his belief that women were more nurturing than men, he advocated for the feminization of the teacher population.

[305] Both Horace Mann and Catharine Beecher are remembered in history as significant champions for the use of "normal schools" (teacher colleges) to prepare public educators for their duties and standardize their educational approaches.

[306] Fred Leonard, *Guide to the History of Physical Education* (Philadelphia, PA: Lea & Febiger, 1923).

[307] Mark Hopkins, as quoted by Leonard, 247.

While gymnastics seemed in vogue on many college campuses in the 1820s, it was viewed as an extracurricular activity. Participation often waned over time, with President Jasper Adams of Charleston College reporting in 1830 that the gymnasium equipment at his college built a few years earlier had fallen into ruin after neglect and infrequent use.[308] Wider student body participation would not appear in the U.S. until 1859 and the creation of collegiate sports. It could be said that Partridge was a generation ahead of his time in identifying a need for a universal educational requirement for fitness training fully incorporated within the college curriculum.

The Greco-Roman traditions, which were in political and philosophical vogue around Partridge's lifetime, stressed the concurrent training of the mind and body. This simultaneous training can be easily seen in accounts of the great Greek philosopher Socrates, often hailed as the founder of Western philosophy. Johann Jakob Brucker's anthology, a text used in the instruction of Partridge and later in his academies, recalls Socrates as

> [I]ntroducing a method of philosophising, which was happily calculated to improve the human mind and to cherish the virtues of social life, is solely to be ascribed to Socrates; a man, whose penetrating judgement, exalted views, and liberal spirit, united with exemplary integrity, and purity of manners, have justly entitled him to that distinction, which by the unanimous suffrage of antiquity he has obtained the first place among philosophers.[309]

In the work of *Xenophon's Memorabilia of Socrates*, Socrates reports the need to train both the body and soul (morals),

> I see those not exercising the bodies not able to do the works of the body, thus also those not exercising the soul I see not able to do the works of the soul; for neither are they able to do what things they ought, nor to refrain from what things they ought.[310]

Several of Plato's dialogues of Socrates are set in a gymnasium.[311] In historical context, gymnasia would have been fields or areas reserved for the

---

308 Leonard, 247–48.

309 Johann Jakob Brucker, *The History of Philosophy: From the Earliest Times to the Beginnings of the Present Century*, vol. I (Dublin: P. Wogan, P. Byrne, A. Grueber, W. M'Kenzie, J. Moore, J. Jones, R. McAllister, W. Jones, J. Rice, J. White, and G. Draper, 1792), 162–63.

310 Xenophon, *The First Part of Xenophon's Memorabilia of Socrates: With a Literal Interlinear Translation* (London: John Taylor, 1831), 26.

311 Heather L. Reid, "Plato's Gymnastic Dialogues," in *Athletics, Gymnastics, and Agōn in Plato*, ed. Heather L. Reid, Mark Ralkowski, and Colleen P. Zoller (Fonte Aretusa: Parnassos Press, 2020), 15–30.

use of physical training. This training would have been conducted in the nude (γυμνός/gymnós), lending to the naming of the area. Various orators, intellects, and philosophers seeking an audience would often frequent these locations for potential students.[312] As a result, gymnasia became associated with the cultivation of both the mind and body.

The concern for both the capabilities of mind and body are seen within Partridge's proposed plan for primary and secondary education:

> At this early age, it is dangerous to trust [young boys] much in the streets, where they are liable to personal injury, and where their morals are constantly exposed to be contaminated, from frequent exhibitions of vice in its grosser as well as more fascinating forms. Under these circumstances they are deprived of other manly and healthful exercises so absolutely necessary for developing and maturing the physical and more energies of youth, and without which, they grow up puny and debilitated, incapable of either mental or physical exertion.[313]

For Plato, this concurrent training environment created the perfect backdrop to explore Socrates' concept of excellence (ἀρετή/virtue), with scholar Heather Reid even suggesting that Plato's dialogues serve as a virtual gymnasium for the reader.[314] Plato records Socrates, within the dialogue with Phaedo, asserting, "with good grounds assume that this struggle towards completeness, this progress, this increase in inward excellence, is the destination of rational beings, and consequently is the highest purpose of creation."[315] It is doubtless that Partridge would have been drawn to Socrates' call to physical and mental excellence, with Socrates' notable military service as a Greek hoplite only further elevating the idolized status of this esteemed philosopher.

The ancient Stoics, drawing upon the Socratic ideal, continued to advance the concurrent training of the mind and body. As explained by Gaius Musonius Rufus:

> For obviously the philosopher's body should be well prepared for physical activity, because often virtues make use of this as a necessary instrument for the affairs of life... We use the training in common to both [soul and body] when we discipline ourselves to cold, heat, thirst,

---

[312] Reid, "Plato's Gymnastic Dialogues."

[313] Alden Partridge, "Communication," *National Intelligencer*, 1826, Norwich University Archives and Special Collections.

[314] Reid, "Plato's Gymnastic Dialogues."

[315] William Whewell, *The Platonic Dialogues for English Readers* (Cambridge, MA: MacMillan & Co., 1859), 439.

hunger, meagre rations, hard beds, avoidance of pleasures, and
patience under suffering.[316]

The parallels between the language of Rufus and Partridge are obvious.

The ancient Stoics used physical exertion as a means to practice and expand their physical fitness and psychological resilience. Much like how a muscle can grow stronger with use, the mind can grow its ability to remain in a calm emotional state if repeatedly exposed to voluntary hardship. Epictetus used the analogy of "Hard Winter Training." In the era of ancient Greece, warfare would stop during the winter months. The soldier who continued to train during these months would be far better prepared for combat in the spring than the soldier who did no training during this period. Thus, the Stoics believed in exposing themselves to voluntary hardships during times of relative comfort in life, so they would be fully prepared to meet adversity when it comes. As explained by Epictetus in this analogy:

> And yet a bull doesn't become a bull all at once, any more than a man acquire nobility of mind all at once; no, he must undergo hard winter training, and so make himself ready, rather than hurl himself without proper thought into what is inappropriate for him.[317]

Modern scholars such as William Irvine have further divided this theory into two parts. The first part is "Stoic adventures," the practice of engaging in activities that are likely to generate hardship or setbacks. Irvine provides the example of trying to climb a mountain in order to "be surprised by unpleasant experiences."[318]

The second part is "Stoic Toughening Training," the engagement of an activity to purposely cause hardship.[319] Irvine provides examples of fasting (to engage hunger) or direct exposure to known fears (like public speaking). While the latter example of facing a fear doesn't relate to Partridge's use of physical education, Partridge's engagement in physical education to create physical and mental stress and therefore develop an individual's "constitution" certainly could be defined as Stoic Toughening Training. Most famous for his long-range marches, Partridge employed a range of physical challenges in the

316 Gaius Musonius Rufus, *Musonius Rufus: "The Roman Socrates"*, trans. Cora E. Lutz (New Haven, CT: Yale University Press, 1947), 55.

317 Epictetus, *Discourses, Fragments, Handbook*, ed. Christopher Gill, trans. Robin Hard (Oxford: Oxford University Press, 2014), 9.

318 William Braxton Irvine, *The Stoic Challenge: A Philosopher's Guide to Becoming Tougher, Calmer, and More Resilient* (New York, NY: W.W. Norton & Company, Inc., 2019), 142.

319 Irvine, *The Stoic Challenge*, 141.

development of his cadets to include the hauling of cannons, pulled by hand, up and down hills even in heavy winter snow.[320]

Partridge the Pedestrian

Partridge became famous for his habit of what some today might describe as "extreme" walking. It is hard to fix a date of when Captain Partridge began his long pedestrian adventures, but they were likely the result of his work as an engineer and desire to survey the heights of local mountain ranges. It is interesting to ponder if Thomas Jefferson had any influence on Partridge, as Jefferson was a powerful advocate for walking. Writing to his nephew, Jefferson once remarked, "There is no habit you will value so much as that of walking far without fatigue."[321]

There is a record of Partridge taking USMA cadets into the Catskill Mountains in 1810.[322] In the 1826 catalogue for the military academy at Middleton, Partridge reports his cadets being able to "walk with facility 40 miles per day."[323] Partridge often walked great distances on multiple-day excursions while ascending and descending mountains.[324]

To put the distance in context with modern military requirements, the U.S. Army Expert Infantry Badge (EIB) requires only a 12-mile foot march.[325] The Marine Corps' boot camp "Crucible" requirement of 40 miles of foot marching over two days is the closest modern training standards to Partridge's

---

[320] John A. Guidotti, "Controversy: The Legacy of Alden Partridge," 1991, 6–7, Eisenhower Leader Development Program Papers, United States Military Academy Library.

[321] Jefferson strongly advocated walking (with a rifle) as a primary form of exercise in a letter to his nephew. It is also notable that he lists a number of Stoic authors of books he believes appropriate in a man's education within this letter. Thomas Jefferson, From Thomas Jefferson to Peter Carr, 19 August 1785, National Archives.

[322] Partridge presented the finding of research on "meteorological observations, [and] barometrical measurements of the heights of the Catskill Mountains" to the Military Philosophical Society in 1810 and later published the works in the *Home Journal and Citizen Soldier* (Philadelphia, PA) of January 1844, as reported in Sidney Forman, "The United States Military Philosophical Society, 1802–1813: Scientia in Bello Pax," *William and Mary Quarterly* 2, no. 3 (1945): 281.

[323] Partridge, Catalogue of the Officers and Cadets, 1826, p. 39.

[324] Mark Bushnell, "#15 Alden Partridge: The First Extreme Hiker," in Norwich Bicentennial, *200 Things about Norwich*, 19.

[325] Department of the Army, *USAIS Pamphlet 350-6: Expert Infantryman Badge* (Fort Moore, GA: Department of the Army United States Army Infantry School, 2023).

practices.[326] Both of these examples fall short of the distances covered by Partridge. Norwich University revived the tradition of long pedestrian marches with their cadets in 2009 with the establishment of the "Legacy March" of 50 miles over the course of three days.[327]

The great distances Partridge could cover during his pedestrian adventures are made even more remarkable when considering he didn't travel over flat terrain. Partridge climbed mountains across New England to become the first "peakbagger" before it became a popular hobby.[328] Partridge's extreme walking would inspire James P. Taylor to build the Vermont Long Trail in the early 1900s, which would inspire Benton MacKaye to create the Appalachian Trail that now runs from Georgia to Maine.[329] Thus, Partridge left a lasting legacy of hiking and mountain climbing on the Republic.

It may be impossible to determine what caused Partridge to begin his famous pedestrian journeys. It is probable that Partridge was exposed to the works of Lewis Lochée, master of the Royal Military Academy at Little Chelsea, while a cadet/professor at the USMA. Lochée specifically recommended "long marches" and "climbing steep and difficult ascents" in his *Essay on Military Education* (see appendix I).[330] Like other military sources of the era, Lochée references ancient Stoic philosophers concerning educational practices. While the USMA has this work within their library, the lack of library inventories during the tenure of Partridge makes it impossible to confirm the connection.

Partridge's pedestrian adventures resembled the ancient Stoics in a couple of ways. First, military marching is used as a means of developing strength and endurance. Plutarch recalled Cato the Younger engaging in walking in all environments to build his endurance:

> [Cato] built up his body by vigorous exercises, accustoming himself to endure both heat and snow with uncovered head, and to journey on foot at all seasons, without a vehicle. Those of his friends who went

---

[326] Charles C. Krulak, "The Crucible: Building Warriors for the 21st Century," in *The Legacy of Belleau Wood: 100 Years of Making Marines and Winning Battles, an Anthology*, ed. Paul Westermeyer and Breanne Robertson (Quantico, VA: History Division, United States Marine Corps, 2018), 313–18.

[327] Norwich University Alumni & Family, "NU Club of the Upper Valley Legacy March Breakfast," 2019. However, the current total distances covered by Norwich cadets and the speed of movement fall far short of the practices under Partridge.

[328] Laura Waterman, Guy Waterman, and Bill McKibben, *The Green Guide to Low-Impact Hiking and Camping* (New York, NY: Countryman Press, 2016).

[329] Bushnell, "#15 Alden Partridge: The First Extreme Hiker."

[330] Lewis Lochée, *An Essay on Military Education*, 2nd ed. (London: T. Cadell, 1776).

abroad with him used horses, and Cato would often join each of them in turn and converse with him, although he walked and they rode.[331] Similarly, for health, Partridge advised to:

> [W]alk at least ten miles each day, at a rate of four miles per hour;—about three to four times each year shoulder your knapsack, and, with your barometer, &c. ascend to the summits of our principal mountains, and determine the altitudes, walking from thirty to eighty miles per day, according as you can bear the fatigue—do all these, and I will insure you firm and vigorous constitutions.[332]

Partridge clearly used his pedestrian adventures in a fashion consistent with Stoic philosophy. The purpose of his marches was exposure to fatigue to build physical fitness while advancing high levels of acclimatization to the environment. Similar to the recommendations of ancient Stoic Seneca, these walks would ensure frequent exposure to nature.

Cato, being a Roman praetor and military commander who stood against Julius Caesar's overthrow of the Roman Republic, would likely be a hero to anyone who supported republicanism like Partridge. It is highly likely that Partridge had been exposed to the story of Cato through either Plutarch or the popular play by Joseph Addison.[333]

Further evidence that Partridge drew on the ancient Stoic tradition is found in a letter to the editors of the *National Intelligencer* newspaper. Partridge wrote the following passage about Greek military history in relation to the practice of climbing mountains to take geographical measurements:

> I read in ancient history that the Spartan soldiers marched from Sparta to Marathon said to be 150 miles—in those days—carrying their armor and provisions—the whole weighing about 70 pounds—that the Athenians made a formal march of 40 miles immediately after fighting the Battle of Marathon, in order to defend Athens from a naval attack of the Persians—that Euclid of Megara was in the practice of walking 40 miles in a night in order to visit his friend Socrates, and that Socrates, himself, the great philosopher of Greece, was found in the ranks, fighting the battle of his country, and cheerfully enduring all the hardships, fatigues and privation of the most hardy soldier, I feel

---

[331] Plutarch, "The Life of Cato the Younger," in *Parallel Lives*, trans. Bernadotte Perrin, Cambridge, MA: Harvard University Press, 1919), VIII:248.

[332] Alden Partridge, "Walking," *Journal of Health* II, no. 8 (December 22, 1830): 122.

[333] Partridge's students would have also been exposed to Cato through Addison's play, performed at the American Literary, Scientific and Military Academy some year between 1825–28. List of Student Actors and the Roles They Are Playing in Dramatic Productions, 1825–1828, Norwich University Archives and Special Collections.

my own inferiority and am similarly impressed with the constitution of modern degeneracy.[334, 335]

While this article was written in 1829 after the development of much of his educational theories, Partridge connects his walking to both Sparta and Socrates (being the founder of the Socratic school of philosophy that Stoicism developed from).

Partridge had knowledge of the exploits of Roman emperor Hadrian. Walking up to 20 miles a day in his armor, the Stoic-influenced Emperor Hadrian may have provided an example for military leaders.[336] Zeno of Citium, the founder of Stoicism, was well known for giving lectures while walking on the *Stoa Poikile* (Ancient Greek: ἡ ποικίλη στοά) or "Painted Porch" of Athens' marketplace. (As a result, the philosophy developed the name of Stoicism.)[337] Partridge was exposed to the ideas of Zeno through at least Plutarch.

Both Partridge and Zeno would share the intent of keeping their students moving to prevent them from lazily lounging while listening to instruction.[338] In this way, Partridge could have been claimed not only to teach Stoic tenants (through exposure to hardship) but also to provide the instruction in a method commonly employed by the Stoics (walking). Partridge objected to the contemporary use of gymnastics because of its incompatibility with lectures;

> I would now, Gentlemen, seriously ask whether the Exercise of walking which invigorates the physical Energies and [illegible] so much to the preservation of health—and which enables us to perform Scientific Excursions and Examine Countries in the best possible manner and at the least expense—to climb mountains and endure the fatigues of military services, should our Country require of us such service—[illegible] fine—which will enable us to discharge all the active duties of life better and more efficiently—be not much more advantageous and appropriate for the youth of our literary Institutions generally, than those Exercises (improperly) called Gymnastics, which are dangerous in the performance, are not as healthy as walking, have

---

[334] Alden Partridge, Letter from Alden Partridge to Gales & Seaton of the *National Intelligencer* newspaper, 3 February 1829, 11–12.

[335] Thank you, Dr. Samuel Limneos, for your help deciphering Partridge's handwriting.

[336] Aelius Spartianus, "The Life of Hadrian: Part I," in *Historia Augusta*, vol. I, trans. David Magie (Cambridge, MA: Harvard University Press, 1921), X.

[337] Donald J. Robertson, "How to Walk Like a Stoic," Stoicism—Philosophy as a Way of Life, March 1, 2020 (Medium).

[338] Donald Robertson, "The Art of Stoic Walking," *Journal of the Stoic Gym* 1, no. 8 (August 2019): 6.

no lecturing to -improve the mind, or induce manliness of deportment, and which, finally, can be applied to no useful purpose in the discharge of the various duties of life.[339]

It may have been only natural for Partridge to default back to the teaching methods of the ancient Stoic philosophers of walking while lecturing to instruct his students. Sadly, this may have led to the loss of contents of many of Partridge's military lectures, as he presented these lectures from memory and not a recorded script.

Experiential Learning

Partridge was one of the first educators to engage in what is now the acceptable practice of "field trips."[340] In this fashion, Partridge combined physical activities with experiential learning. As explained by Partridge:

The pupils should frequently be taken on pedestrian excursions into the country; be habituated to endure fatigue, to climb mountains, and to determine their altitudes by means of the barometer as well as by trigonometry. Those excursions, while they would learn them to walk, (which I estimate an important part of education,) and render them vigorous and healthy, would also prepare them for becoming men of practical science generally, and would further confer on them a correct *coup d'œil* so essentially necessary for military and civil engineers, for surveyors, for travelers, &c., and which can never be acquired otherwise than by practice.[341]

Here it can be seen how Partridge may be drawing on the ancient traditions of Sparta or the Stoics to use walking as a voluntary form of discomfort. This form of Stoic Toughening Training would also carry the extra advantages of experiential learning opportunities for his cadets. Soldiers must have a mastery of the skill of marching and this simply cannot be taught within a classroom. Mastery of this fundamental skill requires a significant amount of experience gained through marching. It is also noted that foot marches are not a singular task but are among a host of tasks required of soldiers; "A well-conducted march is a medium for developing and demonstrating the many indefinable attributes of a good soldier, a good leader, and a good unit."[342]

In 1822, cadets at Norwich University relayed the painful learning lessons shared among novice soldiers:

---

[339] Alden Partridge, Letter from Alden Partridge to Gales & Seaton of the *National Intelligencer* newspaper, 9–10.

[340] Norwich University, "History of NU."

[341] Alden Partridge, Lecture on Education, 1825, The British Library, 10.

[342] Department of the Army, *Field Manual FM 21-18: Foot Marches* (Washington, D.C.: Government Printing Office, 1950), 7.

At this hour of the day, the sun poured forth its burning rays, which together with the dust of the road, that now in clouds covered us, produced in some degree the effect of fatigue, and many began to complain of blistered feet, a consequent evil which the inexperienced pedestrian has always to suffer in the outset.[343]

The journal goes on to not only display how Partridge's cadets participated in these Stoic toughening training exercises, but they also explicitly understood their purpose. From his Class of 1822,

Among other benefits which we had realized in our tour, the happy effects of industrious and well regulated habits, which often form the man anew, had now been experienced. During this our short excursion, a test, in some degree, of the force of habit was afforded, and the lesson was a lesson of improvement. Many of our fellow Cadets, with their equipage for a burthen, their arms and accoutrements, scarce advanced to the age of 14 years, were at no time the sufferers of any degree of inconvenience from the travel of one day and another in succession. With an atmosphere heated almost to an insufferable degree, a road sometimes deep in sand, and then rough and mountainous, many of our younger brothers were unwilling to acknowledge their inability to vie with any in our corps.[344]

From the Class of 1824, "These are productive of the highest benefits to the members of the Institution, in as much as they tend, by a moderate share of fatigue, measureably [sic] to invigorate the constitution."[345] While these passages do not specifically reference Spartan/Stoic philosophy or materials, it clearly displays that cadets understood the philosophical "why" of the activities in which they were engaging.

Partridge's view on physical education was influenced heavily by the Western philosophic and military tradition. Partridge drew inspiration from the Spartan agoge and the philosophy of Socrates and the Stoics. The emphasis of transitioning from learning to doing can be seen within philosophers back to Socrates. Within Stoic philosophy, "Virtue is divided into two parts, *contemplation* and *action*. The one is delivered by institution, the other by admonition: one part of virtue consists in discipline, the other in exercise: for

---

[343] American Literary, Scientific and Military Academy, *A Journal of an Excursion Made by the Corps of Cadets, of the American Literary, Scientific and Military Academy, under Capt. Alden Partridge. June, 1822* (Concord, NH: Hill and Moore, 1822), 13.

[344] American Literary, Scientific and Military Academy, *Journal of an Excursion*, 37.

[345] Joseph Dana Allen, *A Journal of an Excursion, Made by the Corps of Cadets of the A.L.S. & M. Academy, Norwich, Vt. under Command of Capt. A. Partridge, June, 1824* (Windsor, VT: Simeon Ide, 1824), 8.

we must first learn, and then practice."[346] Partridge's pioneering use of field trips to allow his students to practically apply the instruction they were learning from the classroom ultimately led the way for greater use of experiential learning in American education.

Collective Improvement

While Partridge addresses the improvements encountered on a personal level, it is undoubtable that collective improvements would also have occurred within the student body. These collective improvements were similar to those noted in the "Crucible" event in the Marine Corps Recruit Training introduced by General Charles C. Krulak in the late 20th century. Like Partridge, Krulak exposed Marine recruits to shared hardship. However, Krulak's intent was not specifically individual improvement but improvement of the team. The "Crucible" provided Marine recruits with greater insights on the "value of working together in a common cause to overcome the most arduous tasks and conditions."[347] Undergoing this experience, marine recruits "reinforc[es] the teamwork and values that allowed them to prevail in times of duress and hardship."[348] Partridge's cadets had similar collective reactions to their long pedestrian journeys. These marches would have provided individuals a chance to improve upon their own skills while also allowing cadets to develop their leadership abilities by sharing their knowledge and encouraging their comrades.

3. Idle Time / 6. Prescribed Length of Time for Completion

Partridge's third and sixth objections to the state of education concern the use of time. For this reason, these two points will be examined together. The efficient and productive use of time is a sentiment shared by both Partridge and Stoic philosophers.

The amount of idle time within contemporary educational practices was a concern to Partridge. As he explains,

> A third defect in our system is, the amount of idle time allowed the students; that portion of the day during which they are actually engaged in study and recitations, under the eye of their instructors, comprises but a small portion of the whole; during the remainder, those that are disposed to study, will improve at their rooms, while those who are not so disposed, will not only not improve, but will be

---

[346] Lucius Annaeus Seneca, *Seneca's Morals of a Happy Life, Benefits, Anger and Clemency*, trans. Roger L'Estrange, new ed. (Chicago, IL: Belford, Clarke & Co., 1881), 140.
[347] Krulak, "The Crucible," 316.
[348] Krulak, 317.

very likely to engage in practices injurious to their constitutions and destructive to their morals. If this vacant time could be employed in duties and exercises, which, while they amuse and improve the mind, would at the same time invigorate the body and confirm the constitution, it would certainly be a great point gained.[349]

Partridge was concerned that excessive free time may lead cadets to engage in vice. Partridge believed that this time could be better used in building both the body and mind of the student. He continues to address his concerns with the pacing of education in his sixth objection to education:

A sixth defect is the prescribing the length of time for completing, as it is termed, a course of education. By these means, the good scholar is placed nearly on a level with the sluggard, for whatever may be his exertions, he can gain nothing in respect to time, and the latter has, in consequence of this, less stimulus for exertion.[350]

Partridge's concern with the appropriate use of time is visible, asserting the standard length of study was a "sluggard's" pace. Partridge wanted to create an environment where students came, demonstrated proficiency, and were promptly sent to work in their profession. The desire for accelerated program completion can be traced back to Partridge's time at West Point. He details this desire in a letter to General Joseph G. Swift on March 5, 1817.[351] It is unfortunate that the USMA never adapted this enlightened approach to education, as the institution continues to operate on a four-year model,[352] first established in July 1816,[353] with notable exceptions only in times of major war.

Partridge instituted a timesaving approach within his academies. As seen in the following passage from the 1821 prospectus:

Every student will be allowed to progress as rapidly as he possibly can, regard being had to a thorough understanding of the branches in which he engages. No young man of talents and industry will be retarded in his progress, for the purpose of being kept in a class with

[349] Partridge, "Captain Partridge's Lecture on Education," 266.

[350] Partridge, 269.

[351] Letter from Alden Partridge to General Joseph G. Swift, March 3, 1817, National Archives, as quoted by Webb, *Captain Alden Partridge*, 101.

[352] "Academic Program." United States Military Academy West Point.

[353] Adoption of a four-year model approved by Secretary of War William H. Crawford. Barnard, *Military Schools and Courses of Instruction*.

those of a contrary disposition. Any period from three months to six years may, doubtless, be usefully employed at the institution.[354] This approach would benefit from a host of educational advantages (further discussed below).

The ancient Stoic philosophers also stressed the importance of appropriately using time. They understood that life was finite and could be easily wasted. According to Seneca, "It's not that we have so little time, but that we waste much of it. [...] So it is: the life we are given isn't short but we make it so; we're not ill provided but we are wasteful of life."[355] Diognetus took great care to teach a young Marcus Aurelius to avoid quail fighting, what Donald Robertson suggests is an equivalent to modern video games and other popular amusements that were merely trivial matters that waste valuable time.[356] Marcus Aurelius advocated meditation on the following if an individual found it difficult to start the day:

> In the morning, when you find yourself unwilling to rise, have this thought at hand: I arise to the proper business of man, and shall I repine at setting about that work for which I was born and brought into the world? Am I equipped for nothing but to lie among the bed-clothes and keep warm?[357]

With every passing moment drawing an individual closer to an inevitable death, Stoics viewed procrastination as a waste of life. They wished the wise would come to guard their time as much as they did the rest of their property. As stated by Seneca, "Men are thrifty in guarding their private property, but as soon as it comes to wasting time, they are most extravagant with the one commodity for which it's respectable to be greedy."[358]

Partridge kept his students on a rigorous schedule to avoid the waste of time and tendency for vice that he feared. Each day, he allotted a healthy eight hours for sleep. Another eight hours would be dedicated to study and recitation. Two hours were dedicated to military drills or other physical

---

[354] Alden Partridge, Catalogue of the Officers and Cadets of the American Literary, Scientific and Military Academy. Norwich, Vermont, August, 1821, p. 13, Norwich University Archives and Special Collections.

[355] Lucius Annaeus Seneca, "On the Shortness of Life," in *Hardship and Happiness*, trans. Elaine Fantham et al. (Chicago, IL: University of Chicago Press, 2014), 1.3–4.

[356] Donald Robertson, *How to Think Like a Roman Emperor: The Stoic Philosophy of Marcus Aurelius* (New York, NY: St. Martin's Press, 2019).

[357] Marcus Aurelius Antoninius, *The Meditations of the Emperor Marcus Aurelius Antoninus: a New Rendering Based on the Foulis Translation of 1742*, trans. George W. Chrystal (Edinburgh: Otto Schulze & Co., 1902), Book V, 1.

[358] Seneca, "On the Shortness of Life," 3.1.

exercises. Three hours were allotted for meals and duties. The final three hours were "devoted, in due proportion, to practical agricultural and scientific pursuits and duties, and in attending lectures on the various subjects before mentioned."[359]

On Sunday, religious service was mandatory, followed by study;[360] "every Cadet will remain at his own room. The reading and study of the Holy Scriptures, and also of other christian [sic] and moral books, is earnestly and urgently enjoined upon all."[361] With nearly every waking moment focused on self-improvement, Partridge's schedule would be very similar to the monastic life of a religious order. Partridge's approach seemed to have its desired effect with Harmon recording the students of Norwich, "More than all other students we have ever had—they were—time conscious; they had to make good and get going."[362]

Ultimately, Partridge's focus on productive use of time enabled students to move through the curriculum as fast as their learning acquisition allowed, which greatly benefited the student and society. Students who could accelerate their learning would have their interval motivation rewarded through cost-savings, as Partridge's tuition was charged by the time students remained in his institution. Those who could learn faster would not only see the overall cost of their education lowered but could also start their productive lives months or years earlier than other educational models.

4. Excess of Money / Negative Influence of Luxury

Partridge's fourth objection to the contemporary education model was that some students had an excess of money. While Partridge focused his attention on money, the actual issue he raises is the negative impact of the luxury that access to excessive funds might cause. As recorded by Partridge,

> A fourth defect is, the allowing to students, especially to those of the wealthier class, too much money, thereby inducing habits of dissipation and extravagance, highly injurious to themselves, and also to the seminaries of which they are members.[363]

Just as the Spartans would have those attending the agoge possess little belongings, the Stoics believe that wealth was created by having the least

---

[359] Partridge, "Captain Partridge's Lecture on Education," 278.

[360] Alden Partridge, Internal Regulations for the Military Academy at West Point, State of New York, Approximately 1814–1817, p. 5, Norwich University Archives and Special Collections.

[361] Partridge, Catalogue of the Officers and Cadets, August, 1821, p. 15.

[362] Harmon, *Norwich University, Its Founder and His Ideals*, 24.

[363] Partridge, "Captain Partridge's Lecture on Education," 266.

desires. According to Seneca, "It is not the man who has too little who is poor, but the one who hankers after more."[364]

Partridge felt that an excess in free time combined with money to buy luxuries would likely produce students "corrupt in morals, and instead of going forth into the world to become ornaments in society, they rather are prepared to become nuisances to the same."[365] How excesses in wealth and luxury are best detailed by Partridge in his plan for primary and secondary education (see appendix J):

> Let us suppose a youth eight or ten years of age, furnished with his regular weekly allowance of pocket money, 12, 25, or 50 cents, as the case may be, the fact of his having money in his pocket draws him to same public place, confectioner's shop, grocery, or tavern, for the purpose of spending it. He probably at this early age, will commence with purchasing sugar plums, candy, &c. but soon [discover] that boys a little older than himself, drink lemonade, and he readily imitates their example. Soon lemonade becomes insipid, and punch is substituted in its place. As he grows older, he deems it manly to smoke, and cigars are consequently put in requisition. In the train of these practices, naturally and almost certainly follows profanity, gaming, and retinue of other evils, which ultimately involve the unhappy subject in disgrace and ruin.[366]

In this example, Partridge explains how excesses in funds could draw youths to needlessly buy luxurious indulgences and, as a result, to draw into contact with poor role models that may further accelerate bad behavior.

Partridge advocated for the efficient use of time, as discussed earlier, combined with students being provided with all supplies needed for their education with no extra money provided:

> When youths are sent to a seminary, it is presumed they are sent for the purpose of learning something that is useful, and not to acquire bad habits, or to spend money; they should consequently be furnished with every thing necessary for their comfort, convenience and improvement, but money should in no instance be put into their hands.[367]

---

[364] Lucius Annaeus Seneca, *Letters from a Stoic*, trans. Robin Campbell (Harmondsworth: Penguin Books, 1969), Letter II.

[365] Partridge, "Captain Partridge's Lecture on Education," 267.

[366] Partridge, "Communication."

[367] Partridge, "Communication."

Partridge codified his thoughts in numerous draft manuscripts concerning internal regulations while at the USMA to include severe punishment for those who would make unapproved purchases.[368]

While not all cadets would have had access to excess funds, this prohibition of unauthorized purchases likely resulted in more equality among the cadets. The experiences of a son of a poor farmer and the son of a rich banker at the USMA or Norwich University were largely standardized by prohibiting the wealthier from purchasing additional luxuries.

The "long gray line," a term used to describe the long continuation of United States Military Academy graduates, is one of the most notable features of this theory still in practice today. Partridge, while selecting new cadet uniforms in 1814, selected the color gray over blue partially because of its reduced cost.[369] As Partridge founded other military academies, he changed the uniform color to blue but retained a simple style. After this fashion, the uniform was cheaper than clothes typically worn by civilian students.[370] Partridge led by example in his simple manner of dress. During his time at West Point, he wore his uniform so often that it was believed that he didn't own any civilian attire.[371]

Partridge's avoidance of luxury even extended to sleeping surfaces. With military members unlikely to find feather beds on military campaigns, he banned them within his American Literary, Scientific and Military Academy: "The students will be required to sleep on matrasses, or straw-beds, no feather-beds will be allowed in the establishment."[372] In *Meditations*, Marcus Aurelius reported taking "a liking to the philosopher's pallet and skins."[373] An emperor of Rome, Aurelius was so moved by Stoic Toughening Training that he preferred a camp bed with rough furs more than comforted bedding and soft linen because it would build the "Grecian discipline, belong[ing] to that (philosophical) profession."[374] The use of camp beds was likely another

---

[368] Punishments included reprimand, confinement, suspension, or dismissal. Partridge, Internal Regulations for the Military Academy at West Point, State of New York, Approximately 1814–1817, p. 3.

[369] John Robert Elting, ed., *Military Uniforms in America: Years of Growth, 1796–1851*, vol. II (San Rafael, CA: Presidio Press, 1977), 96.

[370] Alden Partridge, Letter from Alden Partridge to John Wright, 16 February 1819, Norwich University Archives and Special Collections.

[371] Ambrose, *Duty, Honor, Country*, 44.

[372] Partridge, Catalogue of the Officers and Cadets, August, 1821, p. 10.

[373] Antoninius, *Meditations*, I.6.

[374] Antoninius, I.6.

adoption of Spartan culture by the Stoics. (Discussion on Partridge's attire and desire to avoid luxuries will continue in chapter VIII.)

There was a fear that the availability of excess money among cadets would lead to an increased likelihood of gambling and games of chance. Partridge strictly forbade these activities both at the USMA[375] and ALSMA.[376] The Stoics recognized gambling as a form of escapism that distracted individuals from living productive and virtuous lives. Marcus Aurelius, in the first chapter of the *Meditations*, thanks his teacher for not encouraging him to support specific charioteers or teams during the circus:

> My First Teachers: Not to support this side or that in chariot-racing, this fighter or that in the games. To put up with discomfort and not make demands. To do my own work, mind my own business, and have no time for slanderers.[377]

The avoidance of gambling also helps to avoid other vices, like cheating, that often accompanies this activity.

With the emphasis on avoiding luxuries among his students, Partridge prepared them well for the lifestyle of military leaders on campaign. By living with little, cadets trained under this method would have a better means of evaluating what was necessary and what was truly a luxury for the military man. This approach parallels the following passage from Seneca: "Until we have begun to go without them, we fail to realize how unnecessary many things are. We've been using them not because we needed them but because we had them."[378]

The purpose of Partridge's restrictions on money directly parallels the intent of the mythical Lycurgus' forbiddance of gold and silver coinage in Sparta. Using iron money prevented the importation of luxury items, negated

---

[375] "Gambling & games of chance of every discription [*sic*] either with or without cards are strictly prohibited." Alden Partridge, Regulations for the Military Academy at West Point during Vacation, 20 December 1813, p. 2, Norwich University Archives and Special Collections; Partridge, Internal Regulations for the Military Academy at West Point, State of New York, Approximately 1814–1817, p. 3.

[376] Rule 8: "Gambling and games of chance of every description, either with or without cards; all profane and vulgar language; every species of irregular or immoral conduct, or of conduce contrary to the true principles of correct military discipline ; all scuffling, and all unnecessary noise, at any time or place, are strictly prohibitd [*sic*]." Partridge, Catalogue of the Officers and Cadets, August, 1821, p. 11.

[377] Marcus Aurelius, *Meditations: a New Translation, with an Introduction, by Gregory Hays*, trans. Gregory Hays (New York, NY: Modern Library, 2002), I.5.

[378] Seneca, *Letters from a Stoic*, Letter CXXIII.

the influence of wealth, and ended lawsuits among the Spartans. As recorded by Plutarch:

> For the iron money could not be carried into the rest of Greece, nor had it any value there, but was rather held in ridicule. It was not possible, therefore, to buy any foreign wares or bric-à-brac; no merchant-seamen brought freight into their harbours; no rhetoric teacher set foot on Laconian soil, no vagabond soothsayer, no keeper of harlots, no gold- or silver-smith, since there was no money there. But luxury, thus gradually deprived of that which stimulated and supported it, died away of itself, and men of large possessions had no advantage over the poor, because their wealth found no public outlet, but had to be stored up at home in idleness.[379]

Xenophon recorded that this iron money was made so impractical that it took a wagon to move even small sums and would be impossible to move without being noticed.[380] Partridge duplicated the effects of the avoidance of private or "hidden" transactions by requiring all cadets to have written preapproval for purchases on the threat of expulsion for failing to do so.[381]

The Spartan prohibition of coins and silver money caused a cultural shift from valuing aesthetics to one that valued utility. In the way, Spartans valued a kothon, a cup made to filter sediment with great utility for soldiers, over a cup of gold.[382] Archeologists continue to debate the actual existence of iron money due to the lack of archeological evidence. The Spartans did adopt the coinage of the countries they interacted with. The iron money of Sparta could theoretically have been lost in time because of rust, or as Mitchell speculates, mentions of such currency in fact refer to Sparta using pig iron, given its high value in the era, as a basis of trade with other nations.[383] Regardless, if Spartan iron money was a historical fact or simply a legend, Partridge would have likely believed the reports of Xenophon and Plutarch. By duplicating these restrictions, Partridge likely sought after the same reported

---

[379] Plutarch, "The Life of Lycurgus," in *Parallel Lives*, vol. I, trans. Bernadotte Perrin (Cambridge, MA: Harvard University Press, 1914), 231.

[380] Xenophon, *The Works of Xenophon*, vol. II, trans. Henry Graham Dakyns (London: Macmillan & Co., 1892).

[381] Rule 16, "16th. Every Cadet is strictly prohibited purchasing any article or articles, or trading in any manner whatever, without a written permission from the Superintendent, and all accounts previous to being settled, must be examined and approved by the Superintendent." Partridge, Catalogue of the Officers and Cadets, August, 1821, p. 11.

[382] Plutarch, "The Life of Cato the Lycurgus," in *Parallel Lives*, I:205–303.

[383] H. Michell, "The Iron Money of Sparta," *Phoenix* 1 (1947): 42.

goals as Lycurgus. Partridge hoped to eventually shift his students' preference for luxury and escapism to utility and productivity.

While not related to money, Partridge had another practice to avoid luxuries worthy of remembrance. Partridge's restriction on bathing hours runs parallel with the desire to avoid luxury and engage in Stoic toughening training. As stated in the ALSMA catalogue,

> Frequent bathing is recommened [*sic*] to all the Cadets, as equally conducive to health and cleanliness. But every one is strictly prohibited going into the water during the warmer part of the day, or at any time after 8 o'clock. A. M. The days and hours for this purpose will be from time to time specially designated by the Superintendent.[384]

By this rule, bathing would be restricted to the colder (uncomfortable) periods of the day. This would duplicate both Spartan[385] and Stoic[386] practices.

Influence of Daniel Defoe

Starting with the Norwich Class of 1826,[387] the avoidance of luxury would have been further reinforced through the use of the first book in Daniel Defoe's *Robinson Crusoe* trilogy within the curriculum.[388] Partridge employed a Spanish translation of this work to aid in foreign language instruction while taking advantage of the use of a Neostoic archetypal story to reinforce this philosophy among his students. This novel captures the first half of the life of a character named Robinson Crusoe as he ventures into the world looking for adventures against the wishes of his father. Eventually Crusoe is marooned on a remote island, has a religious conversion experience, and embraces Neostoic philosophy as he struggles to survive 28 years, largely alone, before his rescue.

---

[384] Rule 17. Partridge, Catalogue of the Officers and Cadets, August, 1821, p. 11.

[385] "With a view to attack luxury still more and remove the thirst for wealth [...] surrendering them to every desire [... including] hot baths." Plutarch, "The Life of Lycurgus," in *Parallel Lives*, I:234.

[386] Seneca encouraged the use of cold baths. Victoria Rimell, "The Best a Man Can Get: Grooming Scipio in Seneca Epistle 86," *Classical Philology* 108, no. 1 (2013): 1–20.

[387] Partridge, Catalogue of the Officers and Cadets, 1826, p. 22.

[388] The full title of the first book in Defoe's trilogy being *The Life and Strange Surprizing Adventures of Robinson Crusoe, of York, Mariner: Who Lived Eight and Twenty Years, All Alone in an Un-Inhabited Island on the Coast of America, Near the Mouth of the Great River of Oroonoque, Having Been Cast on Shore by Shipwreck, Wherein All the Men Perished but Himself: with an Account How He Was at Last as Strangely Deliver'd by Pyrates*. It wasn't possible to determine the exact publication that was used within Partridge's classroom. The books were supplied through Oliver Dudley Cooke & Co. and could have possibly been the *New Robinson Crusoe* abridgement of Defoe's original work written in 1779 by Joachim Heinrich Campe. Cooke, Letter from O. D. Cooke & Co. to Isaac Partridge, Approximately 1825–1829.

Defoe used this narrative to explain Neostoic philosophy and its applications reporting within the third book by reporting, "The Fable is always made for the Moral, not the Moral for the Fable."[389]

In this story, Crusoe clearly values items based on their utility and not traditional societal norms. Coinage has no utility if isolated; thus, a useful tool becomes more valuable than gold. This work would have challenged students to reevaluate value assessments of items they encountered. It is unknown if the USMA cadets were exposed to *Robinson Crusoe* as part of their curriculum in the early years of West Point. However, the USMA Library holds multiple copies of this work of an appropriate age to have been used by cadets during Partridge's tenure. Besides Defoe's *Robinson Crusoe*, the USMA Library also holds copies of Defoe's *An Essay upon Projects* (see appendix K). With the similarities between Partridge's plan for land-grant colleges and Defoe's royal military academy, it suggests Partridge may have been heavily influenced by Defoe.[390]

5. Universal Course of Study

Partridge's fifth objection to American education was the use of a single curriculum for all students. As Partridge explains,

> Every youth, who has any capacity or inclination for the acquirement of knowledge, will have some favorite studies, in which he will be likely to excel. It is certainly then much better that he should be permitted to pursue those, than, that by being forced to attend to others for which he has an aversion, and in which he will never excel,

---

[389] Defoe, *Serious Reflections during the Life and Surprising Adventures of Robinson Crusoe with His Vision of the Angelick World* (Ship and Black-Swan in Pater-nofter-Row: Printed for W. Taylor, 1720), A2.

[390] As explained by Sanftleben, "Just as Justin Morrill's land grant proposal reflected elements of Partridge's philosophy, Partridge's ideas were in turn quite similar to those expressed in Daniel Defoe's *Essay on Projects* (1697/1970). Defoe proposed the King of England establish an academy to teach all who were interested "the necessary arts and exercises to qualify them for the service of their country. ..." He recommended that its structure be "all military" and that students "be disciplined under proper officers." Suggested studies included geometry, astronomy, history, navigation, arithmetic, trigonometry, architecture, surveying, fortification, and entrenching, as well as "all the customs, usages, [and] terms of war." Also, all students would participate in "exercises for the body," especially, swimming, riding, fencing, running, leaping, and wrestling." Kurt Allen Sanftleben, "A Different Drum: The Forgotten Tradition of the Military Academy in American Education" (EdD diss., The College of William & Mary, 1993), 28–29.

or ever make common proficiency, he should finally acquire a dislike to all study.[391]

It is very interesting that Partridge used Blaise Pascal, a Christian apologist and Anti-Stoic, as his example:

The celebrated Pascal, is a striking instance of the absurdity and folly of attempting to force a youth to attend to branches of study, for which he has an utter aversion, to the exclusion of those for which he may possess a particular attachment. Had the father of this eminent man persisted in his absurd foolish course, France would never have seen him, what he subsequently became, one of her brightest ornaments.[392]

Blaise Pascal was a 17th-century child mathematical prodigy who had a great dislike for the traditional study of classical languages. Pascal would go on to create one of the first mechanical calculators. Barnard may provide insight into Partridge's selection of Pascal as an example. Barnard reports, "[F]rom the lateness with which [Partridge] commenced his Latin and his subsequent declarations, his aversion was for the languages."[393] Partridge likely used Pascal as an example because of a shared childhood dislike for foreign languages (although Partridge would overcome this fault in his adulthood).

Partridge seems to advocate for the allowance of extreme specialization in the case of genius. This may display a large degree of tolerance for monomathic individuals who excel in a single field. It is rather interesting as Partridge began his education theory asserting that education wasn't "liberal" enough to prepare students for all the duties of American citizenship.[394] This may suggest that Partridge had a broader range of talent management than would have been the norm either in his era or even today.

Partridge's default curriculum would develop students into highly functional polymaths with skills and knowledge in a wide array of practical and academic fields. As a result, students who progressed through this curriculum would become natural generalists capable of addressing complex problems through interdisciplinary approaches.[395] However, if students displayed no talent, ability, or interest in a subject, they would be encouraged to engage more heavily in areas that held their interest. In this fashion, Partridge's

---

[391] Partridge, "Captain Partridge's Lecture on Education," 268.

[392] Partridge, 268.

[393] Barnard, *Military Schools and Courses of Instruction*, 833.

[394] Partridge, "Captain Partridge's Lecture on Education," 268.

[395] Franklin C. Annis, "Future Army Leaders: Expert Specialists or Master Generalists?" From the Green Notebook, June 5, 2018.

approach would safeguard the intrinsic motivation of students who would be required as citizens to support their continuous self-directed learning.

Conclusion

Partridge's educational theories offer significant advantages in using time, practicality, and maintaining student motivation over the competing educational practices of his day and our contemporary college/university model. Many of his theories were generations ahead of his time. Certainly, notable educational scholars of the 20th century like Knowles, Ryan, and Deci would confirm the validity of many of Partridge's practices. Partridge led the way in experiential learning, introducing his students to a wide range of industrial experiences while exposing his students to the physical fitness and psychological resilience-building fatigue of long marches. He was careful to safeguard the critical aspect of internal motivation in learning. He recognized and adapted to the individual needs of students on a level unseen today.

Partridge's model placed many of the burdens of learning on the instructor to optimize the speed of learning for individual students. This tailored pace of learning magnified the internal motivation of students to progress rapidly through their course of study and thus returned months or years of productivity back to the students, which would not have occurred under an arbitrarily paced educational approach. Partridge built a "complete" educational model for the needs of the citizens of the American Republic. His polymathic approach blended theoretical knowledge and practical application in a scope unseen in our contemporary era. Partridge's Spartan educational environment would have normalized the experiences of students regardless of their social-economic background. The frugality learned from this environment would certainly serve graduates well as they entered private business or military service.

Partridge's educational model presented an ideal mix of subjects and domains within his curriculum. Yet, that curriculum was easily adapted for students who had an unusual disdain for specific subjects to safeguard the critical aspect of internal motivation for learning. In this way, Partridge was well prepared to present a polymathic education to typical students while being able to defend and fully support savants. This approach truly allows for the self-actualization of all students who experienced Partridge's model.

Make a list of problems with our contemporary college and university systems. I would be truly surprised if many of the faults were not noted and addressed within Captain Partridge's *Lecture on Education*. Would we not want to see our colleges and universities control the cost of their tuition and structure the pace of education to meet the unique needs of the individual student? Do we not want to see students taught how to maintain their physical fitness and be well-developed in their psychological resilience? Do we not want to see our students well informed on the system and function of their government so they can fully participate in the democratic process? Would it not be better to educate all students in the experience of a soldier to dissuade the Nation from needless and endless conflicts? Would we not want to see graduates educated in multiple fields and not economically trapped in a single specialty? And yet the answers to resolving these questions have existed largely unused for two centuries. It is past time we remember the wisdom of Captain Alden Partridge and embrace his *American System of Education.*

Chapter VIII: Partridge the Stoic

There was an incredible range of philosophies taught within Alden Partridge's curriculum. From the Scottish common sense realism of Dugald Stewart to the empiricism of the Enlightenment philosopher of John Locke to the utilitarianism of William Paley, students of the American Literary, Scientific and Military Academy would be well versed in a range of philosophies.[396] However, there is one ethical tradition that is especially important to highlight due to its close connection with the Western military tradition, modern democracies, and mental well-being. This critical ethical tradition comes from Stoic philosophy.

One may ask, "Was Captain Alden Partridge a Stoic?" I would have to answer emphatically yes. His actions and behaviors are in keeping with the finest aspects of the Stoic/Neostoic tradition. However, simple written proof of such an assertion may be hard to demonstrate. In all my research, I did not find any record of Partridge directly referring to himself as a Stoic. There is no known direct statement within his writings calling his students to be "Stoics."

However, it is important to remember the historical record is not complete. While we have many publications from Partridge, we do not have transcripts of his military lectures nor the various instructions he would have provided upon the march. While we have the benefit of many correspondences written to Partridge in archives, such as the Kreitzberg Library at Norwich University, most of the private correspondence penned by Partridge has been lost to time. Because of the cultural zeitgeist of the era and the American Founders' great interest in Stoic/Neostoic philosophy, direct references to Stoicism may have been obscured with references towards American ideals that contained Stoic principles.

It would be hard to imagine that someone who railed against the tyrant Julius Caesar for the overthrow of the Roman Republic[397] would not have praised the great Stoic Marcus Porcius Cato (Cato the Younger) who stood in her defense. It is hard to imagine that an officer presenting a lecture on Greek

---

[396] Partridge, Catalogue of the Officers and Cadets, 1826, pp. 21–22.
[397] Partridge, Letter to the Public, 2.

and Roman tactics[398] would not praise the great Stoic general Publius Cornelius Scipio Africanus, as Scipio was a greater strategist than Napoleon Bonaparte.[399]

Stoicism in many ways runs contrary to our present-day philosophical zeitgeist that poststructuralism (postmodernism) has shaped. Stoicism rejects the "importance of materialism, celebrity, self-esteem and victimology"[400] that is associated with the current philosophical worldview. To the military profession, the increased focus on personal responsibility and the rejection of collectivism may provide significant advantages in training and ensuring soldiers operate within the ethical bounds established by the law of war. It is also important to note that while Stoic philosophy reached its height in the Roman era, it was hardly universal in application. The use of Stoic philosophy was largely limited to the Roman elite. While Partridge's curriculum contains a large degree of instruction in Stoic ethics, it is likely only a minority of students would fully incorporate Stoic ethics within their personal philosophies upon leaving his academies.

This chapter will examine Partridge's life, behavior, and curriculum to demonstrate his multiple shared philosophical claims with the Stoic tradition (specifically Stoic ethics).[401] To prevent repetitive language, the term *stoa mementi* is used to describe shared philosophical claims of ancient Stoics that appear in later philosophical schools and, in some cases, the personal philosophies of those who may lack any formal instruction in Stoic philosophy.[402] Furthermore, this chapter will provide the various evolutions of

---

[398] Partridge, Catalogue of the Officers and Cadets, 1826, p. 19.

[399] B. H. Liddell Hart, *Greater than Napoleon: Scipio Africanus* (Edinburgh: William Blackwood & Sons Ltd., 1930).

[400] Michael Evans, "Stoicism and the Profession of Arms," Quadrant Online, April 25, 2018, para. 12.

[401] The three major branches of philosophy are ethics (study of moral principles), epistemology (study of knowledge and how it is acquired), and metaphysics (studies of the features of existences and what is "real"). Stoic philosophy is often presented as a holistic philosophy. However, philosopher Julia Annas suggested Cicero demonstrated that Stoic ethics can be presented and understood independently from Stoic metaphysics and logic. This may have allowed the perpetuation of Stoic ethics through time, even after ancient Stoic metaphysics and epistemology had been abandoned. Julia Annas, "Ethics in Stoic Philosophy," *Phronesis* 52, no. 1 (2007): 58–87.

[402] I would like to thank Professor Jennifer Baker for the suggestion to create such a term. The term was developed by borrowing the *stoa* ("stand" or "porch" of the Stoa Poikile in which the philosophy took its name) and the Latin word *mementi* (meaning "to remember"). Thus, the literal translation is "Remembrance of the Stand." I believe this term works well for its purpose and is also reminiscent of another Stoic concept: of *memento mori* (the mediation on death) to inspire better action in life.

the Stoic philosophy and related definitions. In this review, it will be demonstrated that Partridge looked like a Stoic, behaved like a Stoic, and included Stoic principles within his educational theories and curriculum. Even if there is no record of Partridge naming himself a Stoic, there is more than sufficient evidence that I can safely apply this label to him. The structure of this chapter may be useful in evaluating if others (including ourselves) carry sufficient *stoa mementi* to also be named as "Stoics."

Correctly defining the terms connected with Stoicism will be critical during this review. Like many ancient Greek terms, its meaning has drifted with time. Furthermore, various communities have defined terms differently for unique ends. As a result, an imprecise use of this term can often lead to confusion. The following definitions will seek to examine nuances in the various evolutions within the Stoic tradition. Many of the connections between Partridge's theories and lifestyle to the various evolutions within Stoic traditions will be highlighted.

Partridge's Education, Behavior, Curriculum, & Legacy

In an exercise of abductive reasoning, the general education, behavior, curriculum, and legacy of Alden Partridge will be examined to determine the presence and degree of *stoa mementi*. As poet James Whitcomb Riley once asserted, "When I see a bird that walks like a duck and swims like a duck and quacks like a duck, I call that bird a duck." Unfortunately, abductive reasoning is limited to producing plausible conclusions (but does not provide verification). However, this exercise will give some indication if others would label Partridge a Stoic.

Partridge's Education

Exact details of Partridge's early educational experiences are unknown. He was said to have been a voracious reader, reading any book he could find. Partridge moved through his village school to a regional school. He had a brief tutorship with a local minister before entering Dartmouth College. At the age of 19, while in attendance at Dartmouth, Partridge referenced the Stoic poet Virgil in one of his assignments. This is the first known concrete connection to the Stoic philosophy:

> [T]he roman empire [...] far exceeded all those that preceded it in grandeur and magnificence as in the number and valour of its inhabitants[. H]ere the arts and sciences were cultivated in the highest

degree and brought as it were almost to perfection[. H]ere lived Virgil the glory of his age and as a poet perhaps unequaled.[403]

This essay was written on the topic of the negative influence of luxury (Stoic philosophers universally advised the avoidance of luxury). It is unfortunate that Dartmouth does not have records of the specific curriculum taught during Partridge's years of attendance, as a fire destroyed much of the college's records in 1834.[404] However, there is a strong possibility that Partridge may have also been introduced to the Stoic-influenced works of the great English poet John Milton during his studies, including Milton's writing on educational theory.

At West Point, Partridge's education in Stoicism would be reinforced through Enfield's translation and abridgement of Brucker's *History of Philosophy*. This book was used as the principal source for philosophical education in the early days of the Academy.[405,406] Jefferson wrote to John Adams several times, corresponding positively on its quality of translation and subject material covered within this work.[407,408] This text emphasized the significance and achievement of the Stoic philosophy:

> [T]he Stoics rose to great distinction among the Grecian, and gave birth to many illustrious philosophers, whose names and doctrines have been transmitted with great respect to the present times. This part of the history of philosophy will, therefore, require a diligent and minute discussion.[409]

With Stoics being presented as the pinnacle of pre-Christian thought, Stoic elements being carried into and reinforced by Christian theology (through mandatory church attendance), and Stoic concepts appearing throughout the ancient works of military history, cadets in the early days of the USMA—

---

[403] Alden Partridge, Essay on Luxury by Alden Partridge, approximately 1804, Norwich University Archives and Special Collections.

[404] Webb, *Captain Alden Partridge*.

[405] Waugh, *West Point*.

[406] Barnard, *Military Schools and Courses of Instruction*.

[407] Thomas Jefferson, Thomas Jefferson to John Adams, 22 August 1813, National Archives.

[408] Thomas Jefferson, To John Adams from Thomas Jefferson, 12 October 1813, National Archives.

[409] William Enfield, *The History of Philosophy from the Earliest Times to the Beginning of the Present Century: Drawn up from Brucker's* Historia Critica Philosophiae, vol. I (Dublin: Printed for P. Wogan, P. Byne, A. Grueber, W. M'Kenzie, J. Moore, J. Jones, R. McAllister, W. Jones, J. Rice, R. White, and G. Draper, 1792).

including Partridge—would have had the importance of Stoic ethics reinforced throughout their studies.

Partridge's Appearance

Partridge's appearance received significant attention within the historical record. Military historian and former West Point public relations officer R. Ernest Dupuy attempted to shame Partridge's fashion sense, asserting, "[Partridge] strutted about [West Point] clad in an ancient blue uniform coat overladen with buttons and lace, and with such unusually widespread tails that it became known as the 'Peacock."[410] However, Partridge was simply wearing the appropriate uniform for engineer officers as prescribed by Colonel Jonathan Williams, chief of engineers, in 1807.[411] As previously mentioned, Partridge's constant appearance in uniform at West Point led many to speculate that he didn't own civilian attire.[412] When designing a new uniform for the USMA cadet corps, Partridge designed a uniform in the current European military fashion of the era. Yet Partridge's design was less decorative than the cadet uniforms of the École Polytechnique, with Partridge selecting the less elegant gray color to further governmental cost savings (as previously discussed in chapter VII).

When Partridge developed a uniform for his American Literary, Scientific and Military Academy, he would convert to blue-dyed cloth for cadet uniforms but retained his simple style. While an engineering captain on active service, Partridge would have worn a gold epaulette on his right shoulder. Yet, no image exists of him wearing epaulettes after leaving the military. A painting from 1826 suggests that Partridge wore a uniform identical to his cadets during the early days of the American Literary, Scientific and Military Academy.[413,414] In a painting published in 1842, an addition of a simple gold bullion bar appears on Partridge's collar, possibly used to distinguish faculty from

---

[410] Ernest Dupuy, "Mutiny at West Point."

[411] William K. Emerson, *Encyclopedia of United States Army Insignia and Uniforms* (Norman, OK: University of Oklahoma Press, 1996), 207.

[412] Ambrose, *Duty, Honor, Country*, 44.

[413] John Holbrook, *Military Tactics*.

[414] As reported by Collette Leonard, a painting of Edwin Ferry Johnson (Class of 1825 and instructor from 1825 to 1829) held within the Norwich collection suggests that other faculty wore uniforms identical to the cadets in the early days of the Academy.

cadets.[415,416] A daguerreotype lithograph dating from sometime after 1839 when the technology was invented shows Partridge in a uniform that has a gold-bullion-trimmed collar with a bar and flourish possibly around a bullet button.[417] I postulate this additional flourish may have been used to distinguish Partridge, serving in the presidential role for his academies, from the rest of the faculty uniforms.[418] While paintings of Partridge in the typical formal civilian attire of the day exist, he was comfortable wearing the simple attire of a soldier all his life, with even his fancier uniform designs being far understated for the military fashion of the era. Partridge's actions mimic those of Socrates, who could wear finer clothes to fancy social functions but typically preferred simple attire.[419] The ancient Stoic Seneca advised simplicity in attire but not to the extreme of estrangement to the larger community, "Inwardly, we ought to be different in all respects, but our exterior should conform to society."[420] Therefore, Partridge's wearing of formal civilian dress should not be seen as a violation of Stoic principles. Like the ancient Stoics wearing undyed cloaks in

---

[415] *To Captain Alden Partridge of the Military Institute, Norwich, Vermont, from Volume 3 of the U.S. Military Magazine*, 1842, hand-colored lithograph on wove paper, 11 1/8 × 9 in., Hood Museum, Dartmouth College, Hanover, NH.

[416] The oldest surviving coatee at the Sullivan Museum and History Center at Norwich University is dated to ca. 1830–43. It is believed to be a cadet uniform, as reported by Collette Leonard. The uniform displays cording ("frogging") in similar color to the coatee that runs from the center of the forward edge of the collar to the bullet button, creating a loop above, below, and behind the button.

[417] Norwich University Archives and Special Collections.

[418] In inquiring to the Norwich community on LinkedIn, Colonel Donald Braman suggested that the flower could be edelweiss (*Leontopodium alpinum*). This flower that grows in the alps became a symbol of daring and devotion. It became a popular military symbol starting in the late 18th century. While the popularity of this flower spiked in 1856, two years after Partridge's death, when Austrian emperor Franz Joseph I presented one to his wife, Sisi. She later wore Edelweiss-themed jewelry in official portraits, resulting in edelweiss becoming a national symbol for Austria. However, it is possible that Partridge would have known of this flower. Botany was taught at Partridge's academy in 1825–26. Botany books of the era described *Leontopodium alpinum* as a six- or seven-petal flower (consistent with the shape in the daguerreotype). With Partridge frequently advertising his academies within German-language newspapers, it is possible he was informed of the flower's symbolism by this community. There would have been a strong connection between the alpine-climbing symbolism of this flower and Partridge's own mountain climbing/pedestrian practices.

[419] Diogenes Laërtius, *The Lives and Opinions of Eminent Philosophers*, trans. C. D. Yonge (London: Henry G. Bohn, 1853).

[420] Lucius Annaeus Seneca, *Moral Epistles*, trans. Richard M. Gummere, vol. I (Cambridge, MA: Harvard University Press, 1917).

the fashion of the Spartans, Partridge's normal attire largely represented a simplified version of the military fashion of the day.

Partridge's Behavior

Partridge displayed many behaviors consistent with Stoic philosophy. These traits appeared within his writings, his educational theories, and his daily practices.

Virtue-Focus

In Partridge's letter to John C. Calhoun concerning the release of Partridge's court-martial transcripts, Partridge appealed on Calhoun's virtue and the memory of future generations to release the transcripts that could have repaired Partridge's reputation.[421] Partridge's focus on virtue can be seen throughout his educational designs. For example, in 1826, Partridge presented a design for academies (covering primary through collegiate education) near large cities designed to protect the virtue of largely unsupervised children. "Under this system the morals of the pupils would be effectually guarded, and, it is believed, might be easily preserved."[422] Partridge's *American System of Education* controlled the use of time and luxury within the academic environment to build the virtuous habits that could carry citizens throughout the remainder of their lives.[423] Captain Partridge asserted, "Few, if any, vicious habits have then been formed, and the morals, under a strict and regular system of discipline, may easily be preserved."[424]

Happiness (Eudaimonia)

Like the Stoics, Partridge would focus his students towards living happy lives. For example, in a notice for parents and guardians, Partridge would write, "That body of students who have least spending money, will ever be found most attentive to their studies, most orderly – most virtuous, and consequently most happy."[425] It should be readily apparent that Partridge was not using the term *happy* in our contemporary meaning of joy or bliss. Partridge was drawing on the ancient Greek concept of *eudaimonia* (Ancient Greek:

---

[421] Partridge to Calhoun, November 17, 1818, as quoted by Webb, *Captain Alden Partridge*.

[422] Alden Partridge, Letter by Alden Partridge, 14 October 1826, p. 6, Norwich University Archives and Special Collections.

[423] Partridge, Lecture on Education (British Library).

[424] Alden Partridge, Prospectus and Internal Regulations of the American Literary, Scientifick and Military Academy; to be Opened at Middletown, in the State of Connecticut in the Month of August, 1825, p. 30, Norwich University Archives and Special Collections.

[425] Alden Partridge, Notice to Parents and Guardians of Students at the Academy, Approximately 1820–1825, p. 1, Norwich University Archives and Special Collections.

εὐδαιμονία), representing the highest form of human good achieved through the practice of virtue, fulfillment of duty, and exercise of ethical wisdom. As recalled by 18th-century French political philosopher Charles Louis de Secondat, Baron de La Brède et de Montesquieu (commonly referred to as simply "Montesquieu"),

> Born for society, [Stoics] all believed that their destiny was to work for it, it was all the less burdensome as their rewards were all within themselves, as, happy in their philosophy alone, it seemed that only the happiness of others could increase their own.[426]

It is a shame that most Americans today do not recognize that this was the concept Thomas Jefferson intended when he wrote within the Declaration of Independence, "We hold these truths to be self-evident, that all men are created equal, that they are endowed by their Creator with certain unalienable Rights, that among these are Life, Liberty and the pursuit of Happiness." Jefferson rejected John Locke's "life, liberty, and property," asserting that property was not the highest calling for Americans but rather the pursuit of moral excellence.[427]

Partridge's Curriculum as Observed by Others

Outside the clearly Stoic-influenced authors within Partridge's curriculum—including Virgil, Cicero, Tacitus, Butler, and Locke[428]—Stoic influences within Partridge's curriculum were clearly recognized through the action and experiences of his cadets and by Partridge's contemporaries. The renowned American politician Levi Woodbury visited Partridge's academy in 1823, while governor of Vermont, and recorded his observations in a letter to Partridge the following April (see appendix M).[429] Woodbury's offered high praise for the Academy.[430] Most interestingly, he reported the following:

---

[426] Anne M. Cohler, Harold S. Stone, and Basia C. Miller, trans., *Montesquieu: The Spirit of the Laws* (Cambridge: Cambridge University Press, 1989), 466.

[427] This is not to say that Jefferson did not support property rights. Jefferson did support property rights; however, a nuanced discussion of his opinion on this subject is outside the scope of this research and too large to be included.

[428] Alden Partridge, Catalogue of the Officers and Cadets of the American Literary, Scientific and Military Academy: To Which Is Subjoined the Prospectus and Internal Regulations of the Institution, & c, & c: Norwich, Vt., November, 1821, p. 12, Norwich University Archives and Special Collections.

[429] Levi Woodbury, Letter from Levi Woodbury to Alden Partridge, 5 April 1824, Norwich University Archives and Special Collections.

[430] The use of several Latin phrases demonstrated Woodbury's high level of classical education.

Without casting any disparagement on the wishes or exertions of other teachers in other institutions it must be admitted, that your course of discipline is in itself more severe and at the same time more attractive; that with you neither wealth nor rank can buy an exemption from obedience & study; that your system tends to inspire pupils with sentiments like those attributed to Alexander of Macedon of its being a "royal virtue to labour"; that you learn them by example as well as precept, the rich and intelligent are born to higher destinies than luxury & indolence – *consumere fruges*;[431] that in aid of the common means of exciting to improvement you bring the strict subordination of military tactics, the pride & vigilance of military office and the strong desire to obtain all those manly accomplishments, which ought to be united in the soldier & scholar.[432]

Present-day audiences may miss the allusion Woodbury made to the *De Fortuna Alexandri* by Plutarch by including Alexander the Great's "royal virtue to labour." In this work, Plutarch examines Alexander of Macedon's action in the context of Stoic ethics. Plutarch displays how Alexander, through action, could implement a cosmopolitan empire that would only be imagined by Zeno of Citium. In the words of Plutarch,

Moreover, the much-admired *Republic* of Zeno, the founder of the Stoic sect, may be summed up in this one main principle: that all the inhabitants of this world of ours should not live differentiated by their respective rules of justice into separate cities and communities, but that we should consider all men to be of one community and one polity, and that we should have a common life and an order common to us all, even as a herd that feeds together and shares the pasturage of a common field. This Zeno wrote, giving shape to a dream or, as it were, shadowy picture of a well-ordered and philosophic commonwealth; but it was Alexander who gave effect to the idea. For Alexander did not follow Aristotle's advice to treat the Greeks as if he were their leader, and other peoples as if he were their master; to have regard for the Greeks as for friends and kindred, but to conduct himself toward other peoples as though they were plants or animals; for to do so would have been to cumber his leadership with numerous battles and banishments and festering seditions. But, as he believed

---

[431] Translates from Latin to "consume crops." Possible reference to Horace's *Epistola*, Book II, line 27: "Nos Numerus sumus et fruges consumere nati," implying a "common sort of men, [...] all but the sage."

[432] Woodbury, Letter from Levi Woodbury to Alden Partridge, 3–4.

that he came as a heaven-sent governor to all, and as a mediator for the whole world, those whom he could not persuade to unite with him, he conquered by force of arms, and he brought together into one body all men everywhere, uniting and mixing in one great loving-cup, as it were, men's lives, their characters, their marriages, their very habits of life. He bade them all consider as their fatherland the whole inhabited earth, as their stronghold and protection his camp, as akin to them all good men, and as foreigners only the wicked; they should not distinguish between Grecian and foreigner by Grecian cloak and targe, or scimitar and jacket; but the distinguishing mark of the Grecian should be seen in virtue, and that of the foreigner in iniquity; clothing and food, marriage and manner of life they should regard as common to all, being blended into one by ties of blood and children.

Plutarch is well known for criticizing aspects of Stoic philosophy.[433] However, Woodbury's allusion to this work could be viewed as the highest praise for those who advance the interest of Stoic ethics. Woodbury was essentially asserting that Partridge's cadets were not only capable Stoic philosophers on an intellectual level, but they were also men of action who could carry this ethical worldview into application throughout American society.

Brigadier General George Douglas Ramsay reflected on his experiences of being challenged by Captain Partridge to pull an 18-pound cannon down to the river and back as part of the instruction at the USMA:

The deeper the snow the greater the fun and enjoyment. The going down to the river was no difficult matter, but retracing our steps through the deep snow was the "hic labor hoc opus est" which tried our metal.[434]

In this passage Ramsay references Virgil's *Aeneid* 6.129, which translates into "that's the task, that's the toil," used in reference to the difficulties of journeying back from the underworld. Certainly, one could imagine returning a cannon, weighing over 4,000 pounds, uphill, through the snow, by hand, as a challenge that would seem as equal as clawing oneself out of hell.

It is significant that Ramsay alluded to the *Aeneid*. The *Aeneid* is filled with Stoic phraseology[435] with Aeneas being an imperfect example of a Stoic

---

[433] Jan Opsomer, "Is Plutarch Really Hostile to the Stoics?," in *From Stoicism to Platonism: The Development of Philosophy, 100 BCE–100 CE*, ed. Troels Engberg-Pedersen (Cambridge: Cambridge University Press, 2021), 296–321.

[434] George D. Ramsay, USMA Class of 1820, as quoted by Guidotti, "Controversy: The Legacy of Alden Partridge," 7.

[435] Mark W. Edwards, "The Expression of Stoic Ideas in the 'Aeneid," *Phoenix* 14, no. 3 (1960): 151.

hero and the ideal Roman citizen.[436] In Ramsay's quotation, a student's direct connection of Partridge's educational practices back to Stoic philosophy can be seen. This reference provides an indication that not only did Partridge's students take part in these Stoic toughening exercises (testing one's mettle or resilience), but the cadets also had an understanding of the philosophy behind them. It also must be noted the "fun and enjoyment" Ramsay recalls in these activities. Even with a task that he figuratively compared to as painful as coming back from the land of the dead, it can be seen how he viewed this activity not as a punishment but a challenge and ultimately took great pride in demonstrating the toughness of his mettle.

Partridge's Legacy

A Norwich graduate, Kemp Russell Blanchard Flint, received his education two generations after Partridge's death. Flint composed "A Norwich Man's Creed" in 1927.[437] This creed was re-titled to the gender-neutral "A Norwich Cadet's Creed" in 1977, three years after women entered the Norwich Corps of Cadets. It reads:

> I believe that the cardinal virtues of the individual are courage, honesty, temperance and wisdom; and that the true measure of success is service rendered—to God, to Country, and to Mankind.
> I believe that the fundamental problem of society is to maintain a free government wherein liberty may be secured through obedience to law, and that a citizen soldiery is the corner stone upon which such a government must rest.
> I believe that real education presupposes a sense of proportion in physical, mental and moral development; and that he alone is educated who has learned the lessons of self-control and open-mindedness.
> I believe in Norwich, my Alma Mater, because within her halls throughout the years these tenets have found expression while men have been taught to be loyal to duly constituted authority in thought and word and deed; to view suffrage as a sacred privilege to be exercised only in accordance with the dictates of conscience; to regard public office as a public trust; and finally to fight, and if need be to die, in defense of the cherished institutions of America.

> —K. R. B. Flint, '03[438]

---

[436] C. M. Bowra, "Aeneas and the Stoic Ideal," *Greece and Rome* 3, no. 7 (1933): 8–21.
[437] "Guidon Reprint Creed as Written for Cadets by Prof. K. R. B. Flint." *Norwich Guidon* XXVI, no. 20 (March 5, 1940): 1, Norwich University Archives and Special Collections.
[438] K. R. B. Flint, "A Norwich Man's Creed," *Norwich Guidon* VI, no. 1 (September 16, 1927): 2, Norwich University Archives and Special Collections.

Multiple *stoa mementi* are found throughout this creed, showing the persistence in instruction of the Stoic ethical tradition past Partridge's lifetime. The creed lists three of the four ancient Stoic cardinal virtues, only replacing *justice* with the subordinate virtue of *honesty*. Stoic elements can be seen in appeals to render service to God, country, and mankind.[439] The Socratic view of education of the whole person is seen within the educational goals of moral, mental, and physical development. The endurance of hardships and even the ultimate sacrifice to live for the good of the community, viewed as a sacred honor in public service, is also in line with Stoic philosophy. Flint's use of "thoughts and words and deeds" also alludes to Saint Thomas Aquinas' *Summa Theologica* as realms possible for humans to sin,[440] with Flint's use implying that Norwich graduates were "sinless" in fulfilling their duty to their community and nation. Aquinas was a theologian heavily influenced by the ancient Stoic Seneca. Another reference to ancient Stoicism appears in the phrase "liberty may be secured through obedience to the law" (this passage is reframed on the Norwich University gates: "Obedience to the Law is Liberty"). This phrase is attributed to the sixth-century Roman philosopher and martyr Anicius Manlius Severinus Boethius, who was heavily influenced by Stoic philosophy.[441] Regardless, if Norwich cadets specifically named themselves as Stoics, their current credo remains heavily infused with *stoa mementi*.

Evolution of Stoic Philosophy

Examinations of Partridge's connection with Stoic ethics will continue through a discussion of the various evolutions of Stoic philosophy. This will include Stoic schools that occurred previously and contemporary to Partridge's lifetime. The modern Stoic movement, which began after the death of Partridge, is also discussed to ensure present-day readers understand how this new evolution of Stoicism relates to previous schools. Definitions for new concepts of natural Stoicism and military Stoicism will also be presented. To fully explore the origins of Stoic philosophy would be at least a book-length project. The abbreviated history that follows will be incomplete but sufficient for the general intent of this chapter. It is assumed that the audience explores

---

[439] "Duties may be divided into three classes, as they respect God, ourselves, and our neighbour." Enfield, *History of Philosophy*, I:371.

[440] "On the contrary, Jerome in commenting on Ezek. 43:23: The human race is subject to three kinds of sin, for when we sin, it is either by thought, or word, or deed." Thomas Aquinas, "Whether Sins Are Fitting Divided into Sins of Thought, Word, and Deed," trans. Laurence Shapecote, Summa Theologica, 2020, ST.I–II.Q72.A7.SC.

[441] John Marenbon, "Anicius Manlius Severinus Boethius," Stanford Encyclopedia of Philosophy, September 21, 2021.

further sources on Stoic philosophy as their motivation directs (see appendix L for a recommended resource list).

Pseudo-Stoicism

The best place to begin the discussion on the evolution of Stoic philosophy is to start with what the philosophy is not. Pseudo-stoicism (sometimes referred to as small "s" stoicism) is defined as "not complaining or showing what you are feeling when you are suffering."[442] It is associated with the unhealthy practice of suppressing emotions. Cornell University philosophy professor Tad Brennan further defined pseudo-stoicism as a "mixture of tough-guy bravado, hypocrisy, and heartlessness [that is] neither personally compelling nor philosophically interesting."[443] The modern Stoic author Donald Robertson reports that this false stoicism is often associated with the idiom of having a "stiff upper lip."[444] The practice of pseudo-stoicism has nothing in common with ancient Stoicism as a philosophy. As explained by Robertson,

> [T]he Stoics class [emotions] as natural, inevitable, and morally "indifferent" — neither good nor bad. Pseudo-stoicism, however, does the opposite. It tries to suppress or conceal these involuntary emotional reactions and treats them as though they were bad, harmful, or even shameful.[445]

Canadian veteran Sheridan Taylor pointed to the "three Johns" archetypes (the actor John Wayne and the fictional characters of John Wick and John Rambo) within popular culture as potential sources of pseudo-stoic behavior within modern service members.[446] Other popular culture figures, such as the character of Spock within Gene Roddenberry's *Star Trek*, may lead the general public to attempt to display purely logical behavior with no signs of human emotions.

The historical record clearly demonstrates that Partridge did not endorse the suppression of emotion and even warned against the damage doing so would cause. In 1822, on announcing the death of a cadet at the

---

[442] "Stoicism Noun," Oxford Advanced American Dictionary.

[443] Tad Brennan, *The Stoic Life: Emotions, Duties, and Fate* (Oxford: Clarendon Press, 2006), 8–9.

[444] Donald J. Robertson, "The Difference between Stoicism and Stoicism," Stoicism — Philosophy as a Way of Life, February 19, 2021 (Medium).

[445] Robertson, "Difference between Stoicism and Stoicism."

[446] Sheridan Taylor and Andrew Carlquist, The Voice of Experience: Retired Master Corporal, Sheridan Taylor, other, in *Before Operational Stress – Military* (Calgary: Wayfound Mental Health Group, 2023).

American Literary, Scientific and Military Academy, Partridge states, "On such an occasion not to weep would be to do violence to the finest and most estimable feelings of our nature."[447] The acknowledgement of emotion endorsed by Partridge is fully compatible with Stoic philosophy.

It is also possible that outside observers may falsely interpret the actions and emotional processing of those who engage in Stoic ethics. This can certainly be true for new military recruits looking at experienced combat veterans, thinking they are fearless or unemotional in the face of adversity. The new recruits see the appearance of reduced fear among these veterans while not understanding that the combat veterans have simply become accustomed to processing emotions such as fear and do not add to their own suffering by focusing on things outside their control. Clinical psychologist Dr. Megan McElheran and registered psychologist Andra Stelnicki have coined the term *functional disconnection/functional reconnection* to describe this learned ability to suppress emotions during times of emergency and reconnect to emotions once returning to safe situations.[448] It would be easy to imagine a young cadet looking up to Partridge on one of the many pedestrian adventures and imagining he was impervious to fatigue. Instead of seeking guidance on how veterans have learned to acknowledge and process these emotions so that they can best remain rational and functional on the battlefield, recruits will often attempt to suppress or "eat" their emotions, leading to negative mental health conditions.

McElheran et al. "posit that the adoption of pseudo-stoic behaviors is implicitly linked to the prevalence of operational stress injuries among military [service members]."[449] Therefore, the explicit instruction in Stoic philosophy and exercises is critical to the development of citizen-soldiers in safeguarding mental health. It is not enough to have a Stoic example, but explicit instruction and discourse on how to truly engage in Stoic ethics must be provided. Unfortunately, the instruction in and exposure to Stoic philosophy has largely

---

[447] Alden Partridge, Records Relating to the Death of Cadet Thomas T. Hubbart on 25 October 1822, p. 5, Norwich University Archives and Special Collections.

[448] While those who work in high-stress environments (military/public service) may naturally discover the ability to emotionally disconnect and reconnect, the ability to functionally reconnect is greatly aided through explicit instruction in Stoic principles. Andrea M. Stelnicki et al., "Evaluation of Before Operational Stress: A Program to Support Mental Health and Proactive Psychological Protection in Public Safety Personnel," *Frontiers in Psychology* 12 (2021).

[449] Megan McElheran et al., "Strengthening the Military Stoic Tradition: Enhancing Resilience in Military Service Members and Public Safety Personnel through Functional Disconnection and Reconnection," *Frontiers in Psychology* 15 (September 23, 2024), 1.

been removed from the curriculum of secondary and higher-education institutions in the late 20th century. Ignorance of this foundational philosophy among the general population has led to the growth in pseudo-stoicism and its related negative mental health effects.

Proto-Stoics

It must be acknowledged that many elements of Stoic philosophy existed prior to the establishment of the formal Stoic school. As noted by classicist A. A. Long, characteristics such as "making the best of things, sticking to a goal through thick and thin, drawing on inner resources of mind and will, raising excellence of character, those were Greek values and no doubt [...] human values long before."[450] While Zeno of Citium is credited as founding ancient Stoicism in the third century B.C., there were several philosophies and key philosophers that Zeno would draw upon to construct his philosophy. These preexisting concepts are often termed *proto-Stoic.*

Many proto-Stoic thinkers were men connected to military service. The epic poet Homer, possibly a military surgeon, was reinterpreted by Stoics to support their philosophy[451] (with Partridge using the *Iliad* within his curriculum).[452] Other influential proto-Stoics include the great Greek philosopher Socrates of Athens (a former Athenian hoplite/heavy infantryman) and the Greek general, philosopher, and historian Xenophon of Athens. Doubtless Partridge would have had an affinity towards these veterans, with accounts of Socrates rescuing Xenophon on the battlefield[453] unlikely to go unnoticed.

Partridge extolled Xenophon for his clear descriptions of military operations, as compared to 19th-century historians.[454] Partridge used Xenophon's *Anabasis* to teach Ancient Greek within his academies. The *Anabasis* detailed the exciting story of a Greek mercenary army attempting to escape from the heart of the Persian Empire after a failed conquest and death of Cyrus the Younger. Within this work, Xenophon displays the paradoxical ability to remain optimistic and provide exceptional motivational speeches to his troops while remaining grounded in the dangerous reality of their situation. Stoics would retain this characteristic now known as the *Stockdale Paradox,*

---

[450] A. A. Long, "Stoicism Ancient and Modern," lecture presented at the Stoicon, September 29, 2018.

[451] A. A. Long, *Stoic Studies* (Berkeley, CA: University of California Press, 2001).

[452] Partridge, Catalogue of the Officers and Cadets, August, 1821, p. 12.

[453] Laërtius, *Lives and Opinions of Eminent Philosophers.*

[454] Partridge, Lecture on Education (British Library).

named after Navy aviator Admiral James Stockdale, who endured over seven years of captivity and torture during the Vietnam War.[455]

Outside of key philosophers influenced through the Greek military traditions, Stoicism formed through interactions and responses to major philosophical schools of the age, to include the Cynics, Skeptics, and Epicureans. Zeno of Citium had studied directly with the Cynic Crates of Thebes prior to beginning his own school.[456,457] Elements of these philosophic traditions were absorbed into and helped shape the Stoic traditions. For example, the concepts later echoed by the Stoic philosophers of individuals suffering more from their perceptions/attitudes/opinions of experiences than from the experiences themselves and how the quality of one's thoughts shapes the quality of one's life is discussed within the first chapter of Plato's *Republic*, which predates the formal Stoic school of Zeno.[458,459] Similarly, Plato's assertion that happiness is founded on virtuous actions and does not require other external conditions (i.e., social status, health, profit) was also carried into—or at least ran parallel with—the Stoic tradition.[460] Later in his career, Partridge would add Plato to the Norwich University curriculum in 1841.[461]

Ancient Stoicism

The understanding of ancient Stoicism was heavily shaped by the Roman Stoa, the last period of ancient Stoicism occurring around 0–200 A.D. Complete works from Roman authors such as Epictetus, Marcus Aurelius, and Seneca serve as the foundation for our general understanding of this philosophical tradition. References to the ideas of the Early Stoa,

---

[455] Jim Collins and James C. Collins, *Good to Great: Why Some Companies Make the Leap... and Others Don't* (London: Random House Business, 2001).

[456] Laërtius, *Lives and Opinions of Eminent Philosophers*.

[457] Academic philosopher John Sellars credits the influence of Crates for the Stoic praise of austere living. Sellars asserts that Stoics should see wealth as a moral indifferent (the opinion of ancient Stoics on wealth varied, yet all would praise the avoidance of luxuries). John Sellars, *Stoicism* (London: Routledge, Taylor & Francis Group, 2014).

[458] Eric Brown, "Socrates in the Stoa," in *A Companion to Socrates*, ed. Sara Ahbel-Rappe and Rachana Kamtekar (Malden, MA: Blackwell Publishing Ltd., 2006), 275–84.

[459] Marcus Aurelius Anderson, "Donald Robertson on Marcus Aurelius the Stoic Emperor, Fatherhood, Errors in ChatGPT, Stoic Meta Lessons, and Why the World Needs Philosophy Now More Than Ever," *Acta Non Verba*, February 14, 2024.

[460] Julia Annas, "Is Plato a Stoic?" *Méthexis* 10, no. 1 (March 30, 1997): 23–38.

[461] Alden Partridge, A Catalogue of the Officers and Cadets of Norwich University, for the Academic Years 1838–39: 39–40: 40–41, p. 17, Norwich University Archives and Special Collections.

approximately 300–100 B.C., and philosophers such as Zeno, Cleanthes, and Chrysippus exist only as fragments or had their ideas partially relayed in later secondary references. With the Middle Stoa representing the shift of Stoicism from the Greek to Roman influence running between 100 B.C. and 1 A.D. that included the notable influences of Cato the Younger and the writings of Cicero and Plutarch.[462]

Losing the materials of the Early Stoa may have been beneficial to the long-term influence of Stoicism on Western society. Some scholars have suggested the radical ideas within Zeno of Citium's *Republic* (including incest and cannibalism) became an embarrassment for later Stoics.[463] While academics debate if Zeno was serious about these suggestions or if they were a holdover from his Cynic training, the tempering of Stoicism to a more conservative form that was compatible with the Roman Republic likely ensured its survival through history. However, the lack of direct sources from Zeno makes it impossible to determine if this assertion is true. Like Socrates, Zeno may have simply been examining each of these topics to determine what social norms were appropriate for the sage.

Ancient Stoicism is a practical philosophy that recognizes that humans do not respond to events and occurrences, but to their opinions of these actions. The ancient Stoic Epictetus would refer to an individual's control over their opinion as *prohairesis* (Ancient Greek: προαίρεσις). Changing the opinions of events could lead to dramatic changes in perceptions, increasing overall happiness and mental health. The goal of Stoic philosophy was not to avoid emotions, but to maximize positive emotions through the control of opinions in relation to virtuous action. Epictetus calls individuals to:

> Remember that it is not he who gives abuse or blows, who affronts, but the view we take of these things as insulting. When, therefore, anyone provokes you, be assured that it is your own opinion which provokes you. Try, therefore, in the first place, not to be bewildered by appearances. For if you once gain time and respite, you will more easily command yourself.[464]

Controlling one's prohairesis is aided by the recognition of a dichotomy of control. Epictetus explains this concept in the opening lines of his handbook:

---

[462] "A Brief History of Stoicism," Stoic Journey, November 9, 2019.

[463] Brian S. Hook, "Oedipus and Thyestes among the Philosophers: Incest and Cannibalism in Plato, Diogenes, and Zeno," *Classical Philology* 100, no. 1 (January 2005): 17–40.

[464] Epictetus, *The Enchiridion*, 2nd ed., trans. Thomas W. Higginson (New York, NY: Liberal Arts Press, 1955), XX.

> There are things which are within our power, and there are things
> which are beyond our power. Within our power are opinion, aim,
> desire, aversion, and, in one word, whatever affairs are our own.
> Beyond our power are body, property, reputation, office, and, in one
> word, whatever are not properly our own affairs.[465]

Stoics believe that only those things within one's control (opinions and
individual actions) should require value judgements in relation to virtue.
Anything outside an individual's control should simply be accepted and the
sage should learn how to be indifferent to these events. In fact, these external
events may even provide greater opportunity to exercise virtue. As Epictetus
recalls,

> Upon every accident, remember to turn toward yourself and inquire
> what faculty you have for its use. If you encounter a handsome person,
> you will find continence the faculty needed; if pain, then fortitude; if
> reviling, then patience. And when thus habituated, the phenomena of
> existence will not overwhelm you.[466]

The Stoic philosopher Agrippinus was a role model in his ability to turn an
indifferent hardship into the opportunity for personal growth and excellence.
As Epictetus recalls,

> [Agrippinus'] character was such […] that when any hardship befell
> him he would compose a eulogy upon it; on fever, if he had a fever; on
> disrepute, if he suffered from disrepute; on exile, if he went into exile.
> And once, he said, when Agrippinus was preparing to take lunch, a
> man brought him word that Nero ordered him into exile; "Very well,"
> said he, "we shall take our lunch in Aricia."[467]

The Stoic concept of dichotomy of control is more of a generalization
than a strict demand. There are many things in life that an individual may have
partial control over. For example, the modern Stoic William Irvine asserts there
is a trichotomy of control, giving the example of the skill in playing a game as
being within a person's control but not always guaranteeing success.[468] The
practice of examining one's opinions and reviewing the general attitude of the
dichotomy of control is often only allowable in moments of reflection. As
explained by Enfield,

> The first impression from the senses produce in the mind an
> involuntary emotion; but a wise man afterwards deliberately examines

---

[465] Epictetus, *The Enchiridion*, I.

[466] Epictetus, X.

[467] Epictetus, *Discourses, Books 3–4. Fragments. The Encheiridion*, trans. W. A. Oldfather
(Cambridge, MA: Harvard University Press, 1928), 465.

[468] Irvine, *The Stoic Challenge.*

them, that he may know whether the[y] be true or false, and assents to or rejects them, as the evidence which offers itself to his understanding appears sufficient or insufficient.[469]

Even in moments of reflection, there is no expressed expectation that all variables within a given scenario are examined to come to a reasonable conclusion if the currently held opinion is appropriate. This practice, due to the limits of human cognition, cannot always be practiced. For example, the best Stoic soldier could not practice the dichotomy of control while engaged in hand-to-hand combat, as his cognitive resources are consumed in tasks related to mere survival. However, he could return to this reflective practice once in a situation of relative safety.

Stoicism allows for rational, virtue-based actions even in extreme environments. Since Stoicism is exercised through voluntary action at the individual level, this philosophy is "a powerful method of reasoning involving the rigorous cultivation of self-command, self-reliance and autonomy."[470] Given the requirement to engage in voluntary action, this philosophy is deeply connected with the concept of individual liberty. However, it also retains a deep focus on civic duty and interest in the support of the larger community. Stoicism supported a cosmopolitan view of the world that imparted value to humans from foreign cities and nations. In Stoic ethics, an individual's affinity (*oikeiosis*, Ancient Greek: οἰκείωσις) or recognition of what "belongs to oneself" should expand outward from the traditional self, household, city, and national loyalties to include considerations for all mankind.[471] The Stoics recognized that the *oikeiosis* of animals extended only to their self-interest (displayed by their desire for self-preservation) but that the *oikeiosis* of mankind, with our rational abilities, should extend beyond oneself and immediate family. Stoic ethics didn't see actions required to sustain one's life and family as inherently selfish but as a natural requirement. However, it would also foster an attitude towards healthy civic life and care for the larger community through the exercise of virtues. In the words of Epictetus, "[Stoics] want at once to live as wise men and benefit mankind!"[472]

The Stoics asserted that there was no moral economy in the world and that life was fundamentally unfair. This philosophy allowed for the possibility of good and honest men to die poor while the corrupt and dishonest died

---

[469] Enfield, *History of Philosophy*, I:346.

[470] Evans, "Stoicism and the Profession of Arms," para. 22.

[471] Will Johncock, "Bringing People Closer: Cicero, Hierocles, and Cosmopolitanism," *Epoché Magazine*, May 30, 2022.

[472] Epictetus, and P. E. Matheson, *Epictetus: The Discourses and Manual* (Oxford: Clarendon Press, 1916), 44.

wealthy. As Evans explains, "Stoicism [is] an unrelenting struggle for virtuous character in a world devoid of fairness."[473] Stoicism focuses an individual to pursue virtue as the *telos* or end goal of life, as virtuous action is a reward in and of itself. John Sellars admires the "reality principle" found within ancient Stoicism. Stoics accepted reality as it is (not how we wish it to be). Stoics identify and act appropriately upon those elements within their control while being indifferent towards those outside their control.[474]

By avoiding vice and pursuing virtue, Stoic individuals could maintain their dignity and self-respect in times of both plenty and hardship. Stoic practices (meditation, self-examination, exercises, and worldview) allowed individuals to identify the influences of negative emotions, like fear or desire, yet remain rational in their thoughts and actions. By doing so, Stoics sought to "create in individuals a citadel of the soul based on the cultivation of rational thought and moral values that foster an indomitable spirit."[475]

Nature

Ancient Stoic philosophy instructed followers to live in accordance with nature, both universal nature and human nature. This required practitioners to observe nature and accept natural processes. As reported by Cicero,

> [T]he Stoics hold, everything that the earth produces is created for man's use; and as men, too, are born for the sake of men, that they may be able mutually to help one another; in this direction we ought to follow Nature as our guide, to contribute to the general good by an interchange of acts of kindness, by giving and receiving, and thus by our skill, our industry, and our talents to cement human society more closely together, man to man.[476]

Stoics would not resist processes such as aging and death, viewing these as simply part of the nature of the world.

To aid in the understanding of nature, Stoics advocated for frequent exposure to nature. According to Seneca, it benefits an individual to take "walks out of doors, that our minds may be energized and refreshed by the

---

[473] Evans, "Stoicism and the Profession of Arms," para. 24.

[474] "The Art of Living: Professor John Sellars on Stoicism as a Medicine for the Mind," Daily Stoic, n.d.

[475] Evans, "Stoicism and the Profession of Arms," para. 62.

[476] Marcus Tullius Cicero, *Cicero de Officiis*, trans. Walter Miller (London: Heinemann, 1913), 24–25.

open air and deep breathing."[477] Certainly, Partridge's frequent pedestrian adventures would duplicate this practice both individually and with his cadets.

Cardinal Virtues

Stoics focused their lives around the four cardinal virtues of prudence (wisdom), justice, courage (manly gallantry), and temperance.[478] Virtues were identified through their contributions to *eudamonia* (human happiness/flourishing). The virtues were the only means of achieving the "honourable the perfect good […] for in these goods all beautiful actions have their accomplishment."[479] The Stoics would assert the failures to achieve happiness resulted from our own failures to fully engage in virtue. As stated by Epictetus,

> Happiness, the effect of virtue, is the mark which God has set up for us to aim at. Our missing it is no work of His; nor so properly anything real, as a mere negative and failure of our own.[480]

Discussion of virtues in Greek philosophy long predates Stoic philosophy. Stoic philosophers did not identify these cardinal virtues but carried these virtues forward from Platonic tradition.[481]

Prudence

Prudence (wisdom) is the "knowledge of what is good, and bad, and indifferent."[482] This virtue requires the use of logic and reason to discern correct actions.

Justice

Justice is the "knowledge of what ought to be chosen, what ought to be avoided, and what is indifferent."[483] This virtue requires actions judged by duty to 'o right in relation to one's community and engage in lawful action even in times of difficulty.

Courage

Courage (manly gallantry) is defined as "the cause of a man's not being alarmed amid dangers and formidable circumstances, but standing firm."[484] This virtue does not demand that individuals are unemotional in hardships;

---

[477] Lucius Annaeus Seneca, *Dialogues and Essays*, trans. John N. Davie (Oxford: Oxford University Press, 2007), 138.

[478] Laertius, *Lives and Opinions of Eminent Philosophers*.

[479] Laertius, 295–96.

[480] Epictetus, *The Enchiridion*, footnote 1.

[481] Laertius, *Lives and Opinions of Eminent Philosophers*.

[482] Laertius, 293.

[483] Laertius, 293.

[484] Laertius, 145.

only that individuals should stand resolute even in the presence of fear or anxiety to meet the challenges of life.

Temperance

Temperance is the virtue of moderation, self-control, and emotional regulation. It is the favoring of long-term goals over instant gratification. Temperance "is one of the speculative virtues, and it happens that good health usually follows it, and is marshalled as it were beside it; in the same way as strength follows the proper structure of an arch."[485]

Diogenes Lærtius, the third century A.D. biographer of philosophers, further explained the Stoic structure of virtues subordinate to the cardinal virtues:

> And subordinate to these, as a kind of species contained in them, are magnanimity, continence, endurance, presence of mind, wisdom in council. And the Stoics define [...] magnanimity as a knowledge of engendering a lofty habit, superior to all such accidents as happen to all men indifferently, whether they be good or bad; continence they consider a disposition which never abandons right reason, or a habit which never yields to pleasure; endurance they call a knowledge or habit by which we understand what we ought to endure, what we ought not, and what is indifferent; presence of mind they define as a habit which is prompt at finding out what is suitable on a sudden emergency; and wisdom in counsel they think a knowledge which leads us to judge what we are to do, and how we are to do it, in order to act becomingly.[486]

The Stoics would focus their lives towards good and beautiful actions exercised through virtues great and small. Actions and behaviors contrary to the virtues would be defined as vices. Like the virtues, the Stoics would also develop a hierarchical structure for the faults of mankind:

> And analogously, of vices too there are some which are primary, and some which are subordinate; as, for instance, folly, and cowardice, and injustice, and intemperance, are among the primary vices; incontinence, slowness, and folly in counsel among the subordinate ones. And the vices are ignorance of those things of which the virtues are the knowledge.[487]

---

[485] Laertius, 292.
[486] Laertius, 293.
[487] Laertius, 293.

Through the engagement in the virtues and avoidance of vice, Stoics could seek to live useful lives through the development of their own character to the benefit of all.

Captain Alden Partridge would embrace the virtues and design his *American System of Education* to support and nurture the virtuous actions of his students. In his 1827 prospectus, Partridge asserted that the educational model would produce "A manly, noble and independent spirit [...] cherished in all the cadets" while "The morals of the cadets will be preserved, and habits of systematic attention, temperance, industry and economy, established and confirmed."[488]

In his 1825 prospectus, Partridge would implore parents to send their sons to his academy before they had learned vice, so they could be indoctrinated with virtue. Partridge would implore parents to be truthful when admitting students with known character faults:

> I would much prefer that the great body of my pupils should enter young, and grow up under my system. The mind and body are then more susceptible of improvement, than at a more advanced period. Few, if any, vicious habits have then been formed, and the morals, under a strict and regular system of discipline, may easily be preserved. It is my fixed determination not, knowingly, to admit any young man of confirmed vicious or dissipated habits, into the institution. I would accordingly recommend to parents and guardians not to send me any of this description, for if they should gain admission, and did not immediately reform, (which seldom occurs when the habits are confirmed) it would only eventuate in their dismission and consequent disgrace. It is much easier to prevent a youth from acquiring bad habits, than to correct them after they are acquired. If parents and guardians will send me their sons and wards free from habits of dissipation, immorality and vice, I will guarantee, so far as human agency will authorize, that they shall be preserved free from such habits, while they remain under my care. Every requisite means will be used to correct the foibles and faults incidental to youth—to accomplish this object no pains will be spared. With their foibles I will bear as much as any person, but with their vices I will make no compromise. For the purpose of enabling me the more readily and the

---

[488] Alden Partridge, Catalogue of the Officers and Cadets: Together with the Prospectus and Internal Regulations of the American Literary, Scientific and Military Academy, at Middletown, Connecticut, 1827, p. 30, Norwich University Archives and Special Collections.

more certainly to accomplish this important object, I must request parents and guardians, if their sons or wards have foibles or faults, frankly to state them to me. On this subject there should be no reserve, as, with such information, I should know much better what course to pursue with them.[489]

Furthermore, students that could not be reformed from vice through instruction and corrective action were frequently dismissed by Partridge. Partridge truly held himself and his students to a strict standard of virtuous action. His education system aimed to produce citizens fully prepared for any duty, economic, civic, or military, that might be asked of them. With the fulfillment of all these responsibilities demanding virtuous action, Partridge was uncompromising on his standards for behavior:

The character and conduct of gentlemen and of soldiers must be considered as inseparable from every member of this institution. Any one, therefore, who shall be guilty of ungentleman like or unmilitary conduct, shall be forthwith dismissed.[490]

Laconophilia (Love of Sparta)

Socrates extolled many elements of the Laconian (Spartan) culture. As a result, it should be of little surprise that Stoic philosophers adopted his Laconophilia. The Stoics would adopt many practices of the Spartan agoge. The agoge, established by the legendary Spartan lawgiver Lycurgus, was a schoolhouse intended to prepare free-born Spartan men for combat. Boys would enter at the age of twelve and remain till the age of thirty. In this schoolhouse, they would be exposed to the elements, being only allowed one cloak per year. They would sleep on reed mats as if on a military campaign.[491] The agoge would expose students to all manner of physical and mental hardships while enforcing an austere lifestyle to ensure Spartan warriors would be the fiercest warriors on the battlefield.

Socrates and the later Stoics, including Marcus Aurelius, would adopt the wearing of cloaks of Spartan fashion (that the Romans would later term *pallium*) in undyed wool.[492] This gave Stoic philosophers the outward appearance of Spartans. Captain Alden Partridge also kept the custom of wearing a simple military uniform (as discussed in chapter VIII). Some Stoics adopted the practice of sleeping on cots covered in animal hides similar to

---

[489] Partridge, Prospectus and Internal Regulations, 1825, p. 30.

[490] Partridge, Catalogue of the Officers and Cadets, August, 1821, p. 15.

[491] Pierre Hadot, *The Inner Citadel: The Meditations of Marcus Aurelius* (Cambridge, MA: Harvard University Press, 2001).

[492] Donald J. Robertson, "Stoicism as a Spartan Philosophy of Life," Stoicism—Philosophy as a Way of Life, October 31, 2019 (Medium).

Spartan camp beds.[493] It may be possible that Partridge was mimicking this behavior when he outlawed the use of featherbeds at his academies. Cadet John Hancock, Jr., reported in 1824 on a pending pedestrian adventure,

> [Partridge] mentioned the march that contemplated, saying he should not have tents, (as was wished by the cadets) he does not think it military to carry them, when it can be avoided. We shall carry blankets upon our backs in knapsacks & sleep on the ground, carry provision sufficient to not be dependent upon the inhabitants as they have been heretofore.[494]

Partridge may have simply been copying the practices of Sparta and the ancient Stoics by ensuring his cadets were well prepared for life through the constant practice of acting and sleeping like soldiers.

The connection between Laconia and the Stoic tradition would be reinforced during the reign of the Spartan king Cleomenes III through his Stoic advisor Sphærus of Borysthenes (who was a direct student of Zeno of Citium). While the reign of Cleomenes was relatively short (235 to 222 B.C.), many of his reforms, returning Laconia to its classical warrior code, remained in place for almost 500 years until the end of Roman Greece. The Stoic influence over Laconia would be captured within the works of Plutarch.[495] As Plutarch recalls,

> It is said also that Cleomenes, whilst a boy, studied philosophy under Sphærus, the Borysthenite, who crossed over to Sparta, and spent some time and trouble in instructing the youth. Sphærus was one of the first of Zeno the Citiean's scholars, and it is likely enough that he admired the manly temper of Cleomenes, and inflamed his generous ambition. The ancient Leonidas, as story tells, being asked what manner of poet he thought Tyrtæus, replied, "Good to whet young men's courage;" for being filled with a divine fury by his poems, they rushed into any danger. And so the Stoic philosophy is a dangerous incentive to strong and fiery dispositions, but where it combines with a grave and gentle temper, is most successful in leading it to its proper good.[496]

Plutarch can be seen using Stoic terms and concepts when recalling the lives of famous Spartans including Lycurgus, Agis, and Cleomenes.[497] Plutarch's *Lives*

---

[493] Hadot, *The Inner Citadel.*

[494] John Hancock, Jr., Journal, 1824, p. 79, Norwich University Archives and Special Collections.

[495] Leonidas Konstantakos, "On Cleomenes and Sphaerus: How Stoic Was the Spartan King?," *Anais de Filosofia Clássica* 7, no. 14 (2013): 74–81.

[496] Plutarch, *Plutarch's Lives*, vol. IV.

[497] Konstantakos, "On Cleomenes and Sphaerus."

would be a consummate part of Partridge's curriculum, carrying this knowledge and likely Laconophilia to the cadets of the American Literary, Scientific and Military Academy.

Some academic members of the modern Stoic community (lacking any connection to military service) downplay the interconnection between Stoic philosophy and Spartan military culture. Many within our contemporary military community insist that this imagined connection is a modern invention driven by pop-cultural interests, possibly related to the Zack Snyder movie *300*. Yet, Cicero clearly states with the *Por L. Murena* that "Cato argues with austerity, and in the character of a Stoic" in behavior and speech derived from "Lacedæmonians, the first institutors of this way of living and talking."[498] The term *laconic*, named for the Spartan homeland, means brevity of speech. Stoics copied this precise and short style of language of the Spartans, as noted by Ralph Waldo Emerson, "Spartans, stoics, heroes, saints, and gods use a short and positive speech."[499] The interconnection within Stoic philosophy and Spartan culture is not revisionist history. Stoicism was forged into a highly practical and functional philosophy because of the influence of Greek military traditions and experiences heavily influenced by Spartan culture.

Several references to Sparta can be found within Captain Partridge's publications. In a letter to the editors of the *National Intelligencer* newspaper in 1829, Partridge recalled Spartans' ability to march great distances under heavy loads.[500] Included in this letter was a description of the marching practices being used at Partridge's American Literary, Scientific and Military Academy. Partridge would also add a reference to the Spartans in his *Lecture on National Defense*:

> Let then military Science be duly cultivated, and military information generally diffused throughout the Republic, and it then may be pronounced safe both from external invasion and internal Insurrection—Sparta without walls was far more formidable than Sparta with walls.[501]

---

[498] Lacedæmonia being another word for Sparta. Marcus Tullius Cicero, *Cicero's Select Orations, Translated into English; with the Original Latin, from the Best Editions, in the Opposite Page; and Notes Historical, Critical, and Explanatory*, trans. William Duncan, A New Edition, Corrected (New Haven, CT: Sidney's Press, 1811), 271–73.

[499] Hugh P. MacDonald, *The Power of Emerson's Wisdom* (New York, NY: Pageant, 1954), 128.

[500] Alden Partridge, Letter from Alden Partridge to Gales & Seaton of the *National Intelligencer* newspaper.

[501] Alden Partridge, Lecture on National Defense.

This quote appears at the end of the lecture, with the reference to Sparta serving as the crescendo of the speech.

Modern historians often apply the label of "Spartan" to Partridge's approach to education and his own personal behavior. It would not be a stretch to claim that Partridge's design of an *American System of Education* should have earned him the title of "America's Lycurgus." Romaneski reports Partridge was "spartan in every aspect of his life."[502] Frequent contemporary references to Partridge and the Spartans can also be found. For example, cadet Isaac E. Morse would report he "acquired [...] the almost Spartan military discipline of [Captain Partridge], of whom it used to be said that there were three things that he hated; a horse, a feather-bed and a woman." (Morse would assert that 'he last of the three was merely slander).[503,504] But just as the ancient Stoics could sometimes be confused for Spartans because of their appearances, we are left to ask if modern scholars have failed to correctly identify Partridge's behaviors and theories as Stoic because of our current disconnection with ancient history and philosophy.[505]

Downfall of Ancient Stoicism

The popularity of ancient Stoic philosophy did not survive the fall of the Roman Empire. Facing competition with the resurrected Platonic and Aristotelian philosophic schools and Christian theology (which absorbed many Stoic principles), ancient Stoic philosophy fell out of popularity. That is not to say that the ancient Stoics were not studied by individuals (i.e., Anicius Manlius Severinus Boethius would be heavily influenced by Stoic logic in the early sixth century A.D.[506]); however, widespread study of Stoic philosophy would not return until the 16th century with the rise of Neostoicism.

---

[502] Romaneski, "Importing Napoleon," 124.

[503] Edward Clarke Morse, *Blood of an Englishman* (Abilene, TX: Jones of Texas, 1943), 88.

[504] It is interesting to speculate on the cause of Partridge's hatred of horses. Was Partridge merely trying to copy the example of Cato the Younger, or was there some other cause? I know of at least one historical record of Partridge riding on horseback, but interestingly, it reports of his hatred for horses; "Saturday April 11th, 1824 [...] capt rode on horse back to day, which was a most remarkable sight as he hates ('tis said) horses & women. the cadets all looked after him with astonishment at a sight so novel it has become a subject of common conversation among the cadets & villagers [...] however he rides remarkeably well, for one who so seldom rides. this is the first time he has rode in any way since he returned from West Point." John Hancock, Jr., Journal, 1824, pp. 63–64, Norwich University Archives and Special Collections.

[505] Franklin C. Annis, "Stoicism for Toughening," Medium, December 27, 2022.

[506] Marenbon, "Anicius Manlius Severinus Boethius."

Neostoicism

Neostoicism is a philosophical movement, beginning in the late Renaissance, that aimed at reconciling ancient Stoic philosophy with Christian theology. The Flemish humanist scholar Justus Lipsius started the movement with the publishing of *De Constantia* (*"On Constancy"*) in 1584, as he combined Christian theology with the teachings of ancient Stoic Lucius Annaeus Seneca and Roman historian Publius Cornelius Tacitus. Other scholars, such as Pierre Charron, Francisco de Quevedo, and Michel Montaigne, continued this research by creating similar combinations of Christian theology with the ideas of other ancient Stoics.[507] Classicist Mark Morford laments, "Justus Lipsius is now little known except to students of Seneca and Tacitus and to intellectual historians of the northern Renaissance."[508] However, Lipsius and Neostoics are well worth remembrance in offering a strong conservative response to Machiavellianism.

To the ancient Stoic virtues of wisdom, justice, courage, and moderation, the Neostoics integrated the Christian virtues of faith, hope, and charity (love). To ensure the compatibility of Stoicism with Christian theology, the Neostoics abandoned the materialism and determinism of ancient Stoicism.[509] Neostoicism is a particularly useful philosophy for leaders; thus, many of the Neostoic works specifically address the correct conduct of princes and government leaders. It is important to note that Neostoicism is a revival of a living philosophy and not merely the resurrection of ancient Stoicism for academic purposes.[510] As reported by Leira, "Dignity, self-restraint and discipline were the recipes for the foreign policy of the prince, while the individual was subordinated to the purposes of the state and taught to control his own life by mastering his emotions."[511]

As described by Oestreich, Neostoicism expected "self-discipline and the extension of the duties of the ruler and the moral education of the army, the officials, and indeed of the whole people, to a life of work, frugality, dutifulness and obedience."[512] Many key elements of Neostoicism would be reinforced through Partridge's curriculum, with the responsibility of being

---

[507] John Sellars, "Neo-Stoicism," *Internet Encyclopedia of Philosophy*.

[508] Mark Morford, *Stoics and Neostoics: Rubens and the Circle of Lipsius*. (Princeton, NJ: Princeton University Press, 2017), xiii.

[509] Sellars, "Neo-Stoicism."

[510] Sellars.

[511] Halvard Leira, "Justus Lipsius, Political Humanism and the Disciplining of 17th Century Statecraft," *Review of International Studies* 34, no. 4 (2008): 669.

[512] Gerhard Oestreich, *Neostoicism and the Early Modern State*, ed. Brigitta Oestreich and Helmut Georg Koenigsberger (Cambridge: Cambridge University Press, 2008), 7.

good leaders and citizens adapted to the specific needs of the American Republic.

Neostoicism became deeply tied to the "Western way of war." Hugo Grotius helped to establish international law using a Neostoic philosophical framework with the publishing of major works such as *De Jure Belli Ac Pacis Libri Tres* (*On the Law of War and Peace: Three Books*). *On the Law of War and Peace* was reportedly carried by Swedish king Gustavus Adolphus during the Thirty Years' War.[513] Adolphus, hailed as the "father of modern warfare," is credited with the development of combined arms battle doctrine that is still employed on battlefields today. Italian-born General Raimondo Montecuccoli expanded on the works of Lipsius, in works on philosophy. The memoirs of Montecuccoli were acquired by the American Founder Arthur Lee in Prussia and carried back to George Washington during the American Revolutionary War.[514] Partridge was likely well aware of these great military thinkers and even possibly read the very copy of Montecuccoli secured by Arthur Lee, as a copy of the appropriate age now rests within the USMA Library.[515]

The rise of Neostoicism was concurrent to the use of firearms and linear tactics in the 16th century. New military technology demanded larger military forces and consolidated nations capable of affording such military formations. Seeing the problems of housing large mercenary armies that often behaved unvirtuously toward citizens, Lipsius advanced the idea of militias formed by virtue-based citizen-soldiers as the primary means of national defense. These forces, raised within local communities and destined to return to these communities, were far more likely to engage virtuously towards the local civilian population. The United States would attempt this approach, writing into the Constitution in Article I, Section 8, Clause 15, [The Congress shall have Power] "To provide for calling forth the Militia to execute the Laws of the Union, suppress Insurrections and repel Invasions." The original intent of the U.S. Constitution was that Congress would only "raise and support

---

[513] Christopher A. Ford, "Preaching Propriety to Princess: Grotius, Lipsius, and Neo-Stoic International Law," *Case Western Reserve Journal of International Law* 28, no. 2 (1996): 313–66.

[514] Arthur Lee, Arthur Lee to Franklin and Silas Deane, 15 June 1777, National Archives.

[515] A special thank you to Professor Vanya Eftimova Bellinger. When I reached out to her trying to discover possible reasons why Carl von Clausewitz's theories on war and Partridge's theories on education were compatible, she relayed that Gerald von Scharnhorst (Clausewitz's mentor) frequently referenced Montecuccoli. This information allowed me to discover a common exposure of Partridge and Clausewitz to this Neostoic military philosopher.

Armies" in times of war and not maintain the large standing army that currently exists in the present day.

Known Influence of Stoicism on the American Founding Fathers

To continue the discussion on the impact of Stoicism and Neostoicism, this section will discuss the impact of this philosophical tradition specifically on the American Founders. The Stoic philosophical tradition deeply influenced many of the prominent philosophers and scientists of the early modern era, including the American Founders.[516] These men were individuals of great education, well-read in ancient Greek and Roman literature. They lived in a time of a great resurgence of Stoic philosophy. They were greatly influenced by Joseph Addison's Neostoic play *Cato, a Tragedy*.[517]

The play recounts the story of Stoic Marcus Porcius Cato Uticensis' (better known as Cato the Younger) fight against and ultimately suicide to spite the tyrant Julius Caesar during the fall of the Roman Republic. This play was performed in 1778 at Valley Forge for George Washington's officers, despite a congressional ban on plays.[518] This play was quoted often in the correspondence of the Founding Fathers, as it was a great inspiration for both patriotism and republicanism in the face of tyranny. The famed ultimatum of Patrick Henry, "Give me Liberty or give me death!" is a paraphrase from Addison's play. Both Samuel Adams and George Washington were commonly referred to as the "American Cato," with King George III being referenced as the tyrant Caesar.[519] This same play would even be performed by students in the early days of the American Literary, Scientific and Military Academy.[520]

While Thomas Jefferson claimed to be an Epicurean,[521] a student of the philosophical school that rivaled Stoicism, later in his life he advanced rules

---

[516] Katerina Ierodiakonou, ed., *Topics in Stoic Philosophy* (Oxford: Clarendon Press, 1999), 3.

[517] Joseph Addison, *Cato, a Tragedy* (Edinburgh: John Wood, 1713).

[518] Rob Hardy, "Cato," George Washington's Mount Vernon.

[519] Rob Goodman and Jimmy Soni, *Rome's Last Citizen: The Life and Legacy of Cato, Mortal Enemy of Caesar* (New York, NY: Saint Martin's Griffin, 2014).

[520] List of Student Actors and the Roles They Are Playing in Dramatic Productions, 1825–1828.

[521] Both Stoicism and Epicureanism are variations of eudaimonism (placing happiness/human flourishing as the highest good) in their ethical approaches. While Epicureanism and Stoicism differ on their approach to interpretations of duty and the role of pleasure, they share more commonalities than differences when comparing them to other philosophies.

for daily living that derived directly from Stoic texts. Additionally, in his writings, Jefferson recommended all the chief surviving texts of Stoicism, including the *Letters* and *Essays* of Seneca, *The Enchiridion* of Epictetus, and the *Meditations* of Marcus Aurelius.[522] In 1803, Jefferson responded to Joseph Priestley's pamphlet *Socrates and Jesus Compared*[523] by penning a suggested syllabus comparing Jesus to other philosophers of antiquity. Jefferson suggests that the "most esteemed" philosophers were Pythagoras, Socrates, Epicurus, Cicero, Epictetus, Seneca, and Antoninus (Marcus Aurelius).[524] The final three of the seven philosophers recommended are ancient Stoic philosophers, with Cicero also capturing significant Stoic materials within his writings and Socrates inspiring the creation of the Stoic school of philosophy.

Jefferson even had the writings of Seneca on his nightstand upon his death.[525] Partridge shared a great many interests with Jefferson, including walking, surveying, and education. It is certain that Jefferson had a significant influence on Partridge, with Partridge "profess[ing] to be a republican of the old Jeffersonian School" in his later political campaigns.[526]

John Adams wrote to Benjamin Rush in admiration of Rome and Sparta.[527] Samuel Adams wrote to John Scollay hoping that Boston would become a "Christian Sparta,"[528] echoing the intent of Neostoic philosophy to combine Stoic philosophy with Christian theology. Several Stoic themes can be seen in the speeches the Founders gave as they rallied the forces to fight against the British. As demonstrated by the following abstract from a speech by Samuel Adams,

> From the day on which an accommodation takes place between England and America on any other terms than as *independent States*, I shall date the ruin of this country. A politic minister will study to lull

---

[522] Donald J. Robertson, "The Stoicism of Thomas Jefferson," Stoicism—Philosophy as a Way of Life, February 8, 2020 (Medium).

[523] Joseph Priestley, *Socrates and Jesus Compared* (Philadelphia, PA: P. Byrne, 1803).

[524] Thomas Jefferson, Doctrines of Jesus Compared with Others, 21 April 1803, National Archives.

[525] Richard Beale Davis, *Intellectual Life in Jefferson's Virginia 1790–1830* (Chapel Hill, NC: University of North Carolina Press, 1964), 118.

[526] Alden Partridge, Nomination of Capt. Partridge for Representative in Congress, and Expression of his Views on Public Matters, Approximately 1838, p. 18, Norwich University Archives and Special Collections.

[527] John Adams, From John Adams to Benjamin Rush, 28 July 1789, National Archives.

[528] Samuel Adams, Epilogue: Securing the Republic: Samuel Adams to John Scollay, 30 Dec. 1780.

us into security by granting us the full extent of our petitions. The warm sunshine of influence would melt down the virtue which the violence of the storm rendered more firm and unyielding. In a state of tranquility, wealth, and luxury, our descendants would forget the arts of war and the noble activity and zeal which made their ancestors invincible. Every art of corruption would be employed to loosen the bond of union which renders our resistance formidable. When the spirit of liberty which now animates our hearts and gives success to our arms is extinct, our numbers will accelerate our ruin, and render us easier victims to tyranny. Ye abandoned minions of an infatuated Ministry,— if peradventure any should yet remain among us,—remember that a Warren and Montgomery are numbered among the dead! Contemplate the mangled bodies of your countrymen and then say what should be the reward of such sacrifices. Bid us and our posterity bow the knee supplicate the friendship, and plough and sow and reap, to glut the avarice of the men who have let loose on us the dogs of war to riot in our blood, and hunt us from the face of the earth! If ye love wealth better than liberty, the tranquillity of servitude than the animating contest of freedom, go from us in peace. We ask not your counsels or arms. Crouch down and lick the hands which feed you. May your chains sit lightly upon you, and may posterity forget that ye were our countrymen.[529]

In this speech, the Stoic elements of enduring hardships, avoidance of luxury, focus on virtue, and demand for justice can be clearly noted.

The Founding Fathers were heavily influenced by Montesquieu's *The Spirit of the Laws*, with this work influencing the design of the three branches of the federal government. Within this text, Montesquieu extolled the Stoics, stating,

There has never been one whose principals were more worthy of men and more appropriate for forming good men than that of the Stoics, and, if I could for a moment cease to think that I am a Christian, I would not be able to keep myself from numbering the destruction of Zeno's sect among the misfortunes of human kind. It exaggerated only those things in which there is greatness: scorn for pleasures and pain.

---

[529] Samuel Adams, *An Oration Delivered at the State-House, in Philadelphia, to a Very Numerous Audience on Thursday the 1st of August, 1776; by Samuel Adams* (Philadelphia, PA: E. Johnson, 1776), 413–14.

It alone knew how to make citizens; it alone made great men; it alone made great emperors.[530]

Doubtlessly, a philosophy receiving such praise would be imitated in the young Republic.

The newly established national capital in Washington, D.C., was built using Greek Revival architecture. The desire to reconnect with the ancient Greek and Roman cultures is seen in the names of the towns and cities being established, Athens, New York; Laconia, New Hampshire; and Sparta, New Jersey, being examples. Even among the names of Partridge's peers and students, ancient Greek and Roman references can be found. Many of the original cadets of Norwich University carried Stoic-inspired names, including Lucius, Seneca, Marcus, and Rufus. These deep interactions with Greek and Roman culture, to include Stoic philosophy, certainly affected the zeitgeist of the era and most certainly impacted the educational theories in the early days of the Republic. With so much interest in reviving and adopting elements of Stoic philosophy within America, many intended references to Stoic ethics may be hidden within the works of the early generations of America, as they explicitly referenced "American" ideals that would have incorporated Stoic philosophy.

Emersonian Idealism

In the 1830s, during Partridge's lifetime, the transcendentalist movement coalesced. The transcendentalist movement consisted of a significant number of disparate philosophers, including Amos Bronson Alcott, William Henry Channing, Elizabeth Peabody, and Margaret Fuller, and among others, the philosophies of Ralph Waldo Emerson and his mentee Henry David Thoreau were significantly affected by Stoic philosophy. Given that the transcendentalist movement captured a range of philosophies, not all being heavily influenced by Stoicism, this section will focus specifically on those who could be classified as Emersonian idealists.

While any examination of Emerson's philosophy is worthy of considerable time and attention, it is sufficient here to claim Emerson's philosophy was essentially a non-theological, restricted variation of Neostoicism. Emerson frequently quoted and referenced ancient Stoic Marcus Aurelius (using the name Antonius). Emerson and Thoreau exhibit multiple *stoa mementi*, including an emphasis on self-reliance/inner strength, concordance with nature, focus on virtue, and acceptance of fate.

Transcendentalism ran parallel with concepts of Unitarianism. Unitarians asserted that truth/divine wisdom could be found in the theological

---

[530] Cohler, Stone, and Miller, *Montesquieu: The Spirit of the Laws*, 465–66.

texts of all religions. Neither Unitarianism nor transcendentalism required any immediate individual or institution in this search for the divine. Emerson explored foreign theological texts, to include the Hindu Bhagavad Gita, searching for truths he believed were universal to mankind. Emerson placed a significant emphasis on self-reliance and the ability of individuals to independently recognize truth, recounting in the essay *Self-Reliance*,

> Trust thyself: every heart vibrates to that iron string. Nothing is at last sacred but the integrity of your own mind. A man should learn to detect and watch that gleam of light which flashes across his mind from within, more than the lustre of the firmament of bards and sages... In every work of genius we recognize our own rejected thoughts: they come back to us with a certain alienated majesty. The power which resides in him is new in nature, and none but he knows what that is which he can do, nor does he know until he has tried.[531]

Emerson's philosophy captured what Alexis de Tocqueville had recognized about Americans during his travels a few years earlier: Americans of the era were more likely to trust their own experiences and intuition instead of defaulting to expertise or abstract ideas.[532] The virtue of self-sufficiency is a *stoa mementi* found within the self-reliant/self-helping drive of Emersonian idealism. It is easy to recognize the self-sufficient goals within Captain Alden Partridge's educational theories that wished to ensure that citizens are "prepared in the best possible manner to discharge correctly the duties of any station in which fortune or inclination may place them."[533]

Ralph Waldo Emerson may provide the strongest evidence that the fulfillment of Captain Alden Partridge's educational model calls American scholars to specifically follow the Stoic philosophical tradition. In 1863, nine years after Partridge's death and in the middle of the American Civil War, Emerson addressed the literary societies of Dartmouth and Waterville Colleges, asserting that college graduates should display the following characteristics:

> So let [the American scholar] habits be formed, and all his economies heroic; no spoiled child, no drone, no Epicure, but a Stoic, formidable, athletic, knowing how to be poor, loving labor, and not flogging his youthful wit with tobacco and wine; treasuring his youth. I wish the youth to be an armed and complete man; no helpless angel to be

---

[531] Ralph Waldo Emerson, *Essays* (Boston, MA: James Munroe and Company, 1840), 54.

[532] Tocqueville, *Democracy in America.*

[533] Partridge, Lecture on Education (British Library), 2.

slapped in the face, but a man dipped in the Styx of human experience, and made invulnerable so, — self-helping.[534]

Within this passage, the multiple *stoa mementi* within Emerson's philosophy are again displayed. The Stoic dependence on inner resources is visible with the call to be "self-helping" and "formidable." "[K]nowing how to be poor" is an appeal to the Stoic acceptance of hardship. Stoic discipline and moderation are found within the call to "lov[e] labor" and avoid "tobacco and wine." The demand for "no spoiled child, no drone, no Epicure, but a Stoic," apart from explicitly calling for the use of Stoic philosophy, demands a life of moral integrity and virtue. Stoic resilience is reflected in the figurative call to be "dipped in the Styx of human experience, and made invulnerable." The Stoic focus on productive use of time is visible in those "treasuring his youth." All these sentiments are also echoed throughout Partridge's educational theories.

In June of that year, Emerson visited the United States Military Academy and was struck by the self-sufficiency of its cadets. Emerson would have witnessed firsthand the influences of military discipline that had been created by numerous practices at the Academy, many of these established by Captain Alden Partridge. While Emerson never directly referenced Partridge in these lectures, he covers many of the educational principles that Partridge wished were established within American colleges. At least one edition of the *Citizen Soldier* newspaper contains works by both Emerson and Partridge. Given that both men were on a similar speaking circuit in New England, it would not be unreasonable to believe that Emerson was exposed to, or at least knew of, Captain Partridge. More on the possible influence of Partridge on the transcendentalist movement is planned for a later volume.

Post-Partridge Stoic Movements

There are a few Stoic terms and movements that have appeared since the death of Partridge. These more recent schools may influence the understanding of Stoicism among the present-day audience. For the sake of clarity and understanding of the present usage of the term *Stoicism*, I will offer a brief discussion on the modern Stoic movement, and the concept of natural and military Stoicism will be presented. This discussion may be useful in understanding differences in communities that use the term *Stoicism* and how they might apply this term for different purposes.

---

[534] Ralph Waldo Emerson, *Lectures and Biographical Sketches by Ralph Waldo Emerson* (New York, NY: Houghton, Mifflin, and Co., 1895), 240.

Modern Stoicism

The modern Stoic movement began well after Alden Partridge's lifetime. While there are many different suggestions for the definition of modern Stoicism,[535] for the sake of this research, it is defined as a philosophical movement begun in the 20th century that sought to reconcile Stoic practices with our modern understanding of science, including the field of psychology. This includes works of prominent scholars like Albert Ellis—developer of Rational Emotive Behavior Therapy—and Aaron T. Beck, the "father of cognitive behavioral therapy."[536] With the founding of the non-profit "Modern Stoicism" in 2012,[537] prominent voices inside the Stoic community, including Christopher Gill, Massimo Pigliucci, Donald Robertson, and others, continue to advance the reexamination of ancient Stoic philosophy informed by modern science.

However, many modern Stoic philosophers suffer from the temperamental narrowing associated with the present-day university system, which results in a significant bias within their research. As a result, it is common to find members of this community misrepresenting the connection between the Western military tradition and Stoic philosophy. For example, Massimo Pigliucci underplays the relationship between military service and ancient Stoic philosophy… going so far to assert that military service cannot be reconciled with the virtuous action required of Stoicism.[538] Similar to the challenges Partridge faced in creating a system of education to meet the entire needs of a citizen of the Republic during his lifetime, many academics today continue to assert that "educated" citizens have no need for military knowledge.

Natural Stoicism

While scholars and academics often study philosophical movements that are restricted in their direct relation to a specific time period in which they were used, there is also a thriving community seeking to apply the Stoic philosophical tradition to current problems. As a result, practical philosophers may use the term *Stoicism* in relation to individuals who have discovered the underlying truths of Stoic philosophy by other means than direct instruction in this philosophic tradition. Unsurprisingly, "natural Stoics" are often discovered

---

[535] Gregory Sadler, "Symposium: What Is Modern Stoicism?," Modern Stoicism, July 30, 2017.

[536] Donald Robertson, *The Philosophy of Cognitive-Behavioural Therapy (CBT): Stoic Philosophy as Rational and Cognitive Psychotherapy* (Abingdon, Oxon: Routledge, 2020).

[537] "The Modern Stoicism Team," Modern Stoicism, September 17, 2019.

[538] Massimo Pigliucci, "Stoicism and the Military: An Unvirtuous Coupling," Modern Stoicism, April 22, 2023.

within populations facing extreme environments, from internment camps to terminal diagnoses. Below, a couple of examples of natural Stoics will be explored.

Victor Frankl (1905–1997 A.D.)

Likely the most famous of the natural Stoics would be Viktor Frankl. Frankl would develop a great interest in psychology and psychoanalysis as a teenager. In 1925, he would graduate with a medical doctorate from the University of Vienna. He served as the director of the neurological department at Rothschild Hospital from 1940 to 1942. He had already started the development of his "Logotherapy" (Meaning Therapy) by September 1942, when he was arrested by the Nazi government and taken to the Türkheim concentration camp because of his Jewish heritage. Within the concentration camp, Frankl keenly observed the interaction between the prisoners' attitudes and their survivability.[539] Frankl records,

> We who lived in concentration camps can remember the men who walked through the huts comforting others, giving away their last piece of bread. They may have been few in number, but they offer sufficient proof that everything can be taken from a man but one thing: the last of the human freedoms — to choose one's attitude in any given set of circumstances, to choose one's own way.[540]

This period of harsh captivity provided the ultimate testbed for his theories in preventing and treating mental disease among his fellow prisoners. In 1946, Frankl would publish *Man's Search for Meaning*, a work explaining his "Logotherapy" that he had fully developed and personally tested. Although Frankl lost his pregnant wife, parents, and brother in the war, he retained his optimism for life and became a powerful living example for his own Logotherapy.[541]

Wilko Johnson (1947–2022 A.D.)

The English singer, songwriter, guitarist, and actor Wilko Johnson (July 12, 1947–November 21, 2022) displayed natural Stoicism when facing the end of his life. Johnson was diagnosed with inoperable pancreatic cancer. This event radically changed the way he viewed the purpose of his existence. The

---

[539] Judith Turner, *"Man's Search for Meaning* Teacher's Guide,"* n.d., PenguinRandomhouse.com.

[540] Viktor E. Frankl and Gordon W. Allport, *Man's Search for Meaning*, 3rd ed., trans. Ilse Lasch (New York, NY: Simon & Schuster, 1984), 75.

[541] Turner, *"Man's Search for Meaning* Teacher's Guide."*

modern Stoic Donald Robertson would ask if the documentary *The Ecstasy of Wilko Johnson*[542] was the "greatest documentary on Stoicism."

> *The Ecstasy of Wilko Johnson*, [...] basically allowed Wilko to quote *Hamlet* and *Paradise Lost* a lot, while reflecting on his own mortality. He never mentions Stoicism but there are many quotes such as "What cannot be cured must be endured", which he's drawing from English classics obviously inspired by Stoicism.[543]

Johnson discusses many Stoic themes, such as *memento mori* and *amori fati*, within the documentary without ever naming the formal philosophy. With quotes from various works of Western philosophy that are heavily influenced by the Stoics and Neostoics, Wilko reflects on how he learned to live in the present moment and accept his inevitable death. Ultimately, the 10-month life expectancy given to Johnson in 2014 would be proven wrong. He underwent surgery in 2014 that allowed him to live for another eight years before dying at the age of 75. However, the change in his outlook and philosophy on life was permanently altered by this experience.

While this list of contemporary men whose lives demonstrated the kind of resiliency often associated with Stoicism could be vastly expanded, the examples of Frankl and Johnson display how the title of "Stoic" is often attached to those who find the deep-seated common lessons of the human experience. In this way, the term *Stoicism* is used as a synonym for truth. While many philosophies seek to find the truth of human experience, Stoic philosophy seems to have captured more of this truth than competing philosophies. As a result, when society recognizes the wisdom of those thriving while going through extreme trials, the label of Stoic is often applied.

Another example of natural Stoicism within the military community includes Lester Tenney, who endured the Bataan death march and years of imprisonment under the Imperial Japanese military. His story of maintaining a positive attitude in this extreme environment is recorded within the book *My Hitch in Hell*. It is not uncommon to see Buddhism related to Stoicism because of their similarities. Similarly, Eastern warrior traditions, such as the Samurai, are sometimes related to the Western Stoic traditions due to shared beliefs and practices.[544] *Natural stoicism* is a useful term and concept for those intent on finding a practical philosophy for their own lives outside the strict scholarly

---

[542] *The Ecstasy of Wilko Johnson*, directed by Julien Temple (BBC Imagine, 2015).
[543] Donald J. Robertson, "The Greatest Documentary on Stoicism?" Stoicism: Philosophy as a Way of Life, March 14, 2023, para. 3 (Substack).
[544] Michael Barr, "Shido and Stoicism: An Exploration of Commonalities in Principles and Practices," Shido and Stoicism, September 19, 2024.

restrictions of the academic study of specific schools and periods of philosophy.

Military Stoicism

As explained by Albert Salomon in the foreword of a translation of Epictetus' *Enchiridion*, the Roman Stoics heavily used military service as a metaphor for life;

> This analogy should be taken seriously. The Roman Stoics coined the formula: *Vivere militare!* (Life is being a soldier.) The student of philosophy is a private, the advancing Stoic is a non-commissioned officer, and the philosopher is the combat officer. For this reason all Roman Stoics apply metaphors and images derived from military life. Apprentice students of Stoicism are described as messengers, as scouts of God, as representatives of divine nature. The advancing student who is close to the goal of being a philosopher has the rank of an officer.[545]

Members of the military community are unique in their use of Stoic philosophy not only to prepare themselves for the inconveniences of the everyday world, but also to develop themselves to handle the extremes of the battlefield. Members of this community are highly motivated to learn the appropriate use of Stoic philosophy and all the advantages of psychological resilience, physical fitness, and enhanced leadership abilities it offers. These individuals often reject the strict academic study of philosophy and embrace only the elements of philosophy that can be proven true through practice, echoing the sentiments of Friedrich Nietzsche, who wrote that "The only criticism of a philosophy which is possible, and which also proves something—that of seeing if one can live by it—has never been taught at the universities: but always criticism of words by words."[546,547] For the military community, philosophy isn't something casually debated. But something that should be fully embodied in everyday thought and action, with the abandonment of all principles not shown practical in the most extreme of environments.

The historian Gerhard Oestreich noted the following characteristics in the writings, actions, and directions of Prussian general Gerhard Johann David von Scharnhorst: "love, trust, respect, reputation, friendly persuasion, gentle

---

[545] Epictetus, *The Enchiridion*.

[546] Friedrich Nietzsche, *Schopenhauer as Educator*, trans. James W. Hillesheim and Malcolm R. Simpson (Regnery/Gateway, Inc., 1980), 99.

[547] The assertion here concerns individual members of Western militaries. It is acknowledged that current bureaucratic management styles and active endorsements of philosophies hostile to Stoic philosophy by military organizations present significant barriers to the widespread adoption of Stoic-based ethics.

procedure, justice, development of physical, mental and spiritual qualities, powers and virtues."[548] Scharnhorst was a Prussian contemporary to Partridge and drew on much of the same Neostoic materials to advance the military capabilities of Prussia. A quick review of these materials will reveal all the features of quality military leaders. Carl von Clausewitz, a student of Scharnhorst and one of the most influential military philosophers of the 20th and 21st centuries, captured many Stoic themes in his work, *Vom Kriege (On War)*. For example, Clausewitz records the Stoic ideal of staying rational even when experiencing emotional distress:

> We therefore say once more a strong mind is not one that is merely susceptible of strong excitement, but one which can maintain its serenity under the most powerful excitement, so that, in spite of the storm in the breast, the perception and judgement can act with perfect freedom, like the needle of the compass in the storm-tossed ship.[549]

There is some debate over whether Clausewitz was drawing directly from the Stoic tradition in building his strategic framework or if the Stoic influence occurred primarily through the instruction under Scharnhorst.[550] Oestreich suggested that Clausewitz may have even been exposed to a new translation of Lipsius.[551]

While there has been a continual focus on these characteristics within military leaders, the knowledge of their direct connection to Stoic philosophy among the larger military community has been largely lost. One interesting echo of Neostoic ideals occurred within General Charles C. Krulak's "Strategic Corporal" concept. Krulak described a vision of well-trained Marine junior leaders being able to rationally assess their environment and make decisions on the use of force in keeping with larger strategic goals. These values (virtue)-driven Marines would have been well prepared through difficult and realistic training while being provided with a host of historical examples of exemplary Marine leaders to help guide their actions.[552] While Krulak never mentioned Stoicism nor Neostoicism within his concept, the parallels with the major tenants of this philosophical tradition are striking.

---

[548] Oestreich, *Neostoicism and the Early Modern State*, 87.

[549] Carl von Clausewitz, *On War*, vol. I, trans. Frederic Natusch Maude (London: Kegan Paul, Trench, Trübner & Company, 1908), 60.

[550] I would like to thank Professor Vanya Eftimova Bellinger for pointing out Scharnhorst's affinity towards the theories of Raimondo Montecuccoli. Montecuccoli, a 17th-century military theorist, was heavily influenced by the works of Justus Lipsius.

[551] Oestreich, *Neostoicism and the Early Modern State*.

[552] Franklin C. Annis, "Krulak Revisited: The Three-Block War, Strategic Corporals, and the Future Battlefield," Modern War Institute West Point, February 3, 2020.

One noteworthy recent example of a military Stoic is Vice Admiral James Bond Stockdale. Stockdale, a Douglas A-4 Skyhawk pilot during the Vietnam War, who had been exposed to the writings of the ancient Stoic Epictetus. Stockdale's aircraft was shot down in 1965. He would endure seven and a half years as a prisoner of war (POW) in the Hỏa Lò Prison (also known as the "Hanoi Hilton"). He shared his knowledge of Stoic philosophy with fellow POWs. This resulted 'n the prisoners held with Stockdale experiencing lower rates of post-traumatic stress disorder than the average Vietnam veteran.[553] The *Stockdale Paradox*, named in Stockdale's honor, is the ability to maintain optimism about the future even when evaluating the brutal reality of the current situation. Stockdale demonstrated the true grit of an individual that would accept nothing less than doing one's duty and having faith that your nation would ultimately come to their rescue.

The academic community frequently criticized Stockdale for his imprecise and amateur understanding of Stoicism. Similar to how Oestreich has been criticizing for interpreting Neostoicism through an ideological lens and in light of his own significant life events,[554] military Stoics are frequently criticized for their non-academic approaches to philosophy. However, philosophers from Emerson to Nietzsche would suggest that all philosophies are at least part autobiographical. Academic philosophers do not avoid the influence of ideology and personal bias within their own research. As a result, this criticism may ring hollow. In the end, there are likely too many differences between the intended uses of philosophy between academics and military members to find themselves typically in concordance with each other. Frequent conflicts between academic philosophy and the military community attempting to use Stoic philosophy are not surprising.

Shifting Philosophy through Time and Community Usage

In the various definitions above, changes in Stoic philosophy have been displayed in both the terms of time (evolution of Stoic schools) and uses (academic vs. practical). A semantic argument certainly could question if future philosophic schools should be called Stoics or merely Stoic-influenced if they continue to perpetuate elements of Stoic philosophy. Another largely semantic

---

[553] Peter Fretwell and Taylor B. Kiland, "Review of Leadership Lessons from Hanoi Hilton: Vice Admiral James Stockdale's Principles Can Inspire Any Organization's Leaders," *Proceedings* 135, no. 11 (November 2009): 1281.

[554] Alberto Clerici, "*In publicis malis.* Justus Lipsius and the 'Double Face' of Neostoicism in the European Wars of Religion," in *Crisis and Renewal in the History of European Political Thought*, ed. Cesare Cuttica, László Kontler, and Clara Maier (Boston, MA: Brill, 2021), 262.

argument would question the degree of ancient Stoic tenants needed to be sustained in future philosophical schools to use the name Stoic.

If I were developing a test to determine if someone should be named a Stoic, I would look for four major elements. The first would be a focus on virtue as the sole good. Second would be the examination and correction of opinions to align value judgements with the ability to engage in virtue. Third would be a voluntary engagement in physical and mental hardship (especially in nature). Finally, would be the focus of duty to the community as a requirement for human happiness. Except for the second criterion, Partridge meets these requirements. As previously noted, he likely engaged in the control of opinions and instructed his students to do so as well, but direct evidence of such action has been lost to time. Given Emerson expressed so many of Partridge's opinions on education while specifically naming Stoic philosophy as an ideal philosophical model, I have little doubt that Partridge would have concurred with Emerson's opinion.

In the end, attempting to retroactively identify and name the philosophical school individuals may have carried may be an exercise in futility. For example, it makes little difference in the end if George Washington is labeled an Enlightenment thinker versus a Neostoic. These terms are useful only in conveying the meaning of his actions and thoughts. It is acknowledged that academic philosophers, because of the restrained study of philosophical schools restricted to specific historical time periods, are unlikely to classify Partridge as Stoic. However, for those who use the term more loosely with a focus on practical philosophy, Partridge in thought and word and deed bares the mark of a Stoic.

Return to the Stoa

Instruction in Stoic philosophy was largely removed from American secondary and higher education in the 1960s. This has created a general ignorance of the true tenants of this philosophy among the U.S. population. There are valid reasons to return to the explicit instruction of philosophy and attempt to remove this widespread ignorance. For example, clinical psychologists have recently demonstrated statistically significant improvements in mental health outcomes when Stoic ethics are taught within psychological resiliency programs.[555,556] It is past time for the U.S. population to again recognize the importance of educational instruction in this time-tested system

---

555 Stelnicki et al., "Evaluation of Before Operational Stress."
556 Gabriela Ioachim et al., "Evaluating the Before Operational Stress Program: Comparing In-Person and Virtual Delivery," *Frontiers in Psychology* 15 (July 25, 2024).

of ethics that was valued by seminal figures in the Western canon. While it is doubtful that every student that is exposed to Stoic principles in education would fully integrate this ethical system into their own personal philosophies, knowledge of these principles would better prepare citizens to fulfill their duties in peace and war. Stoic-educated citizens would at least understand Stoic ethics that could be employed in times of hardship, even if they employed other philosophies in times of comfort. Recognizing the value of Stoic practices and key texts, Partridge integrated these resources throughout his curriculum and instruction. Returning to the *American System of Education* would go far in creating more resilient and virtue-driven citizens for the Republic.

Conclusion

This chapter covered the definitions and evolution of ancient Stoic philosophy and Stoic-influenced philosophical schools. Included in this discussion were the multiple connections found with Captain Alden Partridge and an explanation on how diverse communities made different use of Stoicism and Stoic identity. The influence of Stoic and Neostoic philosophers on the American Founders was also presented to include a discussion of how this philosophy affected the zeitgeist of the early Republic. The Stoic philosophy perpetuated through Alden Partridge's education, appearance, and behaviors was presented, as well as examples of how students and outside observers connected Partridge's educational practices with Stoic philosophy during his lifetime and beyond.

In admitting my personal bias as a military Stoic with a personal philosophy close to Emersonian idealism, I have no issue with naming Alden Partridge a Stoic. The dozens of connections between Partridge's ideas and Stoic ethics make me confident in this assertion. However, it is acknowledged that academic philosophers that use a more stringent approach to studying philosophies restricted to specific time periods may reject this assessment. As previously noted, there are no known sources of Partridge specifically claiming to be, or directing his students to be, Stoic. Until further evidence supporting or refuting if Partridge personally identifying with the Stoic philosophy comes to light, the debate on whether to identify Partridge as a Stoic will likely continue.

Before concluding the discussion on Partridge's connection to the Stoic tradition, I present what is probably the greatest proof Partridge was a Stoic: he suffered the public doom of one. Ironically, Partridge may have missed a powerful warning about his own fate within one of the key texts he used in his academies. A footnote within William Duncan's translation of Cicero's orations recalls the ill fortune of Quintus Aelius Tubero in the eyes of the people of Rome caused by his Stoic behavior at the funeral of Scipio Africanus:

> [It was the same from the study of Tubero] Cicero here ridicules the doctrine of the Stoics, shows the absurdities into which it may betray a man and paints the ill consequences that often arise from it. [Quintus Aelius] Tubero, of whom he speaks here had professed himself a Stoic and resolved to regulate his conduct by the tenets of that sect. Accordingly, in an entertainment he gave the Roman people on occasion of the death of the great Scipio Africanus he made use of plain wooden beds, goat skin covers, and earthen dishes. But this ill-timed parsimony was so displeasing to the Roman people that when he afterwards stood for the prætorship they refused him their suffrages though a man of illustrious birth and the most distinguished virtue.[557]

Is there a passage more fitting for the legacy of Partridge and his Stoic behavior? Even when Partridge had built an ideal model for educating a complete virtue-driven citizen worthy of the Republic, few would find the lifestyle required appealing. Being a virtuous man with a sufficient plan for American education was not enough to guarantee his acceptance among the masses.

---

[557] Cicero, *Cicero's Select Orations*, 272.

Chapter IX: Was Partridge a "Classical" Educator?

There is a common question if Alden Partridge should be considered a "classical" educator in the present-day understanding of the term. This question is likely a false dichotomy, as Partridge's curriculum carried many aspects of both classical and modern utilitarian education. However, with educational reformists like Jeremy Wayne currently calling for a return to classical education and offering a classical education-based standardized testing for primary/secondary schools and college entrance exams,[558] it may be valuable to define classical education and clarify how Partridge's *American System of Education* would relate to this movement.

First, it must be acknowledged that there is no singular definition of classical education. However, it is normally related to the seven liberal arts of the *trivium* (grammar, logic, rhetoric)[559] and the *quadrivium* (astronomy, mathematics, geometry, and music).[560] The central intent of this training was to produce ideal citizens (known as *paideia* (παιδεία) in ancient Greek and *humanitas* in Latin). Instruction in poetry, philosophy, and gymnastics (or other fitness training) normally accompanying the curriculum. These subjects are often taught alongside Christian theology and may require students to read the ancient texts in their original language (Ancient Greek and/or Latin). Classical education is often described as teaching the virtuous and beautiful; this idea is an echo of Plato's vision of education,

> [I]t touches the heart, and penetrates into the recesses of the soul and fills it with harmony and moulds it to grace, and gives to the young character an instinctive unreasoning love for the good and beautiful, even before the boy can reason about such things, so that later on, when reason comes, he salutes her as a friend with whom knowledge has long made him familiar.[561]

---

[558] "Alternative College Entrance Exam," Classic Learning Test, June 21, 2023.

[559] A Renaissance term derived from Martianus Capella's work *De nuptiis Philologiae et Mercurii* (*On the Marriage of Philology and Mercury*).

[560] Term attributed Roman senator and historian Anicius Manlius Severinus Boethius.

[561] Plato, *The Republic of Plato: Books I–V*, trans. T. Herbert Warren (New York, NY: MacMillian and Co., 1892), xlvii.

While utilitarian education is heavily focused on skills and knowledge required for employment, classical education has a broad focus on liberal arts to build the character and virtue of students. As stated by T. Herbert Warren,

> [A] classical education—essentially liberal in that it was not intended that any one should get a living by it, and that it contemplated no profession except that of a gentleman who might be called on perhaps to take public office, or to become a soldier: and, essentially classical, in that it trained the intellect mainly in literature, and that literature a selection from old sources.[562]

There have been considerable drifts in methods of classical education occurring since its medieval origins. In the medieval era, it wouldn't be uncommon for a student to complete the *trivium* (primary education) and the *quadrivium* (secondary education) and enter an apprenticeship (tertiary education) by the age of 14. Keep in mind that college/university in the medieval origin referred to the community of scholars and not the schoolhouse or campus as we would understand it today.[563] Thus, what present-day citizens think of college/university (a location with a group of scholars) as tertiary education; would have historically been tutorship with scholars in whatever space was available in the medieval era. In our contemporary education system, it would commonly be expected that students would complete secondary education around the age of 18. However, in 1947, the novelist and poet Dorothy Sayers suggested a model of classical education correlated with human development that would allow students to complete the *trivium* and *quadrivium* by the age of 16.[564]

Evaluating Partridge as a Classical Educator

While not a universal requirement in our contemporary environment, one of the primary markers of classical education is the requirement to read source materials in their original language (Ancient Greek and Latin). Many individuals may misjudge Partridge's statements on ancient languages by not fully understanding the specific context that he is referencing. Partridge wrote educational theories while in two distinct environments (U.S. Army education and civilian institutions using the *American System of Education*). In both environments, students may have had previously had training in ancient foreign languages or could receive this instruction through other means in the future.

---

[562] Plato, *The Republic of Plato*, 56.

[563] Even the concept of what the American college experience should be was in flux well into the 20th century.

[564] Dorothy L. Sayers, *The Lost Tools of Learning* (London: Methuen, 1948).

Three of Partridge's statements on ancient languages will be examined in the hopes of providing a better context of his intended meaning.

Partridge advanced the inclusion of only the modern foreign language of French within the curriculum of the USMA.[565] Partridge likely viewed graduation from the USMA as a terminal degree for military officers. While it was likely many of the cadets attending West Point would have previously been exposed to ancient languages, the need to learn French was of practical importance because it; was the universal language of the era. It is also noted the intent of this specific proposal was to reduce the four-year curriculum at the USMA to a shorter training program. This would have shifted USMA certification of completion away from a college equivalent. Therefore, one could speculate that in this instance Partridge was not supporting the classical educational approach, and actually intended to push the USMA outside a tertiary educational construct.

It is easy to perceive Partridge's criticism of the contemporary model of secondary education and colleges in his *Lecture on Education* as hostile towards classical education. His first objection is the study of Latin and Greek are not of sufficient practicality to the American citizen. However, in this same publication he references the requirement to study ancient and modern history, with a focus on military education. Listing the classical military historians such as "Xenophon, Thucydides, Polybius and Cæsar,"[566] we see Partridge believed it was important to include authors normally associated with classical education presumably through English translations. Many see this method as an acceptable practice in the current classical education movement (as the content of classical works are more important than form).

In a *National Intelligencer* article published in 1826, Partridge suggested a model for primary and secondary education that would restrict the study of "Latin and Greek, for such as are to attend to the classics."[567] In this model, only a select few students who progressed to college/university for non-scientific degrees would benefit from classical education. However, Partridge's model for primary/secondary education carries the same military education and discipline as his plan for higher education. It is likely that his students would have been exposed to many of the classical thinkers through lectures and translated texts related to military tactics and campaigns. One could speculate that this would have been a limited classical education in practice.

---

[565] Alden Partridge, Observations on the Reorganization of the Military Academy at West Point, New York, 12 January 1821, Norwich University Archives and Special Collections.
[566] Partridge, "Captain Partridge's Lecture on Education," 274.
[567] Partridge, "Communication."

Partridge's 1841 Memorandum to Congress on the establishment of land-grant colleges, asserts, "the study of the ancient and modern languages, (not excepting the English language [...]) to be left optional with the student, or their parents or guardians."[568] However, we must understand that Partridge was here addressing tertiary education. It would be likely that students attending land-grant colleges would have already been exposed to ancient language instruction through churches and religious schools. Therefore, it would be logical that students and parents be allowed to determine if previous instruction in ancient languages was sufficient for the future needs of the students or if additional instruction was required. As with the *Lecture on Education,* Partridge includes the requirement for students to attend "An extensive course of history, ancient, and modern."[569] Again, this can be interpreted as Partridge's willingness to allow students to use translations to access classical education materials.

Given the studies of these ancient languages were an option, it can be assumed that courses in these languages would have been readily available. This would imply support for learning classical education materials in their original ancient language. In Partridge's *Lecture on Education,* he used Pascal—a mathematical genius that rejected the study of language in his childhood—as a prime example of how to adjust educational programs for savants to safeguard the critical aspect of internal motivation in learning.[570] Unfortunately, there aren't sufficient materials in the historical record to clearly define how many students Partridge would have opted out of ancient language instruction. This could be interpreted as Partridge allowing for a rare exception to his permitting a majority of students to have little or no knowledge of ancient languages.

Examining the curriculum Partridge developed for the American Literary, Scientific and Military Academy (now Norwich University), the option for the study of ancient and modern languages to include Latin, Greek, Hebrew, French,[571] and in later years Spanish[572] are available. This provides some indication that while Partridge may not have forced students to take ancient language courses, it was his preference that they gain this knowledge. It

---

[568] Partridge, Memorial of Alden Partridge, praying Congress to adopt measures with a view to the establishment of a general system of education, 5.

[569] Partridge, 5.

[570] Partridge, "Captain Partridge's Lecture on Education."

[571] Alden Partridge, Prospectus of the American Literary, Scientific and Military Academy in Norwich, Vermont, 1820, Norwich University Archives and Special Collections.

[572] Alden Partridge, A Catalogue of the Officers and Students of Norwich University for the Academic Year, 1835–6, Norwich University Archives and Special Collections.

is also important to note the intent of Norwich University was to admit students as young as eleven years old[573] (an age normally associated with the *quadrivium*). Therefore, this requirement would be appropriate for students completing what present-day audiences would understand as secondary and tertiary schooling. This example demonstrates the difficulty of trying to assess Partridge as a classical educator, as educational models of his era were less standardized than pedagogies of our time. Partridge modified his education programs to optimize the experiences of individual students. This included using his institution as college preparatory programs.[574] It was only after the departure of Partridge as president that Norwich University in 1848 raised the admissions age to fourteen and formed a collegiate department that would place the institution clearly in the tertiary education (college/university) category as it would be known by present-day audiences.[575] By 1852, Norwich University would require enrollees to be fourteen years of age and able to "sustain a satisfactory examination in […] Latin and Greek Grammar; Latin Reader; Sallust; Cicero, select orations; Virgil; Jacob's Greek Reader, and Homer's Iliad, two Books, or an equivalent,"[576] essentially confirming the completion of the *quadrivium* and clearly identifying the university as part of the classical education model for its collegiate program (Artium Baccalaureus).

Conclusion

The one clear instance where Partridge did not support classical education and ancient languages is when, with the goal of cost savings, he proposed the USMA be converted from an educational institute to a training ground. In Partridge's *Lecture on Education* and his *Memorial to Congress*, it is clear that Partridge desired students to learn from classical education materials. In his recommendation for primary/secondary education, Partridge restricted the study of classics to students who intended to progress to study the classics in college. However, given the military aspect of this model, it is highly likely that Partridge would have still introduced his students to classical historical works

---

[573] Partridge, Prospectus of the American Literary, Scientific and Military Academy in Norwich, Vermont, 1820.

[574] This occurred prior to Partridge's institutions gaining degree-granting privileges and in cases of students intending to progress to other educational institutions but requiring further academic preparation.

[575] Norwich University, Catalogue of the Corporation, Officers, and Cadets of Norwich University, for the Academical Year 1848–9, Norwich University Archives and Special Collections.

[576] Norwich University, Catalogue of the Corporation, Officers and Cadets of Norwich University, for the Academical Year 1852–'53, p. 14, Norwich University Archives and Special Collections.

concerning military history and tactics, either through lectures or translated texts. In the curriculum Partridge designed for what became Norwich University, the ancient languages and texts of classical education are retained. Therefore, I will assert for the sake of present-day audience, Partridge should be considered within the scope of the classical education movement and a valuable resource for reevaluating our current models of education. While Partridge's *American System of Education* offers far more value and practicality than strict classical education models, Partridge certainly does not detract from wisdom that can be learned from the ages.

Chapter X: Partridge and Land-Grant Colleges

Alden Partridge was the first to provide a detailed petition to Congress proposing land grants to support higher education.[577] On January 21, 1841, Partridge's memorial on a *System of National Education* was read in the House of Representatives.[578] It was unfortunate that Partridge's attempts to establish a system of land-grant colleges was a failure. However, his efforts paved the way for multiple others to continue making proposals, Justin Morrill succeeded in 1862 with a plan that carries forward many elements of Partridge's original proposal. Unfortunately, the 1862 Morrill Land-Grant Act is now viewed as a requirement for a series of monomathic degree programs rather than one with a polymathic approach providing instructs in the most important and useful academic fields and practical skills (to include military tactics).

This chapter will explore Partridge's original polymathic design for the land-grant colleges and how the original intent of the plan was lost in the land-grant complex that resulted. To aid in this evaluation, Captain Partridge's and Edmund Burke's 1839 plan to reorganize the militia (see appendix N) will also be examined. This plan provides further details on the intended manning and student population of land-grant colleges. Student to professor ratios will be examined to determine if these staffing levels were viable for quality instruction. This will be followed with a discussion of how Partridge supported proposed congressional action, believing it as constitutional. This includes the federal government's requirement to provide for the common defense, support the general welfare of the Nation, and to secure to ourselves and our posterity the blessing of liberty. Brief thoughts on why Partridge's efforts in 1841 failed will also be presented. Finally, a brief discussion of the purpose, constitutionality, and ultimate credit for the 1862 Morrill Land-Grant Act will be provided.

---

[577] Kathryn Lindsay Anderson Wade, "The Intent and Fulfillment of the Morrill Act of 1862: A Review of the History of Auburn University and the University of Georgia" (Master's thesis, Auburn University, 2005).

[578] Partridge, Memorial of Alden Partridge, praying Congress to adopt measures with a view to the establishment of a general system of education, 1–8.

Partridge's & Burke's Manning and Student Population Ratio

In 1839, Partridge and Edmund Burke authored a proposal to reorganize the militia intended to demonstrate that 600,000 citizen soldiers (and related instruction) could be maintained for less tax outlay than the 25,000 soldier standing army of this day.[579] Partridge and Burke suggested the establishment of 75 seminaries[580] where the "number of graduates each year would be one thousand eight hundred and seventy-five."[581] These would be considered exceptionally small educational institutions by our present-day standards, each having an enrollment of approximately 100 students. This is roughly equivalent to the contemporary enrollment at Norwich University at the time.[582]

The "one professor of tactics, and one professor of military and civil engineering,"[583] paid for at the federal government's (U.S. Department of the Treasury's) expense, could have aptly provided instruction for this 100-student body in all the related military and engineering subjects specified in Partridge's 1841 memorial to Congress. The difficulty of educating these students would have been further reduced through the use of mixed-age classrooms (commonly employed by Partridge) with upper-class students partially responsible for the instruction of junior students. Further educational support would be provided by the militia, with students receiving additional instruction during their required musters and training encampments. States could have essentially cloned the success of the model practiced at Norwich University through the state-level support for five or six additional professors/instructors. With two seminaries authorized for Vermont, and the allowance for existing educational institutions to be converted into supported seminaries, one could speculate that Partridge hoped for two professorships at Norwich University to become federally funded (possibly Partridge's own position and that of H. Villiers Morris, the professor of engineering).

---

[579] Partridge and Burke, Memorial of Alden Partridge and Edmund Burke, in behalf of the State Military Convention of Vermont.

[580] While the term *seminary* is now associated with institutions for religious instructors, for those seeking entry into priesthood (or ministry), during Partridge's era it would have been associated with other educational institutions.

[581] Partridge and Burke, Memorial of Alden Partridge and Edmund Burke, in behalf of the State Military Convention of Vermont, 15.

[582] The student population of Norwich University was listed at "119" in the 1841 prospectus. Partridge, Catalogue of the Officers and Cadets of Norwich University, for the Academic Years 1838–39: 39–40: 40–41, p. 12.

[583] Partridge and Burke, Memorial of Alden Partridge and Edmund Burke, in behalf of the State Military Convention of Vermont, 14.

Partridge and Burke suggested a regressive model for authorizing institutions in their plan to reorganize the militia,

> That the number of such seminaries, to which each State shall be entitled shall be according to the following ratio of population, viz: each State the population of which does not exceed 250,000, shall be entitled to one seminary; each State, the population of which exceeds 250,000, but does not exceed 600,000, shall be entitled to two seminaries; each State the population of which exceeds 600,000, but does not exceed 1,000,000, shall be entitled to three seminaries; each State, the population of which exceeds 1,000,000, but does not exceed 1,500,000 shall be entitled to four seminaries, and each State whose population exceeds 1,500,000 shall be entitled to five seminaries.[584]

Of the 26 states, one district, and three territories surveyed in the 1840 census, only New York and Pennsylvania had populations exceeding 1,500,000, with the largest being New York.[585]

The 1840 Census reports the national population of free white males between the age of "15 & under 20" at 756,106.[586] While this isn't the exact demographic for those entering the obligation of militia service or higher education, it can provide some insights on the proportion of graduates of the proposed seminaries within the militia. During the first four years of operations (assuming the graduation of a single class of 1,875 students for the 75 institutions), the proportion of regular militiamen of this age group to a seminary graduate would be around 403 to one. However, within eight years of operation (five graduating classes to cover every year of this age group) this ratio would drop to 81 to one. This allowed for a seminary educated lieutenant in every militia company.

Exact ratios varied among the states based on population and the authorized number of seminaries. In New York, the most populous state with 130,904 free white men "15 & under 20",[587] the ratio after four years was close to 1,047 militiamen within this age group for every seminary graduate. After eight years of operation, this ratio dropped to 209 to one. It is inferred that the responsibility of reducing this ratio in the highly populated states would be left to the state's legislature to establish additional state-funded institutions or

---

[584] Partridge and Burke, Memorial of Alden Partridge and Edmund Burke, in behalf of the State Military Convention of Vermont, 14.

[585] United States Census Office, *Compendium of the Enumeration of the Inhabitants and Statistics of the United States: As Obtained at the Department of State, from the Returns of the Sixth Census, by Counties and Principal Towns* (Washington, D.C.: Blair & Rives, 1841).

[586] United States Census Office, *Compendium*, 100.

[587] United States Census Office, 20.

provide educated officers through other means (i.e. private educational institutions possibly modeled after Norwich University).

If Partridge's and Burke's 1939 plan to reorganize the militia had been accepted, and all the current and future states executed this plan to the fullest, there would be 250 seminaries within the United States producing 6,250 graduates per year (roughly equivalent of how many officers the current U.S. Army commissions per year).[588] Had this plan been further amended with the population growth of the nation allowing for one seminary for every 300,000 citizens, the country could now have over, 1,100 seminaries producing over 27,000 professionally educated militiamen per year. If this program was to operate for 35 years (covering the entire population of those within the unorganized militia ages 17 to 45), a population of over 891,000 professionally trained militiamen would exist in the general population at the direct cost of ~2,230 professors during this period. This population would rival the size of the present-day strength of the Active Army.

In Partridge's 1841 proposal for land-grant colleges, Partridge expanded the number of recommended institution to 80 higher seminaries (collegiate-level) and 160 secondary (secondary-level)[589] seminaries "so that the great body of the people might receive its benefits."[590] Assuming Partridge intended the 100-student body design for the higher and primary seminaries asserted in his and Burke's plan to reorganize the militia, the results of his 1841 memorial would provide over 6,000 graduates per year for the improvement of the militia (2,000 of the higher seminaries and 4,000 from the secondary seminaries).

It is unfortunate that Partridge did not provide the student body/yearly graduate calculations within his 1841 memorial to determine the number of required seminaries based on population. Had he done so, it may have aided Morrill in building these calculations into the 1862 Land-Grant Act, which might have allowed the program to be scalable as the nation's population grew. While the Morrill Land-Grant Act of 1862 provided

---

[588] Partridge and Burke, Memorial of Alden Partridge and Edmund Burke, in behalf of the State Military Convention of Vermont.

[589] We would have no contemporary equivalent to what Partridge may be suggesting here. This suggested level of education would be above the level of common schools and below the level of bachelor education. If it included all the instruction directed by Partridge, it could be viewed as something between our present-day high school and instruction across multiple disciplines currently taught at humanities and technical community colleges.

[590] Partridge, Memorial of Alden Partridge, praying Congress to adopt measures with a view to the establishment of a general system of education, 5.

proportionality in the amount of land that would be granted based on the number of representatives and senators in Congress (the number of representatives being influenced by state population), the act was limited to a three-year window of action and does not provide any specifications on professorships (specifically in military tactics) that would be provided at the expense of the federal government.[591] This leaves the present-day audience with questions concerning Partridge's opinions on possible size limitations of universities. William Emerson noted in the 1820s that the 1,500-student population of the University of Göttingen, Germany, was "an obstacle to [student] improvement."[592] We can only speculate if Partridge would have supported increasing the staffing as student populations grew. Would he support land-grant colleges growing to student populations in the thousands (or the present-day tens of thousands), or would he have supported increasing the number of educational institutions? This is a difficult question to answer when the largest student bodies of American universities of the era did not exceed 500 students.

While the original tactics professor to student (1:100) ratio within Partridge's and Burke's proposal to reorganize the militia was practicable, the ratios of tactics professors to students quickly became untenable within a few decades after the passage of the 1862 Morrill Land-Grant Act. Student protests against tactics training caused many universities to drop mandatory instruction in military science. Even the famed hero of Gettysburg, Joshua Chamberlain, could not keep students from protesting against tactics training while President at Bowdoin College.[593] While modern readers can express empathy for the rejection of military training for students who witnessed the bloodiest American war in history, this rejection of military knowledge did not prevent the United States from entering further military conflicts. In 1891, Second Lieutenant John J. "Black Jack" Pershing famously became the single professor of tactics at the University of Nebraska at Lincoln. Upon arriving, he found 90 students in the cadet corps. Pershing quickly built up this population to 350

---

[591] "Morrill Act (1862)."

[592] William Emerson, as quoted by Robert D. Richardson, *Emerson: The Mind on Fire* (Berkeley, CA: University of California Press, 2015), 49.

[593] Alice Rains Trulock, *In the Hands of Providence: Joshua L. Chamberlain and the American Civil War* (Chapel Hill, NC: The University of North Carolina Press, 2013).

cadets or ~88 percent of the student population.[594,595] As land-grant colleges grew in post-Civil War America, military support for universal military instruction soon became inadequate. While the National Defense Act of 1916 increased the number of military personnel to provide for instruction of the military arts and sciences, "detailing not less than one such officer or noncommissioned officer to each five hundred students under military instruction," this level of staffing would still be insufficient for universal instruction.[596] It is not uncommon today for land-grant colleges to expose less than one in 100 students to any military arts or science courses. With this level of military education participation, there is a real question about the constitutional foundation for the Morrill Land-Grant Act, as Congress possesses no enumerated powers regarding education.

Constitutional Justification

Partridge was careful to justify the constitutionality for the creation of the proposed seminaries within his 1841 memorial and his and Burke's 1839 plan to reorganize the militia. In both memorials, the federal government's responsibility to *"provide for the common defense"* and Article I, Section 8, Clause 16, was used as the justification for congressional action. This enumerated power states that Congress has the power:

> To provide for organizing, arming, and disciplining, the Militia, and for governing such Part of them as may be employed in the Service of the United States, reserving to the States respectively, the Appointment of the Officers, and the Authority of training the Militia according to the discipline prescribed by Congress.

Rhetorically Partridge asked, "can any plan be devised that would more efficiently accomplish this great object [to discipline the militia] than the proposed system of education?"[597] In this context, the primary role of these land-grant colleges should be viewed as supporting the militia. As Partridge would state, "we shall never have a well-disciplined and efficient militia, unless

---

[594] "Pershing at UNL," Nebraska U: A Collaborative History from the Archives of the University of Nebraska–Lincoln.

[595] "The Rise of the University of Nebraska: And the Celebration of the Silver Anniversary," Nebraska U: A Collaborative History from the Archives of the University of Nebraska–Lincoln.

[596] National Defense Act, approved June 3, 1916, approved April 17, 1918, U.S. Congress, *U.S. Statutes at Large*, vol. 40, *65th Congress* (Washington, D.C.: Government Printing Office, 1919), 531–32, quoted on 532, Law Library of Congress.

[597] Partridge, Memorial of Alden Partridge, praying Congress to adopt measures with a view to the establishment of a general system of education, 6.

it be based on a system of correct military instruction."[598] While there would be a host of advantages for expanding state-sponsored education, these advantages must be viewed as secondary to the primary goal of educating men capable of leading the militia. Excusing students from military training/education would invalidate Partridge's primary constitutional justification for land-grant colleges.

A second constitutional justification for land-grant colleges provided in Partridge's 1841 memorial found in the Constitution's preamble with the federal government is empowered to *"promote the general welfare."* Partridge rightfully points out this clause "is liable to abuse," and by the 20th century it would certainly be distorted from its original meaning (the language first appearing in the Articles of Confederation). Partridge provides a "clear and definite meaning" that he asserts was originally intended by the American Founders, "that Congress can constitutionally exercise any power that would not infringe on the rights and immunities reserved to the States and the people, or otherwise contravene any of the provisions of the constitution."[599] In order for Congress to operate within the scope of the definition provided by Partridge, support for land-grant colleges could only be *offered* to the states. Individual state legislatures would have to accept the offer, and they had the option to refuse, thus preserving the rights of the states. Furthermore, funding for these institutions would be proportional to the states. "[P]ublic funds, (the common property of all,)" distributed proportionally would benefit the *general welfare*, and not disproportionally benefit one state at the expense of others.[600]

Partridge provides a third constitutional justification in this 1841 memorial also drawn from the preamble of the Constitution, to *"secure to ourselves and our posterity the blessing of liberty."*[601] Partridge's proposed institutions would provide instruction on individual rights, the functioning of the U.S. government, knowledge of the interrelation between the United States and other countries, and the military arts and sciences required to defend the Republic. In Partridge's own words,

> [W]hat constitutes the safest basis on which the liberties of a free people can rest[?] This your memorialist believes to be the intelligence of the great body of the people, which enables them to know their rights, and their ability to defend those rights, both against external and internal invasions.[602]

---

[598] Partridge, 6.

[599] Partridge, 7.

[600] Partridge, 7.

[601] Partridge, 7.

[602] Partridge, 7.

Failing to provide instruction on the rights and liberties of Americans threatens their loss through public ignorance. Failure to educate the citizenry in military arts and sciences threatens the loss of rights through a lack of the ability to resist internal tyranny or external conquest. Partridge would assert his concept of land-grant colleges was in keeping with the genuine spirit of the Constitution.

Not only did Partridge justify the constitutionality of his proposed institutions, he highlighted how of the United States Military Academy failed to comply with the same constitutional elements. The USMA was not providing officers for the militia (the primary constitutional ground force). Therefore, it was not providing for the *common defense*. The USMA was operated at the disproportionate benefit of New York at the expense of other states and therefore not beneficial to the *general welfare*. Finally, the USMA was much too small to educate the mass of the citizenry of the United States and therefore could only partially support the prosperity of the Nation.

While it may be acknowledged that the USMA was a needed institution for national defense in the early days of the Republic, this institute was developed and continues imperfectly to the spirit of the American Republic. By the 1840s, with Partridge's demonstrated success at Norwich University and a plan to spread liberal arts, humanities, scientific, and military education widely across the United States, the USMA became functionally obsolete. With the adoption of the Morrill Land-Grant Act and the ability to produce military officers at the state level, one might ask why the USMA was maintained. This question will be explored in more detail in a later volume.

Polymathic Intent to Monomathic Execution

One of the major faults in the language of the 1862 Morrill Land-Grant Act is it allowed Partridge's original polymathic education design (all students being exposed to the wide range of subject areas) to be interpreted as a requirement for a series of independent academic programs. The imperfect language reads,

> each State which may take and claim the benefit of this act, to the endowment, support, and maintenance of at least one college where the leading object shall be, without excluding other scientific and classical studies, and including military tactics, to teach such branches of learning as are related to agriculture and the mechanic arts, in such manner as the legislatures of the States may respectively prescribe, in

order to promote the liberal and practical education of the industrial classes in the several pursuits and professions in life.[603]

As seen within the text, the requirement to provide instruction in scientific, classical studies, military tactics, agriculture, and mechanic[al] arts is placed on the institution, however there is no specification concerning what instruction individual students should receive. The educational institutions created through or supported by the Morrill Land-Grant Act of 1862 enjoy a wide interpretation of the Act's purpose. As explained by Kathryn Lindsay Anderson Wade, in a master's thesis for Auburn University,

> Some [institutions] focused on agricultural education, some concentrated on technical education, and some directed their attention to scientific experimentation and research. Still others accepted the funding but focused on liberal arts rather than agriculture and the mechanic arts.[604]

Many would see the requirement for instruction for military tactics included solely because of the ongoing American Civil War, and believed it appropriate to abandon the military focus after 1865.

The Morrill Land-Grant Act's vague language concerning military education (lacking Partridge's detailed guidance on curriculum and instruction techniques) and the general apathy of the academic community towards military education likely condemned the "military tactics" instruction clause to failure. In the words of General John Fraser,

> The man in charge [of military tactics instruction] was left upon his own to invent a program as best he knew. No outline of study were available, no equipment, no uniforms. As a result, the "teaching" consisted mostly of drill. In many of the colleges it was required of all male students. In some, the military commandant was placed in charge of all student discipline. The faculties were described as tolerant of, rather than sympathetic to, the military effort. There is little reason why one should expect more, in light of the almost total indifference of the Army.[605]

What little was provided in military tactics instruction would fall far short of the comprehensive educational program originally envisioned by Partridge.

In a 1923 federal ruling regarding compulsory military training for land-grant college students, the passage with the Morrill Act "in such manner

---

[603] "Morrill Act (1862)," sec. 4.

[604] Wade, "The Intent and Fulfillment of the Morrill Act of 1862," 9.

[605] General John Fraser, as quoted by Edward Danforth Eddy, Jr., *Colleges for Our Land and Time: The Land-Grant Idea in American Education* (New York, NY: Harper & Brothers, 1957), 65.

as the *legislatures* of the States may, respectively, prescribe" was used to justify the legal opinion that the Act didn't direct universal military training;

> Military training, according to the Federal Law, is clearly placed in the same category as the other branches of learning which are named. Instruction in military tactics is obviously a requirement on the States, as are the other branches which are mentioned. It does not appear, however, from the Federal legislation that instruction in military tactics is any more obligatory on the individual student than is instruction in agriculture or mechanic arts.[606]

This non-compulsory understanding for land-grant colleges was further reinforced in 1930 with a ruling of the attorney general stating,

> I therefore advise you that you are justified in considering that an agricultural college which offers a proper, substantial course in military tactics complies sufficiently with the requirements as to military tactics in the act of July 2, 1862, and the other acts above mentioned, even though students at that institution are not compelled to take that course.[607]

With these legal rulings, present-day land-grant colleges view having a Reserve Officers' Training Corps (ROTC) as fulfilling their legal requirements with no concern over how few students receive training under these programs. The failures of the Morrill Land-Grant Act to ultimately provide universal, or at least widespread, military education might be summed up by Captain McArthur's comment on the matter; "We learn so slowly and forget so quickly."[608]

In both Partridge's 1841 and his and Burke's 1839 memorials to Congress, it was made clear that instruction was to occur in a polymathic fashion. As seen in Partridge's 1841 memorial, students were exposed to the full range of subjects to produce complete citizens capable of safeguarding the Republic,

> Great care should be taken at these institutions to instruct the pupils in the science of government generally, and particularly to make them well acquainted with the principles of our own free institutions, which they should be taught to love, to respect, and to sustain. The whole course of instruction should be such as to

---

[606] Walter J. Greenleaf, *Federal Laws and Rulings Affecting Land-Grant Colleges and Universities* (Washington, D.C.: U.S. Department of the Interior, Office of Education, 1930), 11.

[607] Greenleaf, *Federal Laws and Rulings*, 12.

[608] John C. McArthur, "Civil Institutions and Military Instruction," *Journal of the United States Infantry Association* VI, no. 1 (July 1909): 861.

form practically useful men, and calculated to develop all those manly, noble, and independent sentiments which ought to characterize every American citizen. Military science and instruction should be common to all. This is the aegis that protects a free people from foreign as well as domestic tyranny.[609]

Partridge clearly advanced the duplication of his *American System of Education* that had proven so successful at Norwich University. Students were intended to experience a broad range of instruction, including military arts and sciences. From Partridge's and Burke's 1839 memorial to reorganize the militia the required inclusion of military education is clearly stated in conjunction with traditional liberal education, "Military science and instruction shall be made an appendage to the usual civil education of the students."[610] Clearly, the 1862 Morrill Land-Grant Act failed to support "liberal and practical education" to the degree that Partridge had envisioned.

Failures in Coalition Building

The first documented records of Partridge proposing land-grant colleges come from the May 9, 1835, entry in the diary of Charles Ellis of the University of Virginia. As Ellis records,

[Partridge] then proposed a scheme of National Education, to be promoted if not established by the General Government, viz: that Congress should form a Committee of Representatives to draw up a system of National Education, predicated on the tone of our Government, and freedom of our Institutions, then that it should be declared to each State, that on condition of her raising one or more colleges, and adopting the plan drawn up by the committee that they would receive so much of the Public Land, sufficient to maintain the Colleges, and thus he said, if the System of Education was rightly drawn up, our citizens would be better instructed in Knowledge, in enjoying as well as maintaining Freedom, better inclined to submit to the Laws of our Country, and Education thus be diffused throughout the whole nation; Congress having nothing to do with the operation of the Institutions, only the appropriating Lands, and the adoption of the regular System, after the whole being under the direction of the States,

---

[609] Partridge, Memorial of Alden Partridge, praying Congress to adopt measures with a view to the establishment of a general system of education, 6.
[610] Partridge and Burke, Memorial of Alden Partridge and Edmund Burke, in behalf of the State Military Convention of Vermont.

singly, and there can be no doubt as to the constitutionality of such scheme since they have already made appropriations to several colleges, in some of [the] Western States.[611]

Four years later, we see Partridge further refine his land-grant college concept in a joint memorial to Congress (written with Edmund Burke) on reorganizing the militia. In this memorial, Partridge advocates the adoption of the *American System of Education* that had proven successful at Norwich University, which was granted a charter by the state of Vermont in 1834. With further improvements to his plan in 1841, and robust constitutional justification for Congress to duplicate the success of Norwich University, one can reasonably ask why Partridge's plan was not actioned?

Professor emeritus of history at Norwich University Gary Lord believes Partridge made an error of timing with his proposals. Partridge's 1841 memorial concerning land-grant colleges was well balance in language and could have received majority support. Unfortunately, he also submitted a second memorial to Congress on the same day concerning commissions of Army officers from sources other than the USMA. This document had a much different tone. While some of the language concerning the USMA cadets compared to Norwich University cadets was true, the language was perceived as the hostile and polarizing "rhetoric of a Jacksonian Democrat."[612] This second memorial likely destroyed any chance to build a meaningful political coalition to advance the concepts of land-grant colleges.

One could also speculate that Partridge's proposed limitations of rank within the peace establishment created a resistance movement against his land-grant proposal. Army offices would have likely used any influence they had on politicians to advocate against Partridge and the restrictions he proposed to put on peacetime career advancement. While proposals for limiting the Active Army in peace to ranks no greater than colonel (see appendix O),[613] or possibly one brigadier general,[614] would have reinforced the primacy of the militia over

---

[611] Charles Ellis, "The Student Diary of Charles Ellis March 10 to June 25, 1835," ed. Ronald B. Head.

[612] Gary Thomas Lord, "Alden Partridge's Proposal for a National System of Education: A Model for the Morrill Land-Grant Act," in *History of Higher Education Annual: 1998: The Land-Grant ACT and American Higher Education: Contexts and Consequences*, vol. 18, ed. Roger L. Geiger (University Park, PA: Pennsylvania State University, 1998), 17.

[613] Partridge et al., "Memorial of a Committee of the Military Convention at Norwich, Vt."

[614] Partridge and Burke, Memorial of Alden Partridge and Edmund Burke, in behalf of the State Military Convention of Vermont.

a peace establishment, it likely created significant hostility from Active Army leadership seeking to protect their rank, pay, and status. With the militia retaining several general officer positions within the states, the peace establishment would have been perceived as existing to support the militia. Active officers advocating for military hostilities would be perceived as speaking out partially in favor of personal gain. It is unknown if this approach could have made the United States a more peaceful nation, but it would have created a system within the Army that would be the opposite of what stands today. Could we imagine an Army led by the National Guard and supported by a subordinate Active Army?

We are left to ponder how the destiny of the United States could have been shaped if Partridge would have only advanced his memorial on land-grant colleges, independent of his other actions to abolish the USMA or limit the rank structure of the Active Army. What would America be today if Congress would have established a national system of colleges modeled on Partridge's *American System of Education*?

Morrill Land-Grant Act

Partridge's death in 1854 prevented him from witnessing congressional action concerning land-grant colleges. It was a close neighbor, Justin Smith Morrill, who eventually championed a bill through Congress and gained the signature of President Abraham Lincoln in 1862. The following section will briefly review if Partridge should be credited for ideas contained within this act, the act's intended purpose, the justification used in passing the act, and will end in discussing if the act was ever really needed.

The 2005 master's thesis by Kathryn Lindsay Anderson Wade entitled "The Intent and Fulfillment of the Morrill Act of 1862: A Review of the History of Auburn University and the University of Georgia" is a rare piece of scholarship exploring the purpose and development of the Morrill Land-Grant Act of 1862 in great detail. Wade painstakingly reviewed the writings of major scholars from the passage of the Land-Grant Act to the time of her writing. Wade found, that while the Morrill Land-Grant Act of 1862 was held as one of the most influential congressional acts related to the fields of education and agriculture in the 19th century, most of the major works from historical scholars on this act only appear in the mid-20th century around the centennial anniversary of the act's passage.[615] Furthermore, Wade found, most of the

---

[615] Wade, "The Intent and Fulfillment of the Morrill Act of 1862."

historical scholarship focused on determining who should receive credit for the act instead of the more valuable examination of the act's purpose.[616]

Recent scholarship surrounding the Morrill Act's sesquicentennial in 2012; duplicated and magnified many of the faults originally found by Wade's earlier study. Sources often simplify the purpose and remove/minimize the requirement for military education and the possible influence of Captain Alden Partridge. This is particularly true in scholarship that uses Edward Eddy's *Colleges for Our Land and Time: The Land-Grant Idea in American Education* as the sole primary source. Eddy's only reference to Partridge concerns his appointment as professor at West Point. Eddy falsely suggests the military training requirement was only added in light of the ongoing Civil War and that "no previous examples of military training in civilian institutions" predate the act, totally ignoring Partridge's work in founding Norwich University and similarly modeled institutions.[617] Multiple attempts to claim credit for this act exist within major works of academic history. Authors tend to bias their research towards their own institutions. For example, those associated with the University of Illinois tend to emphasize Jonathan Turner's influence[618] while Norwich University highlights Partridge's[619] impact with each source often failing to mention other major influential characters. Omitting Partridge as a significant influence on land-grant colleges may be associated with the lack of secondary scholarly works focused on Partridge's larger contributions to American education theory as first identified by Webb.[620]

While Partridge was the first to present a detailed plan for land-grant colleges to Congress in 1841, it must be highlighted that he was not the only scholar advocating for such a system (although he was likely the only one doing so with the primary intent of improving the militia). Other educational leaders that may have influenced Morrill include, but are not limited to, Jonathan B. Turner (involved in founding the University of Illinois), Amos Brown (president of People's College, New York), Simon DeWitt (chancellor of the University of the State of New York), Freeman G. Cary (involved in the establishment of Farmers' College, Cincinnati, Ohio), John S. Skinner (editor of agricultural periodicals) and Thomas G. Clemson (involved in the founding of the Maryland Agricultural College and a former student of Alden Partridge). Arnold Tilden suggests that Morrill ultimately borrowed from each of these

---

[616] Wade.

[617] Eddy, *Colleges for Our Land and Time*, 64.

[618] David R. Wrone, "The People's Choice College: The Movement Behind the Morrill Land Grant Act of 1862," Illinois Periodicals Online.

[619] Norwich Bicentennial, *200 Things about Norwich*.

[620] Webb, *Captain Alden Partridge*.

influences.[621] Multiple scholars has acknowledged this opinion, like Earle D. Ross, supporting Morrill's "generalized synthesis" of ideas from educational influencers of his day.[622] So while Morrill can be credited as the "father" of the Land-Grant Act by his championing of this legislation through Congress, it would not be possible to name a singular conceptual "father" for this act.

In his papers and speeches, Morrill never credited others for the ideas within the 1862 Land-Grant Act, claiming the idea only came to him a year prior to his 1857 proposal. However, Morrill had to have been aware of Partridge's previous proposal and would have been in Congress during Turner's campaigns through the Illinois legislature requesting land grants to support education. Many scholars have noted the similarities in Partridge's and Morrill's proposals. According to Lord, Morrill "never acknowledged Partridge's influence, the close correspondence in their thinking is remarkable and in all probability not accidental."[623] To further establish the likely influence from Partridge onto Morrill, Lord points out that Morrill was friends and business partners with Jedediah Harris, a trustee of Norwich University, and Morrill's close proximity to Norwich University (located 12 miles from Morrill's home town of Strafford, Vermont).[624] Norwich President Charles A. Plumley claimed direct interactions existed between Partridge and Morrill, asserting, "Alden Partridge used to visit Justin Morrill when on his tramps and hiking expeditions and discussed with him his educational theories."[625] Lord also reports of a letter believed to be in Morrill's hand, submitted to *The Spirit of Seventy-Six* Whig newspaper,

> [Partridge's] feelings and politics are those of a selfish anchorite, who lives for himself alone… President Partridge has not that practical experience which should command the suffrage of the voters of this district, nor has he any superior abilities or native which entitle him to their affection. Although a great talker in public, his conversational powers are not above the tap-room celebrity; he is not and never can be a ready debater-though always ready to debate; with always a speech

---

[621] Arnold Tilden, *The Legislation of the Civil-War Period Considered as a Basis of the Agricultural Revolution in the United States* (Los Angeles, CA: University of Southern California Press, 1937).

[622] Earle D. Ross, *Democracy's College: The Land-Grant Movement in the Formative Stage* (Ames, IA: Iowa State College Press), 46.

[623] Lord, "Alden Partridge's Proposal," 11.

[624] Lord, 19.

[625] Charles A. Plumley, as quoted by Alfred Charles True, *A History of Agricultural Education in the United States 1785–1925* (Washington, D.C.: Government Printing Office, 1929), 82–83.

to make on a new subject, I have never known him to make a new speech; his lectures on education have long been a bore, and with new texts would answer for any meridian.[626]

The letter was signed "A. Blacksmith." Morrill frequently references his father's profession, being that of a blacksmith. The exclusion of any acknowledgement of Partridge may have resulted from the conflicts between these two politicians of opposing political parties. Because of the political turmoil during and after the American Civil War, it is possible the Morrill, the Whig/Republican, did not acknowledge Partridge's influence due to his Democratic political alignment. The preponderance of evidence suggests that Partridge was an influence on the Morrill Land-Grant Act. Except for scholarship that fails to mention Partridge entirely, the majority of scholars now claim Partridge was indeed influential on the concepts and language in the 1862 Morrill Land-Grant Act.

Envisioned Purpose of the Morrill Land-Grant Act

The true purpose and intent of the Morrill Land-Grant Act remains uncertain because of vague language. Ross would claim the language was purposely left broad enough to capture all the "varied concepts of industrialists."[627] As the act reads.

[T]he leading object shall be, without excluding other scientific and classical studies, and including military tactics, to teach such branches of learning as are related to agriculture and the mechanic arts, in such manner as the legislatures of the States may respectively prescribe, in order to promote the liberal and practical education of the industrial classes in the several pursuits and professions in life.[628]

However, scholars such as Lang report the broadness of this language caused many to be "unclear about what the legislation actually intended."[629] As a result, the implementation of the act was highly mixed. While some states attempted to create institutions for the instruction of agricultural and mechanical arts, other states used the endowments to support traditional liberal college programs. While present-day land-grant institutions will often report the mission to "teach, research, and service" associated with the Kellogg Commission on the Future of State and Land-Grant Universities, Wade suggests, "Land grant universities are no closer today to understanding the true

---

[626] Lord, "Alden Partridge's Proposal," 19.

[627] Ross, *Democracy's College*, 46.

[628] "Morrill Act (1862)."

[629] Daniel W. Lang, "Amos Brown and the Educational Meaning of the American Agricultural College Act," *History of Education* 31, no. 2 (2002): 140.

intentions of the Morrill Act than they were over a hundred and fifty years ago."[630] With the vague purpose and lack of specific instructions within the 1862 Land-Grant Act, reviewing the transcripts of congressional debates on the bill may provide further details on its true intent.

The discussions surrounding the 1857 land-grant bill are primarily economic in nature. The original bill had no requirements for military instruction. At the time, there was a concern of falling fertility of American soil. There were fears that American farmers could not compete in the market against their European peers unless scientific methods of farming were employed. There was a strong desire for governmental support to drive innovation and cost-savings in agriculture. While these were valid concerns, they were outside the scope of the enumerated powers of the federal government. Ultimately, Morrill's attempt at a land-grant bill in 1857 was considered unconstitutional by Southern Democrats. According to Ross, Democrats believed the act "would be an invasion of the domestic rights of the states."[631] While it progressed through the House and Senate by small margins, the act was vetoed by President James Buchanan.[632]

Morrill reintroduced the bill in 1861 after the start of the American Civil War. He added the requirement for land-grant colleges to provide military education, by adding a mere three words to the bill, "including military tactics." Morrill gave an expanded intent of the Land-Grant Act in his in his introductory speech to the House:

> Something of military instruction has been incorporated in the bill in consequence of the new conviction of its necessity forced upon the attention of the loyal States by the history of the past year. A total unpreparedness presents too many temptations even to a foe otherwise weak. The national school at West Point may suffice for the regular Army in ordinary years of peace, but it is wholly inadequate when a large army is to be suddenly put into service. If we ever expect to reduce the Army to its old dimensions, and again rely upon the volunteer system for defense, each State must have the means within itself to organize and officer its own forces. With such a system as that here offered—nurseries in every State—an efficient force would at all times be ready to support the cause of the nation and secure that wholesome respect which belongs to a people whose power is always to its pretensions. In a free government we have proved,

---

[630] Wade, "The Intent and Fulfillment of the Morrill Act of 1862," 29.

[631] Ross, *Democracy's College*, 57.

[632] Wade, "The Intent and Fulfillment of the Morrill Act of 1862."

notwithstanding some 'in time of temptation fall away,' that patriotism is spontaneous, but doubtless many valuable lives would have been saved in the progress of this plague-spotted rebellion had we not so long assumed that military discipline was also spontaneous. If ever again our legions are summoned to the field, let us show that we are not wholly unprepared. These colleges founded in every State will to some extent guard against the sheer ignorance of all military art which shrouded the country, and especially the North, at the time when the tocsin of war sounded at Fort Sumter.[633]

Here, Morrill is clearly appealing to Congress to pass the bill in relation to their duty to provide training and discipline to the militia. While this intent would have certainly improved the constitutional justification of the act, none of Morrill's expanded language is remembered through time. The judicial branch would not enforce the military training requirements of the law. Support for military training waned by state legislatures, and support from the larger academic community was quickly lost.

Carrying some of the concepts of Partridge's *Lecture on Education*, Morrill further asserted,

If this measure [the Land-Grant Act] had been instituted a quarter of a century ago the absence of all military schooling at the outset of the present rebellion would have been less deplorable in the Northern States. The young men might have had more of fitness for their sphere of duties whether on the farm, in the workshop, or on the battlefield.[634]

However, for as much as Morrill appealed for the need for trained military men during the reintroduction of his bill, the scant amount of modifications to his bill demonstrates his appeal is little more than convenient political rhetoric. The addition of three words to his previous act does not capture the messaging provided in Morrill's speech, nor does it make military education a priority.[635] The bill remained focused on supporting agriculture and the mechanic arts for economic purposes. With the majority of Democrats, who opposed the original bill because of a lack of an associated enumerated power, now absent due to ongoing Civil War, the bill easily passed through the House and Senate.

---

[633] Justin Morrill, as quoted by Ira L. Reeves, *Military Education in the United States* (Burlington, VT: Free Press Printing Co., 1914), 80–81.

[634] As quoted in Alfred C. True, "Military Instruction," in *Proceedings of the Eighteenth Annual Convention of the Association of American Agricultural Colleges and Experiment Stations Held at Des Moines, Iowa, November 1–3, 1904*, ed. A C. True, W. H. Beal, and H. C. White (Washington, D.C.: Government Printing Office, 1904), 91.

[635] True, "Military Instruction," 91.

President Lincoln, who was far less sensitive to constitutional issues than his predecessor, signed the act into law on July 2, 1862. In summary, Tilden asserts the Morrill Land-Grant Act was "not so much as a result of a studied policy of the Congress as in a spirit of generosity and one of unwillingness to worry over details."[636]

Finally, one must ask if the 1862 Morrill Land-Grant Act was ever really needed. Factors that could be evaluated include access to education (enrollment and creation of new educational institutions), impact on the economy, and providing training for the militia. Richard K. Vedder, professor emeritus of economics at Ohio University, challenges many scholarly assertions about the impact of this act. He asserts, "The evidence is pretty clear that the Morrill Act had only marginal impacts on American Higher Education."[637] Vetter noted that American higher education was growing faster prior to the Morrill Land-Grant Act than the first two generations that lived after the act was passed. The adoption of the German university model and its focus on research is ultimately what shaped the direction taken by American universities, not the Morrill Land-Grant Act. The economic output of the United States exceeded Britain's before a large number of students could complete their studies at newly established land-grant colleges. Therefore, the economic impact of the Morrill Land-Grant Act is likely not as significant as claimed by present-day scholars. Even the rate of college enrollment was expanding faster in the two decades prior to the passage of the act vs. the two decades after (240 percent vs. 200 percent). Vedder even argues the Land-Grant Act slowed the rate of establishment of private colleges. The idea that the Morrill Act lowered the cost of education may be ahistorical as there are examples of land-grant colleges being equal or more expensive than comparable private institutions.[638,639] As previously discussed, the Morrill Act never fulfilled Congress's obligation to provide for the training and disciplining of the militia. In the end, the 1862 Morrill Land-Grant Act can be judged as a federal overreach of power that provided for the support and establishment of educational institutions that would have likely been established through state and private action.

---

[636] Tilden, *Legislation of the Civil-War Period*, 70.

[637] Richard K. Vedder, *Restoring the Promise: Higher Education in America* (Oakland, CA: Independent Institute, 2019), 91.

[638] Richard K. Vedder, "The Failure of Federal Higher Education: 'The Morrill Land-Grant Act: Fact and Mythology,'" YouTube, April 9, 2019.

[639] Richard K. Vedder, "The Morrill Act," in *Unprofitable Schooling: Examining Causes of, and Fixes for, America's Broken Ivory Tower*, ed. Todd J. Zywicki and Neal P. McCluskey (Washington, D.C.: Cato Institute, 2019), 31–64.

Conclusion

Captain Alden Partridge was the first to propose a detailed plan for the establishment of land-grant colleges to Congress in 1841. Further details of the function of these institutions are also included in Partridge's and Edmund Burke's 1839 memorial on the reorganization of the militia. In both instances, the intent and constitutional justification for establishing said higher educational institutions was to provide for the training and disciplining of the militia. Partridge's plan provided sufficient resourcing to supply an institutionally educated officer to each militia company, with these officers able to further spread their knowledge throughout the ranks.

The approach used within these institutions would have been polymathic. Each student would have been exposed to a broader scope of subjects and practical exercises than contemporary and present-day liberal colleges. Graduates would have been prepared to lead the militia, begin businesses, or hold governmental offices with each well instructed in the American Constitution and the rights it enshrines. Partridge was not suggesting an unproven educational model, but was suggesting the exportation of the *American System of Education*, demonstrated effectively at Norwich University, across the country.

Justin Morrill successfully championed the Land-Grant Act through Congress and into law in 1862. While Morrill never acknowledged Partridge's influence, the majority of education historians today will acknowledge Partridge as an influence on Morrill's proposal. While Morrill used the ongoing American Civil War and the need for men trained in the military sciences as justification for the legislation, language within the bill alludes to its primary purpose of improving agriculture and mechanical arts for economic improvement. The language concerning military education is a mere three words. Furthermore, there is no language to ensure the student to professor ratio for the military arts and sciences would be practicable. Without Partridge's detailed educational model and plan of instruction, Morrill's Land-Grant Act never fulfilled the congressional obligation to train and discipline the militia. Today, it is believed less than one in 100 students in land-grant colleges/universities are exposed to instruction in military arts and sciences. Some scholars question the need for the Morrill Land-Grant Act as there is sufficient evidence to suggest the same or greater economic and scientific achievements could have been made without the involvement of the federal government.

Over 180 years after Partridge's 1841 proposal to Congress to provide a system of education to support the militia, we are still lacking such a system. At the present-day, there is no indication that either the government or academic community is interested in designing or sustaining an educational model that would support the use of a militia as the primary ground defense force for the United States. We should be heedful of the words of Sir William F. Butler, "The nation that will insist upon drawing a broad line of demarcation between the fighting man and the thinking man is liable to find its fighting done by fools and its thinking by cowards."[640] Instead of universally calling upon college students to build upon their courage and resilience through military education, we currently condemn them into ignorance, pusillanimity, and unenlightened self-interest. Failure to provide for this education has largely doomed the country to depend on a standing army, feared by the American Founders, that contributes to the growth of the military-industrial complex, drives the Nation further into debt, and has led to needless military involvement at a global level.

I have little doubt that Partridge would be appalled by the present state of educational institutions founded under the Morrill Land-Grant Act. Our current system of land-grant colleges/universities fulfills none of the intents laid out in Partridge's *American System of Education*. Present-day graduates are not prepared for military service (unless they formally enter a military service obligation during college). They are not universally educated in the U.S. Constitution (including the rights of citizen), are not universally instructed on foreign relations (including the impact of their voting practices), nor are they universally prepared to be productive and practical members of society fully capable of starting and sustaining business enterprises. When combined with additional problems across the higher-education model of today, like crippling student debt, it is clear that our educational practices are not optimized for citizens of a free republic. Significant changes would be required in our education models to reclaim the Founders' vision enshrined in the U.S. Constitution (to be further discussed in a later volume). It is not enough to have land-grant colleges if the purpose of these institutions was lost in the process of establishing them. One cannot say that Morrill was successful where Partridge had failed, but only that Morrill aimed less precisely and achieved far less than what Partridge had envisioned for a national system of higher education.

---

[640] William F. Butler, *Charles George Gordon* (New York, NY: MacMillan and Co., 1889), 85.

Figure 13—Imagined Appearance of the ALSMA at Middletown, Connecticut If Built to Plan[641]

---

[641] This drawing is based on the plans found in the front matter of John Holbrook's *Military Tactics*. This image contains features of the church and main building (displayed to the right of the church) as they were constructed. The general construction features of these buildings were used to stylize the other buildings displayed in the original plan. While not fully accurate, this image provides an indication of how the ALIMA at Middletown, Connecticut, may have looked if fully constructed to Partridge's vision. (Artist Gala S.)

Appendix A: Morrill Land-Grant Act of 1862[642]

AN ACT Donating Public Lands to the several States and Territories which may provide Colleges for the Benefit of Agriculture and Mechanic Arts.

Be it enacted by the Senate and House of Representatives of the United States of America in Congress assembled, That there be granted to the several States, for the purposes hereinafter mentioned, an amount of public land, to be apportioned to each State a quantity equal to thirty thousand acres for each senator and representative in Congress to which the States are respectively entitled by the apportionment under the census of eighteen hundred and sixty: Provided, That no mineral lands shall be selected or purchased under the provisions of this Act.

SEC. 2. And be it further enacted, That the land aforesaid, after being surveyed, shall be apportioned to the several States in sections or subdivisions of sections, not less than one quarter of a section; and whenever there are public lands in a State subject to sale at private entry at one dollar and twenty-five cents per acre, the quantity to which said State shall be entitled shall be selected from such lands within the limits of such State, and the Secretary of the Interior is hereby directed to issue to each of the States in which there is not the quantity of public lands subject to sale at private entry at one dollar and twenty-five cents per acre, to which said State may be entitled under the provisions of this act, land scrip to the amount in acres for the deficiency of its distributive share: said scrip to be sold by said States and the proceeds thereof applied to the uses and purposes prescribed in this act, and for no other use or purpose whatsoever: Provided, That in no case shall any State to which land scrip may thus be issued be allowed to locate the same within the limits of any other State, or of any Territory of the United States, but their assignees may thus locate said land scrip upon any of the unappropriated lands of the United States subject to sale at private entry at one dollar and twenty-five cents, or less, per acre: And provided, further, That not more than one million acres shall be located by such assignees in any one of the States: And provided, further, That no such location shall be made before one year from the passage of this Act.

---

[642] "Morrill Act (1862)."

SEC. 3. And be it further enacted, That all the expenses of management, superintendence, and taxes from date of selection of said lands, previous to their sales, and all expenses incurred in the management and disbursement of the moneys which may be received therefrom, shall be paid by the States to which they may belong, out of the Treasury of said States, so that the entire proceeds of the sale of said lands shall be applied without any diminution whatever to the purposes hereinafter mentioned.

SEC. 4. And be it further enacted, That all moneys derived from the sale of the lands aforesaid by the States to which the lands are apportioned, and from the sales of land scrip hereinbefore provided for, shall be invested in stocks of the United States, or of the States, or some other safe stocks, yielding not less than five per centum upon the par value of said stocks; and that the moneys so invested shall constitute a perpetual fund, the capital of which shall remain forever undiminished, (except so far as may be provided in section fifth of this act,) and the interest of which shall be inviolably appropriated, by each State which may take and claim the benefit of this act, to the endowment, support, and maintenance of at least one college where the leading object shall be, without excluding other scientific and classical studies, and including military tactics, to teach such branches of learning as are related to agriculture and the mechanic arts, in such manner as the legislatures of the States may respectively prescribe, in order to promote the liberal and practical education of the industrial classes in the several pursuits and professions in life.

SEC. 5. And be it further enacted, That the grant of land and land scrip hereby authorized shall be made on the following conditions, to which, as well as to the provisions hereinbefore contained, the previous assent of the several States shall be signified by legislative acts:

First. If any portion of the fund invested, as provided by the foregoing section, or any portion of the interest thereon, shall, by any action or contingency, be diminished or lost, it shall be replaced by the State to which it belongs, so that the capital of the fund shall remain forever undiminished; and the annual interest shall be regularly applied without diminution to the purposes mentioned in the fourth section of this act, except that a sum, not exceeding ten per centum upon the amount received by any State under the provisions of this act may be expended for the purchase of lands for sites or experimental farms, whenever authorized by the respective legislatures of said States.

Second. No portion of said fund, nor the interest thereon, shall be applied, directly or indirectly, under any pretence whatever, to the purchase, erection, preservation, or repair of any building or buildings.

Third. Any State which may take and claim the benefit of the provisions of this act shall provide, within five years from the time of its acceptance as provided in subdivision seven of this section, at least not less than one college, as described in the fourth section of this act, or the grant to such State shall cease; and said State shall be bound to pay the United States the amount received of any lands previously sold; and that the title to purchasers under the State shall be valid.

Fourth. An annual report shall be made regarding the progress of each college, recording any improvements and experiments made, with their cost and results, and such other matters, including State industrial and economical statistics, as may be supposed useful; one copy of which shall be transmitted by mail [free] by each, to all the other colleges which may be endowed under the provisions of this act, and also one copy to the Secretary of the Interior.

Fifth. When lands shall be selected from those which have been raised to double the minimum price, in consequence of railroad grants, they shall be computed to the States at the maximum price, and the number of acres proportionally diminished.

Sixth. No State while in a condition of rebellion or insurrection against the government of the United States shall be entitled to the benefit of this act.

Seventh. No State shall be entitled to the benefits of this act unless it shall express its acceptance thereof by its legislature within three years from July 23, 1866:

Provided, That when any Territory shall become a State and be admitted into the Union, such new State shall be entitled to the benefits of the said act of July two, eighteen hundred and sixty-two, by expressing the acceptance therein required within three years from the date of its admission into the Union, and providing the college or colleges within five years after such acceptance, as prescribed in this act.

SEC. 6. And be it further enacted, That land scrip issued under the provisions of this act shall not be subject to location until after the first day of January, one thousand eight hundred and sixty-three.

SEC. 7. And be it further enacted, That the land officers shall receive the same fees for locating land scrip issued under the provisions of this act as is now allowed for the location of military bounty land warrants under existing laws: Provided, their maximum compensation shall not be thereby increased.

SEC. 8. And be it further enacted, That the Governors of the several States to which scrip shall be issued under this act shall be required to report annually to Congress all sales made of such scrip until the whole shall be disposed of, the amount received for the same, and what appropriation has been made of the proceeds.

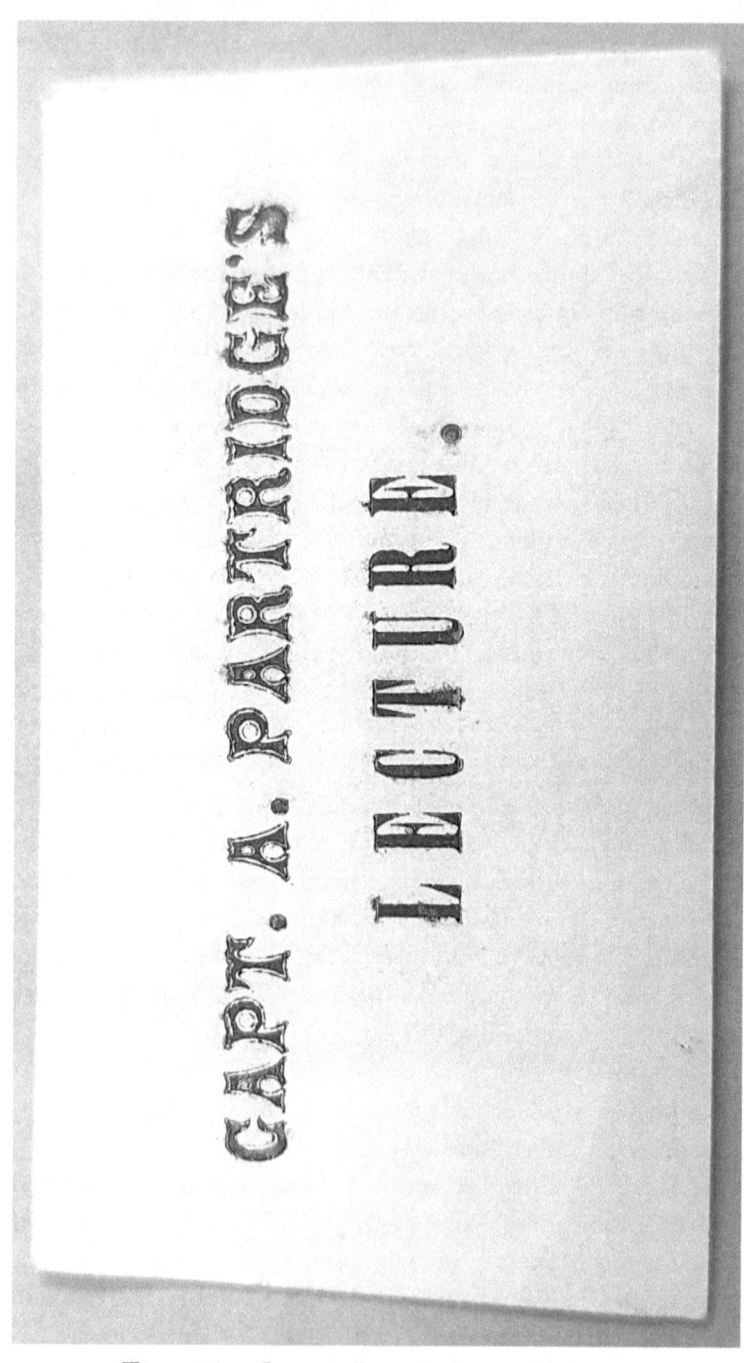

Figure 14—Capt. A. Partridge Lecture Ticket[643]

[643] An example of the tickets for Captain Alden Partridge's public lectures, held within the Rauner Special Collections Library at Dartmouth College.

Appendix B: Guidelines for Establishing Multiple Military Academies[644]

Sir,

I have the honor to inclose [*sic*] to you for consideration, a plan for the establishment in our country, of military academies upon extended scale. I have, for a considerable length of time, been convinced of the insufficiency of the present military academy at this place, to answer all the purposes which ought to be contemplated in the establishment of such an institution. I however did not think it a proper time to bring it forward when the country was involved in war. Peace I considered as more favorable for it. Upon the supposition this plan were adopted, and the academies properly conducted, I think, Sir, we might reasonably calculate upon the following results:—

1st. The diffusion of military science generally throughout our country. By this means, whenever the country becomes again involved in war, it would be provided with all the requisite materials (independent of the existing military establishment) for officering any additional force that might be necessary.[645]

2nd. The furnishing of existing military establishment with officers.

3rd. The furnishing of navy with well-educated midshipmen, who would make scientific officers. This appears to me to be of considerable consequence.

I cannot see why an acquaintance with the theory of his profession, is not as necessary for a naval, as for a land officer. The basis of their education, I consider the same, viz. the mathematics. I also think, that a knowledge of land tactics, or fortifications, of artillery duty on land, of drawing, and of astronomy would be very useful to every naval officer. And finally, I think that educating the officers of the two departments together would have a tendency to cement a friendship between them, which otherwise might not exist. I have discussed the plan with General Swift, who, I believ [*sic*], fully approves it; and I presume, ere this, has addressed you upon this subject. It has also met the approbation of several other well-informed gentlemen. All the members of the academic

---

[644] Partridge, Guidelines for Establishing Multiple Military Academies.
[645] Webb noted that this is the first time a concept like the modern Reserve Officers' Training Corps (ROTC) was presented. Webb, *Captain Alden Partridge*, 207.

staff (now present,) have subscribed to it. The professor of philosophy is absent; but I know his opinion is in favor of it. I would request, Sir, you will have the goodness to examine it, and inform me of your opinion upon the subject. Should it meet your approbation, and the approbation of the other heads of departments, I think there can be little doubt of it being carried into effect.

I have the honor to be, very respectfully,
Sir, Your obed. Servant
A. Partridge
Capt of Engs. U. S. M. A.

Honorable
    A. J. Dallas,
    Acting Secretary of War

[A General Plan for the Establishment of Military Academies]
1st. In addition to the Military Academy at West Point, let there be established two others, one either at, or in the vicinity of the City of Washington, and the other in the vicinity of Pittsburgh, or such other place as shall be deemed most convenient for the Western States.

2nd. At each of the Academies (including the one at West Point) let there be educated at the expense of the Government 150 Cadets—this number to be composed of young men, whose Parents or Guardians are willing and who are themselves desirous of entering the Military Service of their Country.

3rd. In addition to the 150 Cadets just mentioned, let it be allowed to receive at each Academy 250 others; this number to consist of young Gentlemen desirous of receiving a Military Education but who are not disposed to join the Service—these to be under the same regulations and discipline with those in the service of Government, and to pay the Government a stated sum per Annum, as a compensation for the Education.

4th. No pay or subsistence to be allowed those Cadets in the service of Government, but they should be provided with board, an uniform, Swords, books, Stationery, the requisite furniture for their Rooms, in short, with every thing [sic] necessary for their convenience, or the completion of their Education at the Expense of the Government and those Cadets not in service of Government, (that is those who pay for their Education) to be primarily furnished with every thing [sic] at the expense of the Government, in the same manner as the others, for which, and for their education they will afterwards pay a stated sum mentioned in Article 3d.

5th. Let there be allowed at each of the Academies the following Professors and Teachers (viz) a Professor of Natural and Experimental

Philosophy including Astronomy, a Professor of Mathematics, a Professor of Engineering in all its branches, a Professor of Chemistry and Mineralogy, a Professor of History and Geography, a professor of Natural History including Botany, a Professor of Languages, a Professor of Belle Lettres and a Chaplain who should be Professor of Ethics, a Teacher of Drawing and practical Geometry, a teacher of French Language, a Sword Master, and a Riding Master:[646] each of the foregoing Professors and Teachers to have one Assistant.

6th. Let the Professor receive the Pay and Emoluments of Lieutenant Colonels of Engineers, and their Assistants the Pay and Emoluments of 1st Lieutenants; the Sword Master and Riding Master the Pay and Emoluments of Captains of Engineers and their Assistants of 2d Lieutenants.

7th. Let there be to each Academy a permanent Superintendent (who must be a Military Officer) with the Rank, Pay and Emoluments of a Colonel of Engineers, and who will also discharge the duties of Professor of Tactics, the Superintendent to have an Assistant with the Rank, Pay and Emoluments of Captain of Engineers, and who in the absence of the Superintendent should discharge his duties.

8th. Those Cadets in the Service of the Government who should prefer the naval service, could (should they be judged qualified for that service after arriving at a certain state of their education) have their studies directed accordingly, when they had completed their course of studies, they should be put on board their ships as Midshipmen, where they would in a short time learn the practical part of their duty and become qualified for Commissions. By this means the Army and Navy would both be supplied from the Academies with Scientific Officers.

9th. Let there be an Inspector General of Military Academies, with the Rank, Pay and Emoluments of Brigr General, whose duties as well as those of the Superintendents, and all other Academic Officers should be defined by Regulation.

10th. Let the President of the United States be authorized to establish all necessary Regulations, both for the external and internal organization and arrangement of the Academies.

11th. Let the candidates for admission into the Academies be not under the age of thirteen, nor above the age of Eighteen years, the rules of admission as well as the necessary acquirements and qualifications of the Candidates to be prescribed by the President.

---

[646] Webb noted that in the margin but struck out was "or Professor of Architecture and Topography." Webb, *Captain Alden Partridge*, 208.

12th. Let the President be authorized to appoint a stated number of scientific Military Gentlemen who should constitute a board of visitors, of which board the Inspector General should be President, the duties to be prescribed by, and performed under the direction of the President.

13th. Let such number of Musicians be allowed to each Academy as the President shall direct.

14th. Let a Treasurer be authorized to each Academy, who shall discharge his duties under such Rules and Regulation as the President shall direct—and be subject to such penalties, in case of default, as shall be prescribed by law.

15th. Let the Academies be entirely disconnected with any Corps of the Army.

The undersigned members of the Academic Staff of the Military Academy at West Point, fully concur in the foregoing plan for the establishment of military academies, thinking it well calculated for the cultivation and diffusion of military science throughout our country, and would accordingly recommend its adoption.

Signed   And W. Ellicott, Prof. Mats.

Adam Empie, Chap. & Pr.

C. E. Zoeller, teacher of mil. drawing & surveying.

Claudius Berard, teacher of the French language.

John Wright, Lt. Engrs. & Asst. Prof.

W. L. Eveleth, Lt. Engrs. & Asst. Prof. Engrs.

Appendix C: Captain Partridge's Lecture on Education[647]

The Elementary Education of youth, is doubtless, one of the most important subjects which can occupy the attention of an enlightened and free people. It is to the rising generation that we are to look for the future guardians and protectors of the inestimable rights and privileges transmitted to us by the heroes and patriots of the revolution: they are to be future legislators, political economists, and defenders of our country; and on them is to depend, in a very great degree, the future destiny of our mighty republic. It certainly, then, cannot be considered of small importance, that they be prepared, by a proper course of preliminary instruction and the acquirement of virtuous habits, for the correct discharge of the duties indent to their exalted stations; and thereby be enabled to transmit unimpaired to their posterity, the important trust committed to their charge.

I shall define elementary education, in its most perfect state, to be the preparing of a youth in the best possible manner for the correct discharge of the duties of any station in which he may be placed, and consequently, shall consider as most perfect that system which shall be found best calculated to accomplish the object in view. The system of education adopted in the United States appears to me to be defective in many respects; and-

1st. It is not sufficiently practical, nor properly adapted to the various duties an American citizen may be called upon to discharge. Those of our youth who are destined for a liberal education, as it is called, are usually put, at an early age, to the study of the Latin and Greek languages, combining therewith a very slight attention to their own language, the elements of arithmetic, &c.; and after having devoted several years in this way, they are prepared to become members of a college or university.

Here they spend four years for the purpose of acquiring a knowledge of the higher branches of learning; after which, they receive their diplomas, and are supposed to be prepared to enter on the duties of active life. But, I would ask, is this actually the case? Are they prepared in the best possible manner to discharge correctly the duties of any station in which fortune or inclination may place them? Have they been instructed in the science of government generally,

---

[647] Partridge, Lecture on Education (British Library).

and more especially in the principles of our excellent Constitution, and thereby prepared to sit in the legislative councils of the nation? Has their attention been sufficiently directed to those great and important branches of national industry and sources of national wealth—agriculture, commerce, and manufactures? Have they been taught to examine the policy of other nations, and the effect of that policy on the prosperity of their own country? Are they prepared to discharge the duties of civil or military engineers, or to endure fatigue, or to become the defenders of their country's rights, and the avengers of her wrongs, either in the ranks or at the head of her armies? It appears to me not; and if not, then, agreeably to the standard established, their education is so far defective.

2dly. Another defect in the present system, is, the entire neglect, in all our principal seminaries, of physical education, or the due cultivation and improvement of the physical powers of the students.

The great importance and even absolute necessity of a regular and systematic course of exercise for the preservation of health, and confirming and rendering vigorous the constitution, I presume, must be evident to the most superficial observer. It is for want of this, that so many of our most promising youths lose their health by the time they are prepared to enter on the grand theatre of active and useful life, and either prematurely die, or linger out a comparatively useless and miserable existence. That the health of the closest applicant may be preserved, when he is subjected to a regular and systematic course of exercises, I know, from practical experience; and I have no hesitation in asserting, that in nine cases out of ten, it is just as easy for a youth, however hard he may study, to attain the age of manhood, with a firm and vigorous constitution, capable of enduring exposure, hunger and fatigue, as it is to grow up puny and debilitated, incapable of either bodily or mental exertion.

3dly. A third defect in our system is, the amount of idle time allowed the students; that portion of the day during which they are actually engaged in study and recitations, under the eye of their instructors, comprises but a small portion of the whole; during the remainder, those who are disposed to study, will improve at their rooms, while those who are not so disposed, will not only not improve, but will be very likely to engage in practices injurious to their constitutions and destructive to their morals. If this vacant time could be employed in duties and exercises, which, while they amuse and improve the mind, would at the same time invigorate the body and confirm the constitution, it would certainly be a great point gained. That this may be done, I shall attempt in the course of these observations, to show.

4thly. A fourth defect is, the allowing to students, especially to those of the wealthier class, too much money, thereby inducing habits of dissipation and extravagance, highly injurious to themselves, and also to the seminaries of which they are members. I have no hesitation in asserting, that far the greater portion of the irregularities and disorderly proceedings amongst the students of our seminaries, may be traced to this fatal cause. Collect together at any seminary, a large number of youths, of the ages they generally are at our institutions, furnish them with money, and allow them a portion of idle time, and it may be viewed as a miracle, if a large portion of them do not become corrupt in morals, and instead of going forth into the world to become ornaments in society, they rather are prepared to become nuisances to the same. There is in this respect, an immense responsibility resting on parents and guardians, as well as on all others having the care and instruction of youth, of which it appears to me they are not sufficiently aware.

When youths are sent to a seminary, it is presumed they are sent for the purpose of learning something that is useful, and not to acquire bad habits, or to spend money; they should consequently be furnished with every thing [*sic*] necessary for their comfort, convenience and improvement, but money should in no instance be put into their hands. So certainly as they have it, just so certainly will they spend it, and this will, in nine cases out of ten, be done in a manner seriously to injure them, without any corresponding advantage. It frequently draws them into vicious and dissolute company, and induces habits of immorality and vice, which ultimately prove their ruin. The over-weening indulgence of parents, has been the cause of the destruction of the morals and future usefulness of many a promising youth. They may eventually discover their error, but alas, it is often too late to correct it. Much better does that person discharge the duties of a real friend to the thoughtless, unwary youth, who withholds from him the means of indulging in dissipated and vicious courses.

5thly. A fifth defect is the requiring all the students to pursue the same course of studies.

All youth have not the same inclinations, nor the same capacities; one may possess a particular inclination and capacity for the study of the classics, but not for the mathematics and other branches of science; with another it may be the reverse. Now it will be in vain to attempt making a mathematician of the former, or a linguist of the latter. Consequently, all the time that is devoted in this manner will be lost, or something worse than lost. Every youth, who has any capacity or inclination for the acquirement of knowledge, will have some favorite studies, in which he will be likely to excel. It is certainly then much better that he should be permitted to pursue those, than, that by being forced to attend to others for which he has an aversion, and in which he will never excel, or ever make common proficiency, he should finally acquire a dislike to all study. The celebrated Pascal, is a striking instance of the absurdity and folly of attempting to force a youth to attend to branches of study, for which he has an utter aversion, to the exclusion of those for which he may possess a particular attachment. Had the father of this eminent man persisted in his absurd and foolish course, France would never have seen him, what he subsequently became, one of her brightest ornaments.

6thly. A sixth defect is the prescribing the length of time for completing, as it is termed, a course of education. By these means, the good scholar is placed nearly on a level with the sluggard, for whatever may be his exertions, he can gain nothing in respect to time, and the latter has, in consequence of this, less stimulus for exertion. If any thing [*sic*] will induce the indolent student to exert himself, it is the desire to prevent others getting ahead of him. It would be much better to allow each one to progress as rapidly as possible, with a thorough understanding of the subject.

Having thus summarily stated what appear to me the most prominent defects in our present system of elementary education, I will next proceed to point out the remedy for the same. This I shall do by describing the organization, &c. of an institution, such as I would propose.

1st. The organization and discipline should be strictly military.

Under a military system, subordination and discipline are much more easily preserved than under any other. Whenever a youth can be impressed with the true principles and feelings of a soldier, he becomes, as a matter of course, subordinate, honorable, and manly. He disdains subterfuge and prevarication, and all that low cunning, which is but too prevalent. He acts not the part of the assassin, but if he have an enemy, he meets him openly and fairly. Others may boast that they have broken the laws and regulations of the institution of which they are, or have been members, and have escaped detection and punishment, by mean prevarication and falsehood. Not so the real soldier. If he have broken orders and regulations, he will openly acknowledge his error, and reform; but will not boast of having been insubordinate. Those principles, if imbibed and fixed in early youth, will continue to influence his conduct and actions during life; he will be equally observant of the laws of his country, as of the academic regulations under which he has lived; and will become the more estimable citizen in consequence thereof. I shall not pretend, however, that all who wear a military garb, or live, for a time, even under a correct system of military discipline, will be influenced in their conduct by the principles above stated; but if they are not, it only proves that they have previously imbibed erroneous principles, which have become too firmly fixed to be eradicated; or that nature has not formed them with minds capable of soaring above what is low and groveling.

2dly. Military science and instruction should constitute a part of the course of education.

The constitution of the United States has invested the military defense of the country in the great body of the people. By the wise provisions of this instrument, and of the laws made in pursuance thereof, every American citizen, from eighteen to forty-five years of age, unless specially exempted by law, is liable to be called upon for the discharge of military duty—he is emphatically a citizen soldier, and it appears to me perfectly proper that he should be equally prepared by education to discharge, correctly, his duties in either capacity. If we intend to avoid a standing army, (that bane of a republic, and engine of oppression in the hands of despots,) our militia must be patronized and improved, and military information must be disseminated amongst the great mass of the people; when deposited with them, it is in safe hands, and will never be exhibited in practice, except in opposition to the enemies of the country. I am well aware there are amongst us many worthy individuals, who deem the cultivation of military science a sort of heresy, flattering themselves, and endeavoring to induce others to believe, that the time has now arrived, or is very near, when wars are to cease, and universal harmony prevail amongst mankind. But, my fellow-citizens, be not deceived by the syren song of peace, peace, when, in reality, there is no peace, except in a due and constant preparation for war. If we turn our attention to Europe, what do we behold? A league of crowned despots, impiously called holy, wielding a tremendous military force of two millions of mercenaries! Ill-fated Naples, and more ill-fated Spain, have both felt the effects of *their peaceable* dispositions, and were it not for the wide-spreading Atlantic, which the God of nature in his infinite goodness has interposed between us, we also, ere this, should have had a like experience. The principles of liberty are equally obnoxious to them, whether found in Europe, Asia, Africa, or America. If rendering mankind ignorant of the art of war, (as a science,) would prevent wars, then would I unite most cordially with those, usually termed peace-men, for the purpose of destroying every vestige of it. But such, I am confident, would not be the result. Wars amongst nations do not arise because they understand how to conduct them skillfully and on scientific principles; but are induced by the evil propensities and dispositions of mankind. To prevent the effect, the cause must be removed. We may render nations ignorant of the use of the musket and bayonet; we may carry them back, as respects the art of war, to a state of barbarism, or even of savageism, and still wars will exist. So long as mankind possess the dispositions which they now possess, and which they ever have possessed, so long they will fight. To prevent wars, then, the disposition must be changed; no remedy short of this, will be effectual. In proportion as nations are rude and unskilled in the art of war, will their military code be barbarous and unrelenting, their battles sanguinary, and their whole system of warfare, destructive. War, therefore, in such a case, becomes a far greater evil, than it does under an improved and refined system, where battles are won more by skill than by hard fighting, and the laws of war are proportionally ameliorated. What rational man, what friend of mankind, would be willing to exchange the present humane and refined system of warfare, for that practiced by an Attila, a Jenghis Khan, a Tamerlane, or a Mahomet, when hundreds of thousands fell in a single engagement, and when conquest and extermination were synonymous terms. On the principles of humanity, then, it appears to me that, so long as wars do exist, the military art should be improved and refined as much as possible; for, in proportion as this is done, battles will be less sanguinary and destructive, the whole system more humane, and war itself a far less evil. But independent of any connection with the profession of arms, or of any of the foregoing considerations, I consider a scientific knowledge of the military art, as constituting a very important part of the education of every individual engaged in the pursuit of useful knowledge, and this for many reasons; viz.:—

1st. It is of great use in the reading of history, both ancient and modern.

A large portion of history is made up of accounts of military operations, descriptions of battles, sieges, &c. How, I would ask, is the reader to understand this part, if he be ignorant of the organization of armies, of the various systems of military tactics, of the science of fortification, and of the attack and defense of fortified places, both in ancient and modern times? Without such knowledge it is evident he derives, comparatively but little information from a large portion of what he reads.

2d. It is of great importance in the writing of history. I presume it will not be denied, that in order to write well on any subject, it must be understood. How, then, can the historian give a correct and intelligible account of a campaign, battle, or siege, who is not only unacquainted with the principles on which military operations are conducted, but is also ignorant of the technical language necessary for communicating his ideas intelligibly on the subject? This is the principal reason why, as it appears to me, the ancient historians were so much superior to the modern. Many of their best historical writers were military men. Some of them accomplished commanders. The account of military operations by such writers as Xenophon, Thucydides, Polybius and Cæsar, are perfectly clear and intelligible, whereas when attempted by the great body of modern historians, the most we can learn is, that a fortress was besieged and taken, or that a battle was fought and a victory won, but are left in entire ignorance of the principles on which the operations were conducted, or of the reasons why the results were as they were.

3d. It is essentially necessary for the legislator.

The military defense of our country is doubtless one of the most important trusts which is vested by the constitution in the general government, and it is a well known [*sic*] fact, that more money is drawn from the people and disbursed in the military, than in any other department of the government. Now as all must be done under the sanction of the law, I would beg leave to inquire, whether it be not of the greatest importance, that those who are to make such laws should be in every respect well prepared to legislate understandingly on the subject? That there has been, and still is, a want of information on this subject amongst the great body of the members of Congress, I think will be perfectly evident to any one [*sic*] who is competent, and will take the trouble to examine our military legislation since the conclusion of the Revolutionary war. I feel little hesitation in asserting, that from want of this information, more than from any other cause, as much money has been uselessly expended in our military department alone, as would cancel a large portion of the national debt.

4th. It is of great use to the traveler.

Suppose a young man, with the best education he can obtain at any of our colleges or universities, were to visit Europe, where the military constitutes the first class of the community, and where the fortifications constitute the most important appendages to nearly all the principal cities, how much does he observe, which he does not understand? If he attempt a description of the cities, he finds himself embarrassed for want of a knowledge of fortification. If he attempt an investigation of the principles and organization of their institutions, or of their governments, he finds the military so interwoven with them all, that they can not [*sic*] be thoroughly understood without it. In fine, he will return with far less information, than with the aid of a military education he might have derived. As it respects the military exercises, I would observe, that were they of no other use than in preserving the health of students, and confirming in them a good figure and manliness of deportment, I should consider these were ample reasons for introducing them into our seminaries generally; they are better calculated than any others for counteracting the natural habits of students, and can always be attended to, at such times as would otherwise be spent in idleness or useless amusements. Having expressed my views thus fully on this subject, I will next proceed to state more specifically the other branches which I would propose to introduce into a complete course of education: and—

1st. The course of classical and scientific instruction should be as extensive and perfect as at our most approved institutions. The students should be earnestly enjoined and required to derive as much of useful information from the most approved authors, as their time and circumstances would permit.

2d. A due portion of time should be devoted to practical geometrical and other scientific operations in the field. The pupils should frequently be taken on pedestrian excursions into the country; be habituated to endure fatigue, to climb mountains, and to determine their altitudes by means of the barometer as well as by trigonometry. Those excursions, while they would learn them to walk, (which I estimate an important part of education,) and render them vigorous and healthy, would also prepare them for becoming men of practical science generally, and would further confer on them a correct *coup d'œil* so essentially necessary for military and civil engineers, for surveyors, for travelers, &c., and which can never be acquired otherwise than by practice.

3d. Another portion of their time should be devoted to practical agricultural pursuits, gardening, &c.

In a country like ours, which is emphatically agricultural, I presume it will not be doubted, that a practical scientific knowledge of agriculture would constitute an important appendage to the education of every American citizen. Indeed the most certain mode of improving the agriculture of the country will be to make it a branch of elementary education. By these means, it will not only be improved, but also a knowledge of their improvements generally disseminated amongst the great mass of the people.

4th. A further portion of time should be devoted to attending familiar explanatory lectures on the various branches of military science, on the principles and practice of agriculture, commerce and manufactures, on political economy, on the constitution of the United States, and those of the individual states, in which should be pointed out particularly the powers and duties of the general government, and the existing relations between that and the state governments, on the science of government generally. In fine, on all those branches of knowledge which are necessary to enable them to discharge, in the best possible manner, the duties they owe to themselves, to their fellow men, and to their country.

5th. To the institution should be attached a range of mechanics' shops, where those who possess an aptitude and inclination might occasionally employ a leisure hour in learning the use of tools and acquiring a knowledge of some useful mechanic art.

The division of time, each day, I would make as follows, viz.:—

Eight hours to be devoted to study and recitation; eight hours allowed for sleep. Three hours for the regular meals, and such other necessary personal duties as the student may require. Two hours for the military and other exercises, fencing, &c. The remaining three hours to be devoted, in due proportion, to practical agricultural and scientific pursuits and duties, and in attending lectures on the various subjects before mentioned.

Some of the most prominent advantages of the foregoing plan would, in my opinion, be the following; viz.:—

1st. The student would, in the time usually devoted to the acquirement of elementary education, (say six years) acquire, at least, as much, and I think I may venture to say more, of book knowledge, than he would under the present system.

2d. In addition to this, he would go into the world an accomplished soldier, a scientific and practical agriculturist, an expert mechanician, an intelligent merchant, a political economist, legislator and statesman. In fine, he could hardly be placed in any situation, the duties of which he would not be prepared to discharge with honor to himself and advantage to his fellow-citizens and his country.

3d. In addition to the foregoing, he would grow up with habits of industry, economy and morality, and, what is of little less importance, a firm and vigorous constitution; with a head to conceive and an arm to execute—he would emphatically possess a sound mind in a sound body.

I have thus in a summary manner endeavoured to exhibit my views on the important subject of elementary education. I am sensible of my inability to do justice; but if what I have said should be the means of drawing to it the attention of others, abler and more experienced, and who have more leisure than myself, I shall feel that much has been gained. Having devoted about sixteen years of my life to the care and instruction of the youth of our country, and having, under all circumstances, experienced from them a most ardent and devoted friendship, it can hardly be supposed I should feel indifferent to their future welfare and prosperity. I owe them a debt which I can only discharge by continuing zealously to exert myself for their improvement in useful knowledge, and endeavouring to instil [*sic*] into their minds such principles and habits as will render them a blessing to their parents and friends, and an ornament to our beloved country.

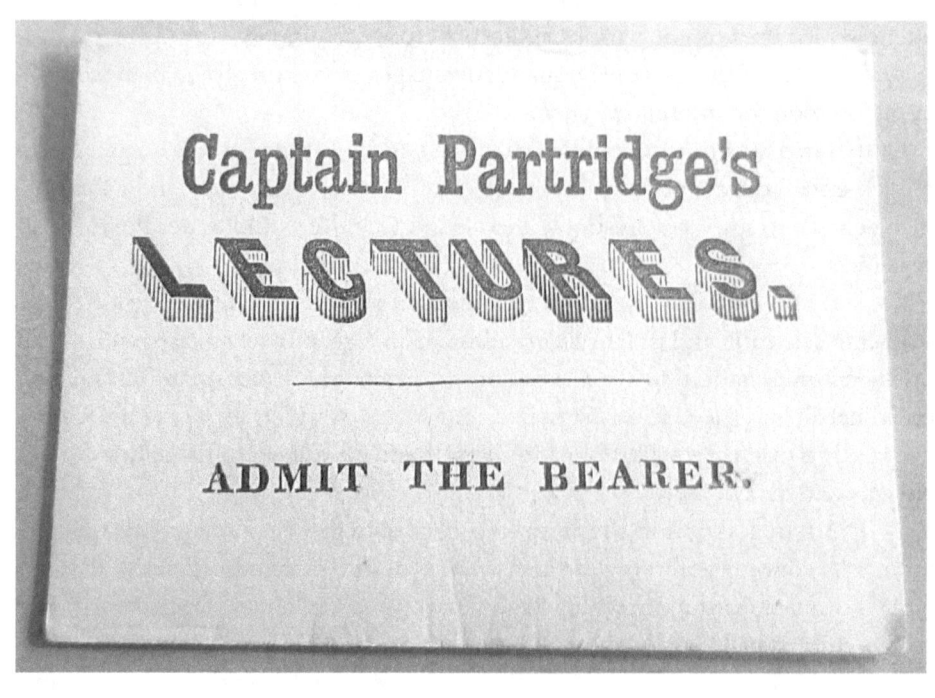

Figure 15—Captain Partridge's Lectures Ticket[648]

[648] An example of the tickets for Captain Alden Partridge's public lectures held within the Rauner Special Collections Library at Dartmouth College.

Appendix D: Lecture on National Defense[649]

Norwich July 18th 1821

Dear Sir

I have read with pleasure, in the Republican of the 2nd Inst, your synopsis of the lecture which I delivered at Windsor, while on a visit to that place with the young gentlemen under my instruction, on the subject of a general system of military defense, adapted particularly to our own country, which while it effectually protects us from foreign aggression would (it is believed) at the same time constitute a most efficient bulwark for our republican Institutions. I have also read several communications, uniting with your own in the request that I would enlarge this lecture and give it to the public. Nothing I can assure both yourself and the other gentlemen making this request, would afford me greater satisfaction than a prompt and full compliance therewith. But the truth is, my constant daily duties are so many and so urgent, as to render it atmost impossible for me to get leisure to answer the letters I receive. Under such circumstances, a few hasty remarks embracing a summary of my ideas on the subject above mentioned, is all I am enabled, at this time, to present to my fellow—Citizens—reserving a more full exposition of the same, to a period of greater leisure should such period ever arrive. Before I proceed however, I would observe, (what is doubtless a well known fact) that there are many Individuals at the present time who believe—I trust conscientiously—that the time is very near when wars and fighting will cease, and that consequently military preparations and the cultivation of military science are unnecessary and ought likewise to cease. That such a time will come, I perhaps as firmly believe as any individual whatever, but that this period is so near as is by some supposed does not appear to me probable. A comparison of the events predicted in the prophecies and revelation, with those which have transpired in the world as recorded in history force upon my mind a conviction, that mankind is doomed to suffer (at least for a time) the evils of wars and of bloodshed, and that consequently that state which intends to maintain its independence free from the encroachments of [avarice?] and ambition, must be prepared to repel force by force. But all other considerations

---

[649] Partridge, Lecture on National Defense.

aside, the present state of the world is such as to afford just cause of alarm to every people not prepared to prostrate themselves at the feet of lawless tyranny. If we turn our eyes to Europe what do we behold? A league of crowned despots impiously called holy—wielding a tremendous military force of two millions of mercenaries, under the [illegible] mark of religion and humanity, aiming a fatal blow at the rights and liberties of mankind. Naples affords a melancholy exhibition of their peaceable and humane intentions, and Spain may be destined, ere long, to share the same fate. And can we reasonably believe that this [unhallowed combination?] these self-constituted umpires of the destinies of nations entertain desires more friendly or less hostile to the principles of our Constitution, than those they have so recently exhibited towards ill-fated Naples? My fellow citizens be not deceived by [hollow?] professions and hypocritical pretensions. Be [illegible] one thing alone is wanting to reduce us to the same degraded condition with the vassals of Europe—and that is the power to do it. But if [true to ourselves?], we may laugh at their threats and set their menaces at defiance. From these considerations (as well as from many others which might be [advanced?]) it appears to me perfectly evident that a competent military defense is absolutely necessary for the safety of every state, and that, consequently, the only thing to be done is, to determine the description of defense which will [conduce?] most effectually to the accomplishment of the object in view. The military defenses of countries—may be considered as of two descriptions; the one, where one portion of the population is equipped with all the habiliments of war, organized into a regular standing army and retained as such in time of peace as well as in time of war. The other where the great body of the people themselves—between certain ages—constitute the military force of the country—not being retained as a military corps in times of peace, but opening this character only, in the emergency of war. The first of these systems evidently contains features highly unfavorable to a republican government. Whenever the military [service?] of a country becomes concentrated in one portion of the population, and that portion furnished with arms and organized into a regular standing army, alienated by feeling, by interest and by profession from the great body of the community, and accustomed to look to its own leaders, and to the executive of the country for its immediate benefactors, I think it may without hesitation be asserted that the liberties of such country— if liberty it [have?] ever enjoyed—are rapidly [illegible] to a close. This conclusion is fully [warranted?] by the [convincing testimonies?] of the histories both of ancient and modern republics. The liberties of Rome were safe while every Roman citizen, considered and felt himself a soldier. But how fatal was the [reversal?], when by the operations of a system organized by Gaius Marius the savior and scourge of Rome, and matured by Julius Caesar, with a view doubtless to the accomplishment of his ultimate object, – the [final?] prostration of the liberties of his country – those noble and patriotic legions, which had so often in times of peril and of danger, proved the shield of their country and the terror of its Enemies, became transformed into mere mercenary bands, alienated from their country, and identified in views in feelings and in interests with their leaders alone. Under the former system she was enabled to set at defiance that Prince of Generals – the great Hannibal himself – though encamped under her very walls at the head of a veteran and victorious Army, while under the latter she fell an almost unresisting victim to the mere advance guard of a Roman army led on by one of her degenerate sons. It being then, as I [conceive?], perfectly evident that the system of defense by means of a standing Army is totally incompatible with the preservation of liberty in any country, I will now proceed to consider the second system mentioned, viz that in which the great body of the people themselves (within certain ages) constitute the military force, and which is so fully recognized in the Constitution of the United States. By the wise provisions of this instrument and of the law made in pursuance thereby the great body of American [citizens?] from 18 to 45 years of age are enrolled and liable to be called into military service, whenever emergencies may require. These compose the grand constitutional military force of the nation—a force identified in views, in feelings and in interest with the great body of the community, and which while it appears an [unpressable barrier?] to foreign invasion will at the same time constitute the surest support to the laws and civil institutions of the country. But this grand constitutional force, in order to answer the purposes for which it was [originally?] intended, must be properly organized and disciplined. For the accomplishment of this latter object, both Congress and the State legislatures have enacted laws requiring the Militia to turn out a given number of times each year, either by companies, battallions or brigades, to improve, as it is termed, in military discipline. But here legislation stops. The manner in which this improvement is to be made is not pointed out. The officers are imperatively required to instruct those under their command, while no provision whatever is made for their own instruction. The officers of the militia being [originally?] taken from the ranks, have consequently enjoyed no better opportunity for becoming acquainted with the duties of their respective stations, than the great body of the privates, unless it has been procured at a considerable sacrifice both of time and money. With what reason then I would I ask can they be required to teach that which they had never had an opportunity of learning themselves? Or if required how can it be expected they will teach it correctly? In the common occupations of life we [calculate?]

differently. If we wish a child to learn the most simple mechanic art, we place it under the care of some person who, in consequence of a course of preliminary instruction, has become master of the same, and is consequently, competent to teach it to others. If [instead?] of pursuing this course we were to place the child under the care of a person, who had enjoyed no means superior to those the child had enjoyed, for becoming versed in the art he was expected to teach, we should be doubtless be [considered?] as going counter, not only to the dictates of reason but of common sense. If then it be absurd to expect that correct instruction can be given in a simple mechanic art, where those required to instruct have themselves, not enjoyed the means of instruction, can it be considered less so to expect that, under similar circumstances, correct instruction can be furnished in the military profession, which embraces within it wide range almost the whole circle of sciences? I shall leave the answer to the sound discretion of every reflecting mind. My conclusion from the foregoing is, that the militia system of the United States is radically defective, and that until it be thoroughly reformed, no real improvement can reasonably be expected under it. This radical defect consists in there being no means of instruction furnished for the officers. It is in vain to suppose that they can instruct, correctly, those under their command, without being themselves, previously, correctly instructed, and it is presuming too much on their zeal and patriotism, to suppose that they will, generally, sacrifice both time and money for acquiring a species of knowledge to be applied to the public good, without any corresponding personal advantages. To remedy the foregoing defect, I now take the liberty to submit, with due deference, the following plan, which [appears?] to me both feasible in its execution, and in every respect calculated to answer the purpose [contemplated?] As the Constitution has invested Congress with full power to provide for organizing and disciplining the militia—let a law be passed embracing the following provisions, viz—

1st Let the United States be divided into military departments—say thirty in number—each of those departments to be wholly comprised within the same state whenever this can be done—

2nd To each of these departments let there be attached a military instructor (under the authority of the United States) who should receive the pay and emoluments of a colonel of infantry, and have the brevet rank of a brigadier general. These Instructors to be gentlemen of established character and reputation, and who have received a regular scientific military education—

3d Let the officers of each brigade of the militia in the United States be required to assemble annually, at stated periods, either in camp or rendezvous, at some central point in the brigade [there?] to remain six days for the purpose of military instruction. Let each Instructor attend in succession at the several camps or places of rendezvous in his department, and devote himself assiduously to the instruction of the officers there assembled. One portion of the day might be appropriated to practical drills, and field evolutions—also to the turning off, mounting and relieving guards and sentinels, while the remainder could be most usefully employed in explaining and illustrating the principles of tactics generally, of artillery, of permanent and field fortification, of the duties of troops in camp and in garrison, and such other branches as time and circumstances might permit—by means of familiar explanatory lectures.

4th Let each officer receive from the government a reasonable allowance for his expenses while attending the instruction, and also while going to, and returning from the camp or rendezvous.

Some of the principal advantages that would result from the adoption of the foregoing plan I conceive would be as follows—viz—

1st The same system of tactics and discipline would pervade the whole mass of the militia—The instructors being imperatively required to adhere to one system. This would be a very important advantage—

2nd By this means the country, in the course of a few years, would be furnished with a well organized military force of at least one million of men, composed of the best materials in the world for soldiers, the whole of which, the officers having been regularly and correctly instructed, might be rendered in the course of a few weeks after being called into service, perfectly competent to the efficient discharge of all the duties of the field. This assertion is not founded upon conjecture. An experience of nearly fifteen years in military instruction has convinced me, that any of our regiments of militia in their present state of discipline, if brought into the field and placed under competent officers, could by three weeks instruction be prepared for discharging all the duties of regular troops. The instruction of the officers then in time of peace becomes an object of great importance. That of the privates is of secondary consideration. There is no difficulty in making soldiers, when officers understand their duty and are disposed to perform it. It may perhaps be objected to the foregoing plan, that the time proposed for the officers [remaining?] in camp or rendezvous is too limited to admit of their deriving much advantage therefrom. In answer to this I would observe, that a due share of experience in this species of instruction has fully convinced me that they would acquire more correct military information in Six days under a competent and systematic Instructor than they usually acquire under the present system during the whole period from eighteen to forty five years of age; and that after attending two or three similar courses the great body of them would be perfectly competent to the correct, efficient, and useful discharge of all the duties of the field. From the best calculation I have been able to make, I feel confident the whole necessary expense of carrying into full and effective operation this plan would not exceed Six hundred thousand dollars—it would probably fall short of that sum. Whether the expense then is to be considered as disproportioned to the object in view and therefore to constitute a barrier to its accomplishment must be decided by the sound discretion of the Representatives of the People—It appears to me however to bear no greater ratio to it than does a grain of sand to the globe we inhabit. The cultivation of military science must also be received as of the first importance in a system of military defense for our country. The plan already detailed is calculated for the general dissemination of practical military information throughout the community, but is not adapted to the investigation of its principles. This can only be done at seminaries where it constitutes a branch of regular attention and study; and where theory and practice can, in due proportion, be combined. At those seminaries would be found our military instructors, our engineers, and our generals, and from them as from so many foci, would all the improvements in the military art, be diffused throughout the country my time does not permit, at present, to detail the general plan on which I would propose those Institutions should be organized—This will be the subject of a future communication. I shall next proceed to make a few observations about fortifications viewed as contributing a part of the system of military defense for the country. And [here?] I regret I cannot consistently, with what I esteem my duty, declare my approbation of the course [pursuing?] in relation to this branch of our national defense. There appears to be a kind of charm in the word fortification which pervades and [powerfully?] influences both the people and their Representatives; and if to this word, the word permanent be attached it becomes almost irresistible. But if I do not much mistake the American people will, ere long, discover that a radical Error has, in this respect, been committed. We construct works of masonry, and to these attach platforms of wood, on which are placed the guns and carriages generally exposed to the weather—and call the whole permanent By a permanent fortification we appear to understand one that is to last forever—unless battered down by an enemies shot. Here we deceive ourselves. Not only the more perishable parts of the works—the platforms and gun-carriages—but even the masonry itself will yield to the dilapidations of time. The water falling on the top of the parapet penetrates the interstices of the masonry—where by its expansions and contractions, in consequence of alternately freezing and thawing, it overcomes the resistance of the cement, the stones or bricks are loosened and displaced, and the result is the work tumbles into ruins. If any one doubt the correctness of this conclusion, I would beg leave to refer him to some of the works in the harbour of New York, and more especially to Fort Columbus on Governor's island. This work is one of the permanent kind—the rampart having a full revetment of masonry faced with free stone with a parapet of brick—and has been finished about fourteen years. This work is now fast verging to ruin; the platforms and other parts of wood being in a state of decay, and the revetment of the rampart and parapet tumbling down. This fortification not withstanding is one of the permanent kind, and in point of materials and workmanship was probably second to few if any in the United States. In adducing this work as an example, I would observe that it is by no means intended to attach the least blame to the distinguished scientific engineer under whose direction it was constructed, nor to the [officers?] or garrison who have subsequently been attached to it, (for as it respects police, I believe it has been uniformly equal to that of any military station in the country) but it is intended solely for the purpose of illustrating the position above stated. Other examples might be adduced proving the same fact, but of these I shall mention only one, and that is the fortifications at Quebec. These works in point of materials and finish are of the most permanent kind; yet when at that place in the summer of 1819, I witnessed the repairs then making on a large portion of the front facing the plains of Abraham, and was informed that similar repairs at different points were necessary every year.

My conclusions from the foregoing are—

1st That fortifications and their appendages however permanently constructed are not exempt from the dilapidations of time, but like every thing else in Nature tend to decay and dissolution—

2nd That for the purpose of keeping them in repair an annual expense will eventually be necessary which will bear a very considerable ratio to the original cost:—otherwise they will in the course of a few years, comparatively speaking, tumble into ruins and become useless.

3d That consequently the American People by the present intended system of fortifying, are taxing themselves with something in the shape of national debt, not bearing an Interest of Six nor Seven, but more probably of twenty or twenty five [percent?], and which if not regularly discharged the loss of the principal itself will be hazarded. But fortifications on the extensive scale we are now constructing them, will not only be found primarily and subsequently a very expensive mode of defense, but will also be ultimately found not to answer the purposes contemplated. What countries in the world are more plentifully provided with strong fortresses than Holland and the Netherlands?—yet with what rapidity were they overrun by the Armies of France under the Command of Generals Pichegrue and Jourdain, in the celebrated campaign of 1794, and [beginning?] of 1795. Why did the French in

1792, with their armies disorganized and their fortresses but indifferently provided with the means of defense repel the formidable invasion headed by the Duke of Brunswick? and why were the allies in 1814 permitted to march to Paris—notwithstanding the fortresses at this latter period were in a much better state of defense than in the former? The answer to all these enquiries is the same: the great body of the people both in Holland and the Netherlands were in a degree indifferent to the contest carrying on between the French and allies—having perhaps a predilection in favor of the former,—and consequently took no active part in. The French also—as a Nation—in 1814 had become tired of war and no longer considered themselves as combatting for freedom and the rights of Man;—and the results in both instances were the same—the armies in the field being defeated. The countries, notwithstanding all the fortifications, were soon subjugated. But in 1792 the state of feeling in France was very different. At that period the energies of the people were roused: they considered that they were contending against tyrants, for liberty and the dearest rights of man, and the result was as might have been expected—the invaders were ignominiously driven from their territory. The total inefficiency of our maritime fortifications for the secure defense of our cities it appears to me was fully proven during the last war. In the harbor of New York alone more than one million of dollars had been expended in constructing permanent works for the defence of the place; and what was the result? Why that in 1814, when an attack was expected the people were obliged to turn out by thousands for weeks in succession, in order to construct works for the defense of those points where they were likely to be assaulted. Every one acquainted with the local situation of New York knows that the enemy by coming down the sound might have landed a force within about 20 miles of the city which in its advance would not have been exposed from a single shot from any of the works in the Hudson. Having possessed himself of Brooklyn Heights, the city, the Navy-yard and the fortifications on Governor's Island would have been at his mercy. Throughout the whole of the last war we do not discover that the British meditated attacks upon any important points with their fleet alone. This appeared to act a Secondary part by means of diversions &c. to the operations of their land forces. Washington was taken by a land attack, and Baltimore and New Orleans were saved by lines temporarily constructed. It appears to me then totally absurd to attempt the defence of a Sea-Board, so extensive as ours, by means of permanent fortifications. However judiciously those works may be placed an active and skillful enemy will always find means to evade them, or to take them in reverse, and in either event to render them equally useless. From the foregoing it may perhaps be inferred that I am opposed to the construction of any permanent fortification whatever on our Sea-Board. This however is by no means the case as will be made to appear hereafter. It is not to fortifications in the abstract, but to the undue and unnecessary extension of them that I am opposed. Having thus summarily but freely expressed my ideas on the present system of extended fortification, viewed as a branch of national defense, I will next take the liberty to submit to the consideration of my fellow citizens a substitute for the same, which, when combined with the cultivation of military science, and the general diffusion of practical military information amongst the militia—agreeably to the plan already stated, I feel confident would afford a more perfect defence, and at the same time be far less expensive than the system now adopted. This plan is as follows (viz)

1st At the most important and exposed points on our sea-board let one or two principal works of the most permanent kind be constructed. These works to be kept in perfect repair, to be plentifully supplied with all the munitions of war, and the guns and carriages well secured from the weather by means of pent houses.

2nd In the vicinity of all the most exposed and vulnerable points on the sea-board let spacious and permanent arsenals be constructed, in which let there be deposited ample supplies of cannon, mortars, gun-carriages, materials for Platforms and other munitions of war, where they would perfectly secure from the weather.

3d In case of war or threatened invasions let temporary works either of earth or wood, be constructed at all the most vulnerable points, which could be readily furnished with cannon, gun-carriages, platforms, and all the necessary implements and munitions from the arsenals in their vicinity

4th As soon as peace is restored these works should be dismantled, and all their apparatus returned to the arsenals whence it was taken. In case of future emergencies they could be restored or others of the same description constructed in their places, which could be supplied from the arsenals in the manner above stated. The efficacy in marine defense, of works of the above description I presume will not be doubted by any scientific military man. Should any one however be disposed to doubt it, I would beg leave to refer him to the defence made by Fort Moultrie, in the harbor of Charleston S. Carolina, when attacked by the British shipping during the Revolutionary War, and also to the defense made by the small fort at Stonington when attacked in a similar manner during the last war. By adopting this system I think the following advantages would result—

1st A more secure defence would be obtained—By knowing the description of force we had to encounter, we should be enabled to construct our temporary works in a manner the best calculated to repel it, and as the gun-carriages, platforms, and implements, when taken from the arsenals, would be sound and in perfect order, we might reasonably calculate they these works would make a more vigorous resistance, than permanent ones which with their apparatus are in a state of partial dilapidation and decay

2nd This System would be much less expensive than the one by permanent fortification. Those temporary works could ordinarily be constructed by the troops with very little if any additional expence; but in cases of pressing emergency the zeal and patriotism of the people might be relied upon with safety, to supply any amount of labor that might be necessary—as was the case at New York in 1814. As it is not proposed they should be retained as military Stations in time of peace, the expence of keeping them in repair would be nothing. As it respects a navy command as constituting a branch of national defense, I have time now only to observe, that I view it as an important appendage to the defence of every commercial nation. The foregoing is a summary, hastily drawn up, of my views on the system of military defense adapted to the United States. It appears to me to be facile in practice, and at the same time well adapted to answer all the purposes contemplated. I am likewise convinced it would be found by experience far more economical than the present system. In presenting this project to the considerations of my fellow citizens, it is by no means my intention to obtrude it upon them, nor to dictate the course they ought to pursue respecting it. The subject is undoubtedly one of the most important that can occupy the attention of the American people and of their Representatives; and while I consider it not only the right but the duty of every American citizen who has bestowed a share of reflection upon this or any other subject of national importance, to lay the results before the public, it is likewise the right and becomes equally the duty both of the people and of the constituted authorities to weigh the subject deliberately, and finally [decide?] it upon the merits of the case alone. Before I conclude I would make one further observation,—and that is—that neither ships of war nor fortifications either temporary or permanent, ought be relied upon as constituting the principal defenses of our country. Bulwarks of earth or of masonry and wooden walls are of themselves very harmless and inoffensive, they become formidable only when bravely and skillfully defended. It is to the people of the country then that we are to look for its only efficient defence,—fortifications and ships are to be considered only as their auxiliaries. Let then military Science be duly cultivated, and military information generally diffused throughout the Republic, and it then may be pronounced safe both from external invasion and internal Insurrection—Sparta without walls was far more formidable than Sparta with walls.

Appendix E: Letter to the Public[650]

To The Public

Having recently [issued?] a prospectus containing a plan of the System of Education which I propose [implementing?] in the Literary, Scientific and Military Academy I am at present engaged in establishing, I deem it a duty which I owe to the public, to explain more fully my [views?] in the establishment of this seminary and also of the principles on which it is to be conducted. [Than?] could well be done in a new [prospective?] notice. In organizing the plan of education for this institution I have taken for my guide, in fact, the Constitution of the United States. By the wise provisions of this instrument and of the laws made in [purpose?] thereof, the grand military defense of our favored country, both against external invasion and internal insurrection, is vested in the great [mass?] of American citizens from eighteen to forty five years of age. These constitute the grand military force of the [Nation?]. A force whose feelings and interests are [illegible] with those of the great body of the People, and which, while it forms an impregnable barrier around the Constitution and liberties of the country, is in no respect dangerous to either. But in order that this [constitutional?] force should [illegible] the purpose [for?] which it was [originally intended?] it must be properly organized, and duly [instructed?] in the elements, at least, of military science and tactics. Hence there arises the [necessity?], in our country, of an [esteemed?] system of military education, and of a general diffusion of military knowledge, (If the great body of American citizens do not feel that they are something [more?]) If [these?] [illegible] [requisites?] be not [attended?] to than [newly universal?] soldiers, [over?] [illegible] will gradually degenerate our militia, So [emphatically styled?] the bulwark of our liberties and independence, will lose their military spirit – will decline and finally be destroyed on their ruins will leaving us a standing army detached by feeling and by interest from the great mass of the People, and when this crisis arrives it will not require the spirit of prophecy to [predict?] our fate, from that of the most celebrated republic of antiquity. The liberties of Rome were safe while military information was generally diffused,

---

[650] Partridge, Letter to the Public.

and every Roman citizen [consigned?] and felt himself a soldier. But how fatal were the results, when, by the operation of a system organized by Gaius Marius, the savior and scourge of Rome, and [matured?] by Julius Caesar, with a view doubtless to the accomplishment of his ulterior object – The final [castration?] of the liberties of his country. Those noble and patriotic legions which had so often, in times of evil and danger, proved the shield of their country and the terror of its enemies, were transformed into mere mercenary bands alienated from their country and [identified?] in [views?] and interests with their [illegible] alone. But Rome, [though?] the most striking is not the only instance in which similar causes have produced like effects. The Republic of Greece [furnish additional?] instances. Can any one believe that if the Greeks in the age of Demosthenes had possessed the same spirit and organization which they did in the age of Pericles and [illegible], or if the [soldiers?] who fought at [Chaeronea?] had been of the same [illegible] with those who fought at Marathon and Thermopylae, they would have so tamely submitted their necks to the yoke of the Macedonian conqueror. From an attentive [consideration?] of these as well as of many other similar instances which might be [added?], I am forced to the conclusion that in every republic the due cultivation of a proper military spirit amongst the great mass of the people, and a general diffusion of military information, are [indispensably?] necessary for the preservation of liberty; and that those republics which neglect these requisites, will eventually be driven to exchange their freedom for a form of government, [becoming?], at [best?], an military despotism. In making these observations however I beg not to be [understood?] as recommending a system of education for our youth [purely?] military. Being far from this: I mean nothing more than that the military should constitute an appendage to their civil education and thereby qualify them for the correct and efficient discharge of their duties as soldiers, when their country requires their services in that capacity. I have not attempted to prove the necessity of a [competent?] military defense to every [State?] which intends to maintain its independence free from the encroachments of surrounding nations. This necessity is too self-evident, I [presume?], to be doubted by any rational [person?]. Should any one however after a thorough and candid examination of the past history and present state of the [world?], be [disposed?] [illegible] to question it, I would waive [endeavoring?] to [convince?] him. Having [thus?] freely expressed my ideas on the importance, in a national point of view, of general diffusion of military information I will proceed to notice more particularly some of the advantages [arising?] from a due cultivation of military science – applicable more especially to young gentlemen destined for a liberal education. It is from this class of citizens that we are to look for a large portion of our statesmen, legislators, historians and travellers. And I would ask, will not the statesman be much better qualified to estimate correctly the military strength and resources of his country, the [legislator?] frame laws on military affairs with a more [perfect adaptation?] to the object in view, the historian to compile with greater judgement and more general utility his narrations of battles and sieges, and the traveller to estimate with greater precision and correctness the real strength and military resources of other nations and therefore to return to his own country with a greater [stock?] of useful information, when [illegible] by a scientific military education, than could possibly be done without it. On this subject I believe there can be but one opinion. A systematic knowledge of fortification and of military tactics is also [indispensably?] necessary for a full and correct understanding of history, a large [portion?] of which is made up of descriptions of battles and sieges; and also, in many cases, of common news paper reading without such knowledge the traveller likewise would find himself much embarrassed were he to attempt a description of almost any of the principal cities on the continent of Europe, of which, generally speaking, the fortifications constitute the most important [appendages?]. A scientific military education I conceive then may [without?] hesitation be pronounced as conferring many and important [advantages?] abstracted from my [illegible] with the military profession without the least disadvantages to [counterbalance?] them. Its importance, when received in connection with the profession of arms, is too evident to need any illustration. The practical military exercises I should consider of sufficient importance to warrant their introduction into seminaries of learning, generally, were they of no other use than to give to the students a good [illegible], a manly and noble demeanor, and, what is of more importance, to [render?] them healthy and vigorous. It is a melancholy fact, that many of our most promising youths by the time they have completed their course of education, and are prepared to enter on the grand theatre of active and useful life, have so completely lost their health, that even if they survive for a time, they are nevertheless [rendered?] in a great measure useless to society. This I conceive is occasioned in a great degree by the want of regular and healthy exercise, and also from the habit that many students, particularly those who are most studious, acquire of leaning over their tables to study. By this means the habit of drawing forward the shoulders is gradually acquired, whereby the chest becomes contracted, the lungs compressed, their motions impeded, the stomach disordered, and disease naturally follows. To counteract this habit the elementary military drill is [particularly?] adapted; for amongst the first habits the well drilled soldier acquires, is that of conveying himself erect, and keeping his shoulders well back. Fencing is also admirably calculated to give a young man the free and perfect use of his limbs, and at the same time to render him healthy vigorous and active. At the present interesting period of internal Improvement, it is believed the department of [natural?] science, applicable to civil engineering, will be of essential importance [this?] is evidently a defect in the general system of education in this country, which our colleges are not [calculated?] to remedy. The course of instruction at those respectable seminaries is [adapted?] more particularly to preparing young Men for the learned professions, but is little calculated to qualify them for practical scientific operations. The department of civil engineering opens a [civil?] field on which enterprising young men, duly qualified, may display their talents with honor to themselves and the greatest advantage to their country. A field equally [civil?] is also opened for the useful display of practical Science, in the immense tracts of unsurveyed lands in our western country. To form accomplished civil engineers and scientific surveyors will be a leading object at this seminary. The military lectures (as stated in the prospectus) are intended for the accommodation of gentlemen who do not wish to go through with a systematic course of military study and instruction. By attending even a single course and making the necessary notes, a person (might acquire sufficient information to) qualify him for a correct and efficient discharge of the duties of the [illegible] field, and would moreover acquire a very competent knowledge of the principles both of permanent and field fortification, of the attack and defense of fortified places, and also of the technical terms used in the various departments of military science. I have introduced the Constitution of the United States into the course of studies, from a conviction that an early and thorough acquaintance with its provisions by the great mass of American citizens is absolutely necessary for the permanent preservation of their liberties. Hence, I would ask, are the People to know whether their rights are, or are not infringed, if they are ignorant of the provisions of that instrument which designates and guarantees those rights; [As?], under our happy form of government, every American citizen, when he arrives at mature age, must exercise in a greater or lesser degree the duties of a statesman, I have thought that a reading room, containing a suitable collection from the most respectable newspapers printed in different parts of the United States, by which the students might become acquainted with the current political and other [pressing?] events of the day in every section of the country, would be a useful and important acquisition to the seminary. I beg however to be understood, that while discussions [grounded?] on politics as a science and tending consequently to the improvement of the mind, will not be [discountenanced?], at proper times, no attempts will ever be made or permitted which may have the [illegible] tendency to bias any student either in favor of or against any political party or sect. I shall be satisfied, I trust, with endeavoring to [instill?] into their minds American principles, such as are contained in the farewell address of the Father of [this?] country, without attempting to make political partizans. The love of country and obedience to its laws will ever be earnestly incubated. A regular attendance [on?] divine [divine?] service will be required, and it will be endeavored [earnestly?] to impress on the minds of the students a due sense of the importance of the sublime truths of Christianity, as contained in the Holy Scriptures; but in this, as in politics, it will be considered enough to attempt in a proper manner, to form Christians without making sectarians. No means will therefore ever be used or permitted, which may have the least tendency to bias any one either in favor of or against any Religious Sect or denomination whatever. On this as well as on all other subjects, the students will be taught candidly to examine, to reason and to think for themselves. The organization of the institution will be strictly military, and the students will be required to attend to all their duties with the same promptness and precision that they would be required of them in active military service. The government will be in a measure parental; The students will all be considered as members of the same family, and I shall ever consider it an imperative duty to bestow upon them all the care and attention which their parents themselves under similar circumstances would bestow between myself and my pupils of all ages, the freest and most unrestrained (personal) intercourse will ever exist. To conduce to their improvement, and to render them contented and happy, will ever be my leading and ultimate object. The [necessary?] annual expenses of a student, whose parents may reside at such a distance as not to be enabled to furnish him with any necessaries from home, will, from the best estimate I can make, be about two hundred and fifty dollars. It was with a view, particularly, to the accommodation of such parents, especially in cases where their sons from youth and inexperience might not be qualified to provide for themselves, that I made the proposition in the prospectus to receive them for a specific per annum fee and furnish them myself. I also offer my services, freely, to [Superintend?] the expenses of all those whose parents or guardians may request it. The institution will ever be open for public inspection, and the parents and guardians of the students, as well as gentlemen of [literature?] and science, generally, are particularly invited to visit it. Having [thus?] freely and fully [endeavored?] to explain to the public the plans of this seminary, and the principles on which it will be conducted, I would beg leave before I [conclude?] to repeat the [assurance?], that no exertions will be wanting to render it an ornament to the country, and in every respect worthy the patronage of the American People.

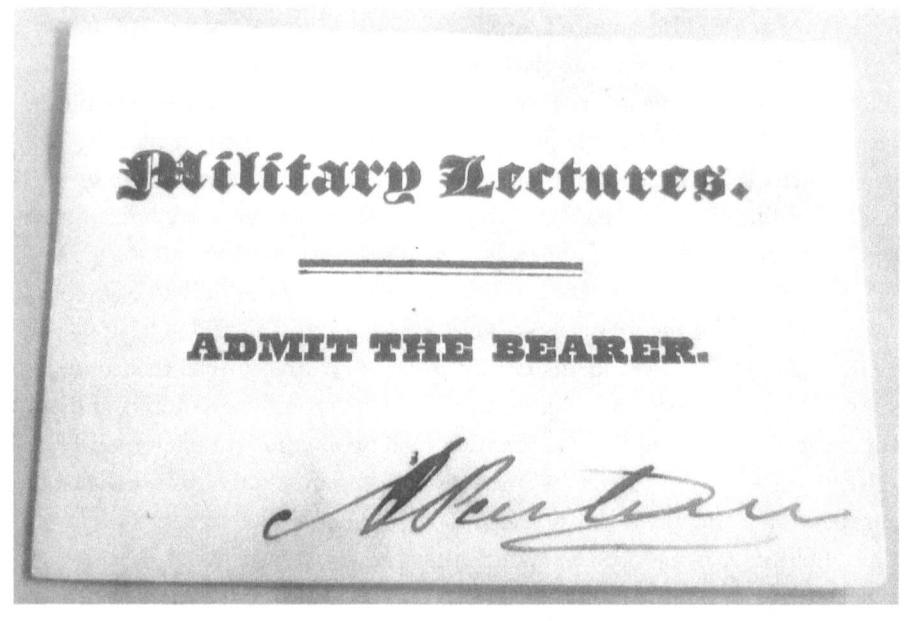

Figure 16—Military Lecture Ticket[651]

[651] An example of the tickets for Captain Alden Partridge's public lectures held within the Rauner Special Collections Library at Dartmouth College. A ticket would "admit a gentleman and lady."

Appendix F: System of National Education[652]

Memorial of Alden Partridge, praying Congress to adopt measures with a view to the establishment of a general system of education for the benefit of the youth of this nation.

January 21, 1841. Read, and laid upon the table.

To the honorable Congress of the United States:

The memorial of Alden Partridge, of Norwich, State of Vermont, Respectfully showeth:

That in all ages of the world education has constituted a great and leading object of attention by all those nations which have been most renowned in the arts, sciences, and arms. How admirably was the system of education adopted by the ancient Persians calculated to develop the physical and moral energies—yea, the whole character of Cyrus, the founder of the second great empire that arose in the world. Amongst the Greeks, their system of education was equally well adapted to develop and improve all the energies of their youth, and to prepare them for the correct and efficient discharge of all the important duties, whether of a civil or military character, devolving upon them as citizens of a free and enlightened republic. While all the faculties of the mind were fully developed and cultivated, under the instruction of a Pythagoras, a Socrates, a Plato, and an Aristotle, the martial and athletic exercises of the gymnasium produced a corresponding effect upon their physical energies. Thus was formed a race of men who have never ceased to command the admiration and esteem of the civilized world. Among the Romans; similar causes produced similar consequences. Their excellent system of republican education developed all the energies of this wonderful people, and prepared them ultimately to sway the sceptre over a large portion of the civilized and barbarian world. It is also well worthy of observation, how admirably the system of education adopted by all these nations was calculated to sustain their civil institutions. Indeed, it constituted the very basis on which those institutions rested. Persia was formidable only so long as she regarded the wise institutions of her ancestors; and Greece and Rome were free and

---

[652] Partridge, Memorial of Alden Partridge, praying Congress to adopt measures with a view to the establishment of a general system of education, 1–8.

independent only so long as the stern principles of their republican system of education were observed.

The systems of education adopted by the nations of modern Europe are well calculated to sustain their civil institutions, whether those institutions be of an aristocratical, monarchical, or more liberal character. The antiquated universities of Oxford and Cambridge, in England, are, in principle, now, precisely what they were one thousand years ago; and only modified, in some degree, in practice, by the stern mandate of popular opinion. The system of education at these institutions is, as it ever has been, purely monastic, and, in every respect, admirably calculated to sustain the union of church and state, as well as all the other aristocratical institutions of the country. Indeed, your memorialist is well convinced that, if those institutions at which the aristocracy of the land are generally educated were abolished, the mighty fabric of spiritual and temporal aristocracy and tyranny in England, would, ere long, be entirely prostrated, and the people restored to the enjoyment of their rights and liberties.

Your memorialist will now proceed to the system of education generally adopted at the colleges and universities in the United States— premising that, by education, he understands the preparing of a young man, in the best possible manner, for the correct and efficient discharge of the duties of any situation in life in which interest or inclination may place him; and that, consequently, that system is best which most effectually accomplishes this object. The question then arises, does the system of education generally adopted at our colleges and universities confer on the youth who resort to them the qualifications above mentioned? Does it qualify them to become efficient and active cultivators of the soil?—that most useful and honorable occupation, by which all others are supported. Does it qualify them for the active duties of the mechanic's shop, or of the counting-house, or other of the more active duties of life? Does it qualify them to stand forth in the hour of danger as the vindicators of their country's honor, or the redressors of her wrongs, at the head of her armies, or in the ranks? Does it confer upon them that expansion of mind, and implant in their bosoms those principles of patriotism, which enable them fully to understand and duly to appreciate the excellencies and importance of our republican, institutions? In fine, is it not better calculated rather to imbue their minds with a knowledge of the aristocratic and monarchical institutions of Europe, and implant in them feelings favorable to those institutions, and thus to send them forth into the world with their moral sense perverted, and their physical energies paralyzed, for want of a judicious system of physical education? If this subject be examined candidly and dispassionately, your memorialist is well convinced that the conclusion must be, that our system of education is not calculated to answer the important purposes contemplated. When our revolutionary fathers severed the bonds which had so long bound them to the mother country, they established our civil institutions upon a broad and liberal basis, but, unfortunately, made no definite provision upon the all-important subject of education. The consequence has been, that the system of education which had been introduced into our colleges and universities, while we were the subjects of a foreign and arbitrary Government, and which, in principle, is the same as prevails in the antiquated universities of Oxford and Cambridge, (the main pillars which support the union of church and state, and the other aristocratical establishments of England,) was retained in our free republic. Your memorialist will grant that those institutions have been somewhat modified in practice, in obedience to the imperative requisitions of public opinion; but again asserts, that they differ not essentially in principle from the antiquated universities of the mother country. And can a system of education, which is so well calculated to sustain the aristocratic institutions of a regal Government, be also congenial to the civil institutions of a republic? Can any one be so blind as not to discover in the prevailing excitements, got up under the auspices of sectarian leaders, persevering efforts for, and a direct tendency to, the union of church and state—the former, however, to exercise the supreme control? Did not the ecclesiastical, boldly attempt to raise itself above the civil power, when the laws of a sovereign State (Georgia) were openly set at defiance, by professed missionaries and followers of the lowly Jesus, whose every-day example was one of implicit obedience to the constituted authorities? Did not professed religious leaders, a few years ago, use every exertion, by means of Sunday-mail memorials, to induce, and even force, Congress to violate the constitution of the United States, by legislating indirectly on the subject of religion? Has not the sacred name of religion been dragged into the arena to aid in fomenting that unhappy and unhallowed excitement, which, under the name of abolitionism, has been, and is now, agitating the country from one extreme to the other, and which, unless hushed by the indignant thunders of the popular voice, will eventually rend asunder our happy Union? Your memorialist believes that all these facts are too notorious to be denied. And are not many of our colleges and universities the very hot-beds in which are brought into active operation those elements of discord, so hostile to the harmony as well as liberties of the people? Is it not notorious, that associations and combinations have been there formed for the purpose of agitating and producing agitation on all those subjects? How much pure religion, so necessary to the well-being of society, has retrograded by these efforts to pervert it to purposes so foreign to its nature, is a subject well worthy the attentive consideration of a Christian and enlightened people.

Should further proof be required of the hostility of these seminaries to the principles of our republican institutions, and also the rights and liberties of the people, your memorialist would beg leave to ask the attention of your honorable body to one or two facts connected with some of the most interesting events in the political history of the United States. The first event to which your memorialist would allude, is the first election of Mr. Jefferson to the presidency. In introducing the name of Mr. Jefferson, your memorialist trusts that no exception will be taken by any party, or the leaders of any party: for all now claim to be his disciples; and each strenuously endeavors to prove that the mantle of this great apostle of liberty has specially descended on him. Now, is there any fact during the whole of our political history better established than that of the bitter hostility manifested by nearly, if not quite, all of our old colleges and universities, to Mr. Jefferson's election? Did not the pulpit and the press, in any degree under their influence, teem with the most bitter, and even envenomed, denunciations against him? Was he not denounced as an infidel, who, if elected, would aim to subvert religion, annihilate the Bible, destroy ministers and churches, and finally sell our liberties and independence to France? And why was this unparalleled persecution raised, and these violent denunciations fulminated, against the immortal author of the declaration of independence? But one answer can be given. He was the advocate and friend of civil and religious liberty, who did not believe in the infallibility of the dogmas of the church, or in the efficacy of legal enactments to sustain true religion. The same rancorous hostility was also manifested towards the successor of Mr. Jefferson in the presidential chair, (Mr. Madison,) and for the same reasons.

During the late war with Great Britain, what course of conduct, your memorialists would ask, did those institutions adopt? Without one exception, so far as the knowledge of your memorialist extends, their united influence was exerted in favor of the enemy, and against their own country. The war was denounced as unholy and unjust, and England was proclaimed to be the bulwark of our religion. During the darkest period of that conflict, when the country was bleeding at every pore, your memorialist witnessed a celebration of one of those treasonable associations, called a Washington Benevolent Society, and heard a reverend professor of one of those institutions, who was invited to officiate on the occasion, invoke the vengeance of Heaven on the rulers and arms of his own country, and its blessings on those of her enemies. From a due and candid consideration of the whole matter, your memorialist is impelled to the conclusion that the system of education generally adopted in the United

States, so far from sustaining the republican institutions of the country, is in direct hostility to the same; and that, consequently, one or the other must ultimately be subverted. If education prevails, then will our free institutions be gradally, but certainly, superseded by the aristocratic ones of England. If, on the other hand, the intelligence of the people, combined with the love of liberty, should prove too strong for the influence of education, the latter must and will be superseded by a system which shall be in accordance with the principles of our civil institutions.

Your memorialist will next proceed to propose a plan, which, if carried into practical effect, would establish a national system of education in the United States, which would be in perfect accordance with the principles of our republican institutions, and which would supersede the present anti republican and monastic system. It is as follows: Let Congress pass a general law, appropriating forty millions of dollars, to be paid by annual instalments, out of the proceeds of the sales of the public lands, for the purposes of education: this money to be distributed among the States, in proportion to their representation on the floor of Congress, in such manner that the smallest States shall have at least one institution, and the largest five. The terms on which the States shall be entitled to receive money to be as follows, viz: That the Legislature of each State shall establish (either by establishing new or remodelling old institutions) such number of seminaries as it shall be entitled to, on the following plan, and embracing the following course of instruction, viz:

1. Every thing of a sectarian character, in religion as well as politics, to be utterly and entirely excluded therefrom.

2. An extensive course of mathematics, theoretical and practical, with their application to civil and military engineering, and geometric operations generally, to the various departments of physical philosophy, and astronomy, navigation, &c.

3. A complete course of physical philosophy, embracing mechanics, hydrostatics and hydraulics, pneumatics, optics, chemistry, magnetism, electricity, natural history, &c.

4. Political economy, embracing the three great departments of national industry—agriculture, manufactures, and commerce; with an examination into their mutual relations, and combined effects upon the welfare and prosperity of a nation. The subjects of currency, monopolies, and labor, would necessarily be involved in the preceding.

5. The science of government generally, in its various forms, with an examination of the objects for which it is professedly instituted; and the consequences which have resulted from it, under its several forms, in different

ages and in different countries. To the foregoing should be added a thorough acquaintance with the constitution of the United States, and the principles of our republican institutions generally, embracing civil administration, &c.

6. An extensive course of history, ancient and modern, of geography, &c.

7. Moral science, mental philosophy, logic, natural and political law, laws of nations, elocution, and an extensive course of ancient and modern literature; the study of the ancient and modern languages, (not excepting the English language, to which all the students should be required to give a full share of attention) to be left optional with the students, or their parents or guardians.

8. Civil engineering, embracing the construction of roads and canals, aqueducts, viaducts, locks, bridges, topographical drawing, &c.

9. Military science and instruction, embracing permanent and field fortification, artillery, the attack and defence of fortified places, the construction of batteries, sea coast and harbor defence, garrison and camp duty; tactics, including the grand and minor tactics, the schools of the soldier, platoon, company, battalion, and the evolutions of the line, military drawing, &c.

10. Architecture.

11. Each student should be allowed to progress as rapidly as possible in his studies, consistently with the thorough understanding of the same, and not be retarded, to be kept in college with such, as might have less capacity, or be less studious than himself.

Your memorialist is well convinced that, under such a system, at the least twice as much useful knowledge would be acquired by any given number of students, in the same time, as is now acquired under the present restrictive system.

12. A course of physical education, which would preserve, the health of the students, render them vigorous and active, and prevent injury to their constitutions, however intense might be their application to study. Regular military exercises, including fencing, &c., would constitute the best system of physical education. These could be attended to at such hours of the day as would otherwise be spent in idleness or useless amusements, for which they would be a pleasing and useful substitute.

Such is the course your memorialist would propose should be adopted for the education of our youth, in all the higher seminaries of the United States. The proposed plan would give about eighty seminaries in the United States; and forty millions of dollars would furnish a fund of five hundred thousand dollars for each seminary. A portion of this (say two hundred thousand dollars) should be appropriated to the erection of buildings, furnishing a library, apparatus, &c.; and the interest of the balance should be applied to the payment of the professors, and to defraying the other current expenses of the institution. By this means, the charges for tuition and other collegiate expenses would be so reduced, that young men of very limited means could enjoy the benefits of the system. But, for the purpose of extending this plan, so that the great body of the people might receive its benefits, it would be necessary to make provision for the establishment of secondary institutions, much upon the plan of the "polytechnic schools" of France. The course of instruction at these institutions, although less extensive, should be conducted on the same principles as in those of the first class. Great care should be taken at these institutions to instruct the pupils in the science of government generally, and particularly to make them well acquainted with the principles of our own free institutions, which they should be taught to love, to respect, and to sustain. The whole course of instruction should be such as to form practically useful men, and calculated to develop all those manly, noble, and independent sentiments which ought to characterize every American citizen. Military science and instruction should be common to all. This is the aegis that protects a free people from foreign as well as domestic tyranny. In order to insure the advantages of the secondary class of instruction to the great body of American youth, who may wish to acquire an education of a higher order than can be obtained at the common schools, your memorialist would propose that there should be twice as many allowed to each State as of the first class; and that an additional fund of twenty millions of dollars should be appropriated, from the proceeds of the sales of the public lands, for their establishment and support. The whole number of these institutions would be about one hundred and sixty; and twenty millions of dollars would furnish a fund of one hundred and twenty-five thousand dollars for the establishment and support of each. A due proportion of this fund should be applied to the construction of the necessary buildings, and the procuring of a library, apparatus, &c.; and the balance to constitute a permanent fund, the interest of which should be applied to the support of instruction, &c.

Having thus given his views on the plan of a national system of education, which he believes to be in perfect accordance with the principles of our republican institutions, your memorialist will proceed to the inquiry whether Congress possesses the constitutional power to do all that is required in this plan. As a preliminary, however, he would observe that he professes to be a strict constructionist of the constitution; or, in other words, a State rights man, agreeably to the old Jeffersonian standard, and should decidedly oppose the exercise of any power by Congress which is not clearly warranted by the provisions of that great charter of our liberties. In the preamble to the constitution of the United States, the great objects for which it was formed are clearly specified; and, among these, the following stand conspicuous, viz:

1. "To provide for the common defence." And your memorialist would ask—can any plan be devised that would more efficiently accomplish this great object than the proposed system of education? By adopting it, military science would be disseminated among the people, throughout every part of our extended republic; and the people themselves would be the depositaries of this knowledge, (so necessary to the protection of their liberties.) as the patriotic framers of the constitution evidently intended they should be. The consequence would be, that we should soon have a well-organized and well disciplined militia, which the founders of our Government have declared to be necessary for the security of a free State. Your memorialist would here repeat what he has often asserted: that we shall never have a well disciplined and efficient militia, unless it be based on a system of correct military instruction. With a well-disciplined force of two millions of citizen soldiers, we might laugh at foreign invasion, and set a world at defiance. Such a force would, constitute the best guaranty of peace, arid render a mercenary army totally unnecessary.

2. "To promote the general welfare." This clause, your memorialist is sensible, is liable to abuse; but still he believes it has a clear and definite meaning, viz: that Congress can constitutionally exercise any power that would not infringe on the rights and immunities reserved to the States and the people, or otherwise contravene any of the provisions of the constitution. Now, your memorialist is not aware that the plan proposed would have any such tendency. Congress would not assume the right to go into the territory of the States and establish seminaries of learning, in defiance of the local authority of said States. The plan merely proposes that Congress offer to the States a portion of the public funds, (the common property of all,) to be appropriated under State authority, for the accomplishment of a great State and national object, agreeably to certain conditions prescribed by Congress. Here is certainly no encroachment on the rights of the States or of the people; nor can your memorialist discover any contravention of either the letter or spirit of the Constitution.

3. "To secure to ourselves and our posterity the blessings of liberty." This is the consummation of the whole matter. To secure liberty to themselves and their posterity was evidently the grand design of the patriotic framers of the Constitution, and of the people in adopting it. Can any system of legislation, them, by Congress, which has a direct tendency to preserve liberty, be unconstitutional? Your memorialist thinks not. Would the proposed system

of education have that tendency? In order to answer this question correctly, it will be necessary to ascertain what constitutes the safest basis on which the liberties of a free people can rest. This your memorialist believes to be the intelligence of the great body of the people, which enables them to know their rights, and their ability to defend those rights, both against external and internal invasions. And would not the proposed system of education make the people well acquainted with their rights? And would it not also, by making them citizen soldiers, enable them effectually to defend those rights? Your memorialist believes there can be no doubt on this subject; and that, consequently, it would have a direct tendency to secure liberty, agreeable to the true construction of the Constitution.

But, should any one still doubt on this subject, your memorialist would observe that Congress has already adopted the principles of the proposed plan, in practice, by repeated acts of legislation, appropriating portions of the public domain for the purposes of education, and even by appropriating millions of dollars of the public revenue, to be drawn directly from the Treasury of the United States, to support the Military Academy at West Point, under the plea that it was necessary to provide for the common defence. Now, all the appropriations hitherto made for purposes of education have been partial, benefiting some States at the expense of the others, and have also aided to sustain the present anti-republican system; while the plan proposed by your memorialist would do equal justice to all the States, and insure the establishment of a system of education in the United States, that would prove the strongest support and safest protection of the liberties of the people. Your memorialist has thus frankly submitted to your honorable body his views on this all-important subject of education; and he is satisfied that none of deeper import to the permanency of our republican institutions can engage the attention of the representatives of a great and enlightened people. Education is more powerful than the lever of Archimedes, either to sustain or to crush the political institutions of every country—just in proportion as it is, in principle and in practice, in accordance with, or adverse to, those institutions. Can the proceeds of the sales of the public lands—the common property of the people of the United States—then, be applied to any object better calculated to provide for the common defence; to promote the general welfare; and, finally, to secure to ourselves, and to our posterity, the blessings of liberty, than to appropriate them in the manner your memorialist has proposed—to the establishment and support of a system of education, truly national in its character, and in strict accordance with our civil and political institutions? Your memorialist believes not; and would, consequently, earnestly but respectfully urge upon your honorable body the passage of a law embracing the provisions recommended in the foregoing memorial.

All of which is respectfully submitted:

A. Partridge

January 13, 1841.

Appendix G: *The Military Academy at West Point, Unmasked*[653]

The Military Academy at West Point, Unmasked: or, Corruption and Military
Despotism Exposed.
By Americanus
Sold at the Bookstore of J. Elliot, Penn Avenue.
1830

TO THE
MEMBERS OF CONGRESS:
GENTLEMEN, I take the liberty of addressing; you, as the constitutional
guardians of the lights and liberties of the people, on a subject which appears
to be fraught with the most important consequences to the future destinies of
our (at present) free and happy republic. The military academy at West Point
has been so long the idol and pet of a certain class of individuals in the United
States, that even to doubt its utility will probably be considered by them as
downright heresy, and to question the correctness of the principles on which it
is conducted and the purity of those who are concerned in its management as
arrogant and presumptuous. But believing, as I conscientiously do, that there is
not on the whole globe an establishment more monarchial, corrupt, and
corrupting than this, the very organization of which is a palpable violation of
the constitution and laws of the country, and its direct tendency to introduce
and build up a privileged order of the very worst class—a military aristocracy—
in the United States, I should feel that I was wanting in my duty to my country
were I to remain longer silent. The military academy was established by a law
passed March 16, 1802. By the provisions of this law the President of the
United States was authorised to establish a corps of Engineers, which corps
was to consist of 16 officers and 4 cadets. This corps was to be stationed at
West Point in the state of New York—was to constitute a military academy,
and its members were subject, at all times, to be ordered on such duties and at
such places as the President of the United States should direct.—Thus it
appears that the military academy at its commencement was simply the corps

---

[653] Partridge, *Military Academy at West Point Unmasked.*

of Engineers, which corps is limited to 20 members, that this corps of Engineers would constitute a very useful appendage to the military establishment was very evident. They were intended to plan and superintend the construction of the necessary fortifications, arsenals, magazines, &c. and when not engaged in any professional duties they were to remain at West Point (their legal head quarters,) and there improve themselves under the direction of the senior officer present, by study and practical exercises. On the 29th April, 1812, a law was passed, entitled an act, making further provision for the corps of Engineers. By the provisions of this act, an addition of several officers was made to the corps of engineers, and of professors and instructors to the military academy. The President of the United States, was also authorized to appoint 250 cadets, who might be attached, at the *discretion* of the President, as students to the military academy, and who were to receive sixteen dollars per month, and two rations per day. Having made these preliminary remarks, I will next state as clearly and precisely as possible, the objections to the institution, under its present organization, and the consequences that must inevitably result therefrom.

1st. The organization of the institution, and the regulations of the War Department, growing out of them, are a palpable violation of the principles of the constitution of the United States. By the provisions of the law of the twenty-ninth of April, 1812, no youth can be admitted into the military academy, who is under fourteen, or above 21 years of age; and by the established regulations, and the usage of the War Department, no person can be commissioned in the military establishment of the U. States, who has not been educated at West Point. Consequently, every American citizen who has attained the age of manhood, is absolutely precluded from holding a commission in the military service of his country, unless he be one of the favored few who have been educated at this national *charity* school, at the public expence. But it may be said that none except those who are educated at this military academy, are qualified for commissioned officers in our military establishment. This however I deny, and assert there are hundreds of young men at this time, in the U. States, who have received a regular, practical and scientific military education at their *own* expence, and who are in every respect as well, yea, better qualified, to discharge the duties of any, or all of the corps of the army, either in peace or in war, than those educated at West Point; because, with equal, if not superior military acquirements, their literary and practical scientific education is far superior. But should one of these apply for a military commission, he would be told there is no room for him—that a chosen few have claims which he cannot urge; that is, they have been supported four years at the expence of the people, at West Point, which gives them a paramount claim to fill all the offices, and thereby thus supported during the rest of their lives. I would now ask, gentlemen, whether this be not a most flagrant violation of the very first principles of our republican constitution, viz: that *offices of honor, trust and emolument, shall be equally opened to all!*

I think there can be but one answer to this question. And, gentlemen, can you, consistent with your duties, as the guardians of the people's rights, allow this monstrous usurpation of power, by an executive department, longer to exist? Will you longer suffer the dearest rights of a vast majority of your fellow citizens, to be thus invaded; yea, trampled upon, and themselves as it were, thrown out of the pale of the constitution? I cannot believe you will.

2nd. The military academy under its present organization has introduced a *military* aristocracy into our republic. What is an aristocracy? I answer, an order or class of individuals in a community who claim and exercise privileges and immunities of which the great body of the people are deprived; and is not this emphatically the case, as it respects those who are admitted as Cadets into the military academy at West Point? Do they not by claiming and enjoying the privilege of filling all the vacancies for officers that occur in our military establishment, claim and enjoy privileges and immunities, of which the great body of the people are deprived? And if so, do they not constitute in the strict sense of the word, an aristocracy, and an aristocracy of the most dangerous kind, a military aristocracy? Can this be permitted longer to exist?

3d. Some of the regulations for the government of the military academy are in direct violation of the law. In the 3d section of the law of 29th April, 1812, it is declared, that the cadets shall be encamped *at least* three months of each year, and taught all the duties of a regular camp. By the 46th article of the regulations for the government of the academy, it is declared, there will be an encampment of the cadets annually, to commence on the first day of July, and end on the 31st of August next, ensuing. The law then, imperatively declares, that there shall be an encampment for three months, and the regulation as imperatively declares it shall be but two months. Is not this a palpable violation of law?

4th. In the 3d section of the law above quoted, it is declared, that the cadets attached to the military academy, shall be arranged into companies of non-commissioned officers and privates, according to the direction of the commandant of Engineers, and be *officered from the same corps* (corps of Engineers) for the purpose of military instruction. From the army register, it however appears, that the instructor of tactics and commandant of the corps of cadets, is an officer of infantry, and that all the other Instructors in the military Department, are either officers of Artillery or Infantry, and that there is not a single officer of Engineers attached to the corps, as a military Instructor. Is not this a palpable violation of the law? And are not the cadets by this arrangement subjected to receive orders from officers who have no legal right to command them, and who, consequently, exercise none but usurped authority over them? There can be but one answer to this question. But I am aware it will be urged by way of palliation, that officers of Engineers cannot be spared from their ordinary duties. Why, then, I would ask, does not the Superintendent do the duties of a military Instructor? His predecessor discharged all those duties personally, besides discharging all the other duties of his station, including those of a regular Professor, and required the aid of no officers of Artillery or Infantry to exercise illegal authority over his pupils. Why, I ask again, does not the present Superintendent do the same? There is a very ready answer to this inquiry—indolence and incompetency! Yes, gentlemen, much as it may surprise you, this Superintendent who has been so much *puffed* by reports of boards of visitors, &c. is totally incompetent to discharge the duties of a practical military instructor to the cadets, and thus Artillery and Infantry officers become necessary to supply his deficiencies. He is incompetent to teach *practically* the elementary duties of the soldier, has never commanded a parade since he has had charge of the academy, and cannot turn off a Corporal's guard correctly. Such are the *practical* military acquirements of the present Superintendent of the United States military academy. But more of this gentleman hereafter.

5th. By the provisions of the law of April 29, 1812, eight instructors are allowed to the military academy, viz: one professor of natural and experimental philosophy, with one assistant; one professor of mathematics, with one assistant; one professor of engineering, with one assistant; one teacher of the French language, and one teacher of drawing. By reference to the army register, however, it appears that instead of 8 instructors (the whole number allowed by law) there are actually employed at the academy about thirty, that is, upwards of twenty more than the law allows. The whole amount of the pay and emoluments of these additional instructors cannot be less than $20,000 per annum.[654] Now, I would inquire, by what means the money to pay these additional instructors, finds its way out of the Treasury? The constitution of the U. States declares that no money shall be drawn from the treasury except in consequence of appropriations made by law; but there could have been no appropriations by law to pay these Instructors, because they themselves are not recognized as Instructors by any law whatever. Still they receive their pay, and the money to pay them is taken from the public Treasury. Is not this a violation of both the constitution and the law? I would also inquire how it is possible that 30 instructors can be diligently and usefully employed in teaching 250

---

[654] Approximately $674,622 adjusted for inflation to 2024 dollars.

youths? Such was not the case previous to the year 1818, when the present *economical* and *improved* system, as it is called in reports of boards of visitors, &c. commenced. Previous to that time, all the duties of instruction were discharged by the instructors allowed by law, with the assistance of the Superintendent, and of four or five of the more advanced cadets, who received no extra compensation for such instruction; when the average number of cadets, for the preceding three years was very nearly, if not quite, as great as the average number for any equal term of time since.

6th. By a regulation of the academy, several of the cadets (I believe six) are selected to discharge the duties of assistant Instructors, and are allowed $10 per month, in addition to their pay and emoluments as cadets.

When this regulation was made, there was no law authorizing such extra allowance; nor, to my knowledge, has any been passed since. I would ask then, is not this also a direct violation of the constitution and law?

7th. During the session of the last congress, an appropriation was asked for by the military committee, of $6,000[655] to erect a proper building for the cadets to exercise under, during cold weather; the appropriation, however, was rejected; but notwithstanding the rejection of the appropriation, the building has been erected, and it is believed was erected at the time the appropriation was asked for. Query—Out of what funds were the expenses paid?

8th. A splendid public house, or tavern, has recently been erected at West Point, on the site of the old barracks which were burnt down. This tavern is kept by a private citizen, who is said to pay rent for the same. Query—Has congress made any appropriation from the public treasury for defraying the expence of putting up this tavern, or public house; and if not, from what funds were the expenses defrayed, and who receives the rent for the same?

9th. In the code of academic regulations we find the following article, viz: "All publications relative to the military academy are strictly prohibited. Any professor, assistant professor, teacher, academic officer, or cadet, therefore, who shall be at all concerned in writing or publishing any article of such character, in any newspaper or pamphlet, or in writing or publishing any hand-bills, shall be dismissed the service, or otherwise severely punished"

The foregoing, gentlemen, is a literal copy of one of the articles of the regulations established for the government of the United States military academy. What a commentary on our boasted freedom of speech? Why declaim against the tyranny of the alien and sedition laws, or against the censorship of the press in despotic governments, while such an article as the

---

[655] Approximately $20,239 adjusted for inflation to 2024 dollars.

foregoing, ten-fold more tyrannical than all of these, is suffered to disgrace the pages of that code, established for the government of our national institution? Is this article constitutional and legal? Or, rather, is it not a libel both on the constitution and laws?

10th. Under the present *improved* system adopted at the military academy, the cadets have repeatedly been subjected to degradation of laboring at the wheel-barrow, as a punishment. Yes, gentlemen, the sons of free-born Americans, sent to the military academy to acquire an education to fit them for stations of honor and trust in the military service of their country, and where they ought to imbibe those refined and noble sentiments, so intimately connected with the character of a soldier, are there subjected to the same degrading punishment (save the ball and chain) that is awarded to private soldiers for the highest offences, as desertion, &c., or to convicts in Europe, who are sentences to work on public highways. And this, in our enlightened republic, in dignified with the name of *discipline*. Suppose this mode of punishment were adopted at our colleges and universities, would it be submitted to? No. Both parents and pupils would revolt at it. Why then is it submitted to, at the military academy? But one answer can be given—it is a *mercenary* disposition. Parents can there get their sons' education at the *public* expense, and to accomplish that object, will allow them to be subjected to degradation. How much superior the situation of that youth, who, while supporting himself by his own honest industry, feels that he is free and independent, to such a state of gaudy slavery.

11th. The 91st article of the rules and articles of war imperatively prohibits courts of enquiry, in all cases, unless ordered by the President of the United States, or demanded by the party accused. The present Superintendent of the military academy, has however, in direct violation of this article, repeatedly ordered courts of inquiry to investigate the conduct of cadets, without such cadets demanding them. These illegal tribunals have administered oaths, and cadets have been obliged to testify before them under oath, or be dismissed from the institution. What, I would ask, are we to expect from a corps of *martial* youths, who witness almost daily, the violations of the laws of their country, by those who are placed over them, and see the solemn forms of justice turned into mockery? I leave the answer to every candid and reflecting mind.

12th. During the year 1827, an individual, not subject to martial law, was seized by military force at West Point, and confined in the guard-house. It is believed that several other instances of a similar kind, have occurred there within the last ten years. For one of these high-handed violations of the laws of his country, the Superintendent of the academy was arrested and tried by the civil authority, and sentenced to pay a fine. And here, I would ask, why there should have been so much excitement and clamour about tyranny and military despotism, because one or two individuals were arrested and confined at New Orleans (as a precautionary measure) when the public safety was threatened by a powerful invading force, and an equal violation of the laws, (by military force) at West Point in the time of profound peace, when no necessity arising from public danger, could be urged in justification of such a measure, should have been passed over in silence, by the constituted authorities?

13th. The pay of the cadets has been repeatedly stopped by the order of the Superintendent of the academy, under the pretext of damages done to books, &c. This is a direct violation of the law, as the pay of no individual in military service can be *legally* stopped, except by sentence of a court martial. Query, what becomes of the money thus detained? I will next proceed, gentlemen, to submit some remarks relative to the expenses of the military academy. But the official report of the commandant of Engineers, submitted to Congress in 1817, it appears, that $115,354.15, had then been expended on the public buildings and their appurtenances. In this report, however, is not included the east barracks, which cost about $45,000, making therefore upwards of $160,000, up to the close of the year 1817.[656] If this be added to the amount expended since that time, on additional buildings and appurtenances, it is believed that the whole will not be less than three hundred thousand dollars. Now I would ask, could economy have been consulted, when $300,000 is expended in preparing an establishment for the accommodation of 250 young men? I have known buildings to have been erected, at the expense of private individuals and under private direction, which afforded ample accommodations for 250 pupils, with the greater part of their instructors, were as permanent and more elegant than those at West Point; and yet, with all their appurtenances— including the grounds attached to them—cost less than $40,000.

The foregoing will illustrate, in a small degree, the *rigid system of economy* pursued at West Point, and about which so many splendid reports have been made. The whole annual expense of this institution to the country, is probably not less than $200,000;[657] the whole annual expense of supporting a cadet there $500, (this it is believed, was the estimate of a military committee several years ago) and consequently every cadet who completes his term of 4 years costs his country each $2,000. But as only about one half of those who are appointed cadets, ever finish their education, and obtain commissions, each officer

---

[656] Approximately $5,396,974 adjusted for inflation to 2024 dollars.
[657] Approximately $6,746,217 adjusted for inflation to 2024 dollars.

obtained from this seminary really costs the country about $4,000.[658] In this economy? or is it not rather *"paying too dear for the whistle?"*[659] It is sometimes urged in favour of this institution, in order to make it palatable to the people, that it is of great use in improving the militia. This I deny. Those cadets who receive commissions in the army, are confined to the regular service, and consequently have no connection with the militia. Those who leave the institution without being commissioned, are but partially instructed themselves, and are generally those who are considered vicious or incompetent.

Now, as the militia claim and exercise the right of choosing their own officers, it is not probable they will select the *refuse* of the military academy for their commanders? It is a reflection on them to assert it. The ideas of discipline imbibed at that institution are also totally abhorrent to the feelings of the great body of the militia and volunteer corps, and would prove a powerful obstacle to their serving under officers educated there, even in time of war. The truth is, that comparatively, very few of those who have left West Point have ever held any commissions in the militia, and if they are never disciplined until it is done through the medium of this institution, it will never be effected.

I have thus, gentlemen, endeavoured in a frank and candid manner to pourtray to you some of the evils and abuses of the military academy; I now seriously ask you, whether you can reconcile it to your consciences, and to the duty you owe to the people, to be annually voting away their hard earnings to support so unconstitutional, so aristocratical, so corrupt, and at the same time, so totally useless an institution as the military academy at West Point, under its present organization.

There is not a single useful object can possibly be accomplished by it now, that cannot be better accomplished under a different organization, which would be entirely free from all the evils attending the present, including even the expense, which, great as it is, I consider among the least of the evils

---

[658] Approximately $134,924 adjusted for inflation to 2024 dollars. The cost to education a single academy graduate in 2024 has increase to ~$450,000.

[659] "Paying too dear for a whistle" is a phrase that originates from Benjamin Franklin. In his childhood, he carelessly purchased a whistle on a whim for several times its actual value. He had significant remorse when his siblings reminded him of all the nice things he could have enjoyed if he had paid an appropriate price for the whistle. This phrase became a mantra for Franklin to avoid unnecessary spending and to ensure appropriate evaluation of value when purchasing. The phrase became well-known after the publication of Franklin's autobiography. Benjamin Franklin, *The Compleated Autobiography by Benjamin Franklin*, ed. Mark Skousen (Washington, D.C.: Regnery Publishing, 2005).

attendant upon it. I will therefore, take the liberty to submit to your consideration, the following plan for its re-organization, viz:

1st. Let the corps of Engineers remain, as fixed by the laws of the 16th of March, 1802, and the 29th of April, 1812.

2nd. Let the whole system of cadets be entirely abolished.

3rd. Let the vacancies for offices, which occur annually in every department of the military establishment, be filled by candidates taken from the several states in proportion to their representation in Congress—reserving a reasonable proportion of such vacancies (say 1-6) to be filled by meritorious non-commissioned officers.

4th. Let all the newly appointed officers be ordered immediately to repair to West Point, where they should be put under such a course of instruction as would prepare them to discharge efficiently, and correctly, the duties of their respective stations in service. This instruction should comprise a course of practical mathematics, of field engineering of infantry and artillery duty, and also the duties of light troops—the duties of camp and garrison, and a complete course of practical military tactics. Now the whole of this course could be completed by men of ordinary capacity and intelligence in six months, when they should be ordered to join their companies, and would be found much better prepared to discharge the duties of their respective stations, than the cadets at the military academy are in 4 years, under the stiff and pedantic system there adopted.

The vacancies occuring in the corps of engineers could be filled by transfer of the must scientific and meritorious officers in the army, or by such other candidates as from their talents, scientific acquirements and other qualifications, should be judged by the President of the United States, best qualified to discharge the duties of engineers.

The members of this corps are, by law, considered learners from the time they enter service, and when not on other duty, they should remain at West Point, and attend to study. &c. There need be very little, if any, additional expense attending this system of instruction, as all, or very nearly all, of the instructors could be furnished from the army.

5th. Let such numbers of officers, non-commissioned officers, and privates as can be spared from each regiment or corps of the army, embracing all the recruits (at least) from the northern department, be sent in rotation to West Point, and constitute a school of practical instruction, let the non-commissioned officers and privates be organized into a brigade, to be officered by the officers on detachment, and by those on new appointments, for the purpose of practical military instruction. Under the present organization of the army, one thousand non-commissioned officers and privates could be detached constantly, including the recruits, for the purpose of instruction, and still leave enough to do all the necessary garrison duty. One third of the officers might also be detached for this purpose, including those on new appointments, the whole would constitute a handsome brigade, and would enable the officers to practice, and become well acquainted with all the evolutions of the line. When one detachment shall have completed its course of instruction, let it be followed by another, and so on in rotation. By this means the whole military establishment would in a short time, be practically well versed in the higher branches of military tactics, and a uniformity of system practically prevail throughout the army.

6th. At the head of this school of practice, let there be placed an officer of experience, and of the first order of talents and of military and scientific acquirements; who, while he is able to explain principles, is also able to give practical instruction. By adopting the foregoing organization, about 200,000 dollars would be saved annually, the military establishment better officered, and all the objections, evils, and abuses, that have been pointed out in the foregoing, avoided.

To the President of the United States.

Sir,

Having in the preceding pages, endeavoured to exhibit to the Representatives of the people, the unconstitutionality, as well as illegality, of many of the provisions in the organization of the United States military academy, and also to pourtray to them, some of the most prominent abuses practised in the institution, I now take the liberty to address to you as chief magistrate of the Union, some further remarks on the same subject, and more especially, in reference to some individuals who have been for several years, and now are, employed about the seminary.

The first individual whom I shall notice, is Lieut. Col. Sylvanus Thayer, the superintendent of the academy. This individual in consequence of his situation, has acquired a fictitious importance and popularity, with a certain class of persons, to which it is believed, neither his merits nor his *real* consequence, aside from circumstances, give him any claims. He has been twice brevetted, this would of course, induce a belief that some very important public services have been performed, but what they are, I believe the public has never been informed. I know that repeated efforts have been made, to impress the public with a belief that the whole system of organization, discipline, and indeed every thing else connected with the military academy, owes its origin to him. I know also, that the commandant of engineers declared in an official report, made several years ago, in substance, that previous to the year 1818, which was shortly after he took charge of the institution, there was little or no organization or system in the seminary—that the cadets were admitted into it, and commissioned from it without any examination. &c. &c. The assertions contained in that report on this subject, I positively deny, and assert that they are untrue in point of fact. Previous to the re-organization of the academy, under the law of the 29th of April 1812, a code of general and internal regulations, and a course of studies suited to the state of the institution at that time, had been drawn up by the superintendent, the immediate predecessor of Lieut. Col. Thayer, which were sanctioned by the proper authority. Under this system, the cadets were regularly examined, previous to being recommended for commissions; and it was under this system that some of the most valuable and distinguished officers, that have ever gone from that institution (several of whom were particularly distinguished in the last war) were formed. During the war, the secretary of war would not allow the cadets to remain long enough at the academy to finish a full course of education, but ordered that whenever they had acquired a practical knowledge of military duty, &c. &c. the names of all such as were of proper age for commissions, should be immediately forwarded to him for that purpose, as their services were wanted in the field. Soon after the close of the war, the then superintendent, drew up a code of general regulations for the government of the institution, under the new organization, which was forwarded to the secretary of war, and after considerable delay, was finally sanctioned by him, in the month of March 1816.

In one of the articles of this code, it was declared that the superintendent in conjuncion with the academic staff, should compile a course of studies, embracing definitively, all the branches of instruction to be pursued at the academy, and which, when approved by the secretary of war, should be considered as comprising a complete course of education at the institution. The then superintendent drew up a course of studies under this regulation, which was agreed to by the academic staff, with but one trifling alteration: it was immediately forwarded to the secretary of war, who approved it (making but one slight alteration) on, or about the 4th of July 1816.

In this course, provision was made from the regular examination of cadets, for admission into the seminary, and also for commissions, and all candidates were regularly examined from that time. As it respects the military discipline of the cadets, it was, at least as perfect previous to the year 1818, as it has ever been since, and it should be obversed, that the greater part of the military instruction, which the cadets have received since the year 1818, was given by those who were *personally* instructed by the superintendent previous to that time.

The present military commandant of the corps, and instructor of infantry tactics, and also the instructor of Artillery, are of that class. However well then the corps may have been instructed in military duty, no credit whatever attaches to the present superintendent, who has never instructed them a moment, either in military duty or any thing else, since he has had charge of the Institution.

It was previous to the year 1818, that a series of experiments on the fire of infantry and artillery, in addition to the regular courses of target practice with cannon and mortars, was made under the personal direction of the then superintendent. Has any thing of the kind been done under the personal direction of the present superintendent? It was previous to the year 1818, that the series of barometrical and thermometrical observations, were carried on by the then superintendent accompanied by the cadets, which resulted in determining the altitudes of all the mountains and principle eminences in the vicinity of West Point; and also of the Catskill mountains, and the highlands of Neversink, above tide-water.

How many mountains and eminences have been measured by the present superintendent; and how much of that kind of practical instruction, so necessary for military men, has be given his pupils? It may be proper here to remark, that previous to year 1818 there were no *mutinies* among the cadets— these belong exclusively to the present *improved* system. It must also be further remarked, that under the present *improved* system, it requires the superintendent and seven others, whose pay and emoluments, cannot be less than $12,000 per annum, to discharge the duties that were formerly discharged by the superintendent alone, whose pay and emoluments never exceeded those of a major. This is a practical illustration of the. present system of *rigid* economy practised in the institution. It appears from the foregoing statements, and I challenge a contradiction of them by the commandant of Engineers who made the report above alluded to, that, the cadets were examined for admission into the military academy and also for commissions as officers, long before the beginning of the year 1818, that the institution was in a perfect state of organization, and the corps of cadets in a high state of military discipline, when the present superintendent took charge of it; and that he consequently had, literally, nothing to do, except to go on with what was already prepared to his hand. I now take the liberty, sir, to submit to your consideration, the following allegations, which I prefer against him, all of which I believe can be clearly proved, viz:—

1st. For the repeated violation of the 91st article of war, by ordering courts of enquiry, directly contrary both to the letter and spirit of the said article, and compelling cadets, under the penalty of dismission, to testify, under oath, before these illegal tribunals.

2d. For making a false report to the War Department, relative to certain occurrences which took place at West Point, on the 29th and 30th of August, 1817. The truth of this allegation will be sustained, by comparing the statements in this report, with his own testimony, and also the testimony of several other witnesses, given before the general court martial assembled at West Point, on, or about the 20th of October, 1817.

3d. For charging double rations, as commanding officer at West Point, and receiving pay for the same, for nearly three weeks, in the months of August and September, 1817; he being at the same time absent from the Post, (it is believed in New York) and another officer actually in command, who charged double rations, and received pay for the same, as commanding officer, during the same time.

4th. For tyranical, unjust, and even barbarous treatment, towards cadets Nathaniel H. Loring, Thomas Ragland, Wilson M. C. Fairfax, Charles R. Vining, and Charles R. Holmes, in the month of November, 1818.

5th. For violation of law, by arbitrarily stopping the pay of cadets in repeated instances.

6th. For violation of law, by confining, or causing to be confined, in the guard house at West Point, one or more persons not subject to martial law.

7th. For neglect of duty, in not attending, personally, to give military and other instruction to the cadets, his pupils, and also, in not attending, personally, to the inspection of their commons, quarters, &c.

8th. For inflicting degrading and unmilitary punishments on the cadets, by putting them to work at the wheel-barrow, &c.

9th. For authorizing or allowing corporeal punishment to be inflicted on soldiers at West Point, in direct violation of the law, which prohibits corporeal punishment in all cases; it is believed that one soldier received two or three hundred lashes, in as many successive days.

The foregoing allegations are preferred against Lieutenant Colonel Sylvanus Thayer, as superintendent of the military academy at West Point. For further particulars respecting him, I would beg leave to refer to the charges and specifications preferred against him, on or about the 5th of December, 1817, which, it is presumed, are on file in the War Department.

The next individual to whom I beg leave to call your attention, is Captain David B. Douglass, professor of Engineering, in the military academy. The conduct of this individual, was partially investigated by a court of enquiry, assembled at West Point, on, or about the 21st of October, 1817—when he was found guilty of disobedience of orders, and of other conduct, which amounted almost, if not quite, to exciting mutiny—for the whole of which, see the proceedings of the court above mentioned. But one of the principal charges against him, the court, for reasons best known to its members, decided they had no power to investigate. The substance of this charge was, that Captain Douglass, then Lieutenant Douglass, defrauded the members of the company of Bombardiers, Sappers, and Miners, of a part of their rations while they were on their march, under his command, from West Point to Buffalo, in 1814.

As this charge was not investigated by the court of Enquiry, I will take the liberty, summarily, to state the facts, as they are contained in an affidavit of several of the members of the company, including the first and second Sergeants.

They state, that on their way to Albany, Captain Douglass sold a barrel of pork belonging to the company for $10; that on the march from Albany to Buffalo, he sold all the small parts of their rations, as candles, vinegar, soap, &c. promising them that they should receive an equivalent in something else.

They also declare, that all they received for the parts of their rations thus sold, including the barrel of pork, was *one small glass of brandy* each. The whole number of non-commissioned Officers, artificers, musicians and privates in the company, was about 70. It must have been very *dear* brandy, if the cost of all they received, was more than one third of the price for which the pork alone was sold.

They also state, that Captain Douglass, while on the march, was guilty of beating and bruising in a most barbarous manner, although corporeal punishment was prohibited by law, one of the privates of the company, and then of tying him to a baggage waggon, and marching him in that situation thro' the village of Cannandaigua, and that the said private was so much injured by the treatment received, as to be unable to march for several days, but was carried in the baggage waggon.

They further state, that the *mighty* offence of which this soldier was guilty, was, *whistling* in a room through which Captain Douglass was passing. Is justice longer to slumber? The foregoing facts can be fully proved by living witnesses.

The third and last personage I shall exhibit, is Claudius Berard, Esq. teacher of the French language at the military academy. This individual's conduct was investigated by the court of Enquiry before mentioned, when he was found guilty of disobedience of orders, and of having been engaged in trafficking with the cadets, his pupils, without the knowledge of the Superintendent, by selling them very cheap watches—six or seven dollar watches—thereby getting away their money for what was of no use to them, (for the watches were generally good for nothing,) and also inducing them to violate the regulations of the Institution, which imperatively prohibited the cadets purchasing any articles without the written permission of the Superintendent. Yes, sir, the man filling the honourable station of an instructor in our boasted National Institution, which has cost the people, since the year 1818, nearly $3,000,000, condescended to trafick with his pupils, in *very cheap watches*; which, if any reliance can be placed on common report, were a gross imposition on their youth and inexperience.

I would now candidly ask you, sir, whether individuals guilty of the acts contained in the foregoing allegations, are the proper ones to hold important stations in the U. States' Military Academy? Is it from men, such as these, that a portion of the American youth, selected to fill stations of honor and trust in the military service of their country, are to imbibe those noble and refined sentiments, those principles of honor, and that respect for its laws, which ought to characterize every American officer? I leave the answer to every honorable and candid mind. I am well aware, however, that it will be said by some, these allegations are not true. I answer, test their truth by an impartial, but thorough investigation. Let such an investigation be had, by a competent and impartial tribunal, and I feel, I hazard little in asserting that more, rather than less, will be proved than I have alledged. I have made the foregoing statements, sir, from a sense of duty, and have perfect confidence that all abuses, so far as depends on the executive, have only to be pointed out to be corrected.

To the people of the United States.

FELLOW-CITIZENS:

I beg leave to call your particular and serious attention to the contents of the preceeding pages, in which you have a deep, a vital interest. You are called upon to pay annually, about $200,000, for supporting the military academy at West Point. And you certainly ought to know, whether the advantages resulting [f]rom this academy to the country, aside from the unconstitutionality and illegality of its organization, are such as to authorise this great expense.

You are told by the advocates of this institution, that it is absolutely necessary your military officers should be educated at your expense; otherwise they would be totally unable to discharge the duties of their stations. In answer to this, I would ask, how much did you pay for the education of your Washington, your Greene, your Putnam, your Warren, your Sullivan, your Knox, your Lincoln, your Wayne, your Starke, your Morgan, your Lee, your Moultrie, your Marion, your Sumter, your Pickens, your Shelby, your Laurens, and a host of other revolutionary worthies? How much did you pay for the education of your Jackson, your Brown, your Pike, your Gaines, your Ripley, your Coffee, your Carroll, your Harrison, and many others who distinguished themselves during the late war? to all those enquiries, you readily answer, nothing.

I ask, do you expect to obtain from the military academy, officers more brave, more patriotic, or more skillful than those who gained your independence, or fought the battles of the late war? If you do, I must say, you greatly deceive yourselves. The *effeminate* and *pedantic* system now practised at West Point, is, the last of all, calculated to form efficient officers for active service.

Youths there learn a number of abstract mathematical propositions, without ever reducing them to practice; this they may do at almost every seminary in the United States. They are also taught a few *stiff* and *pedantic* rules about the arrangement of troops in battle, the planning of campaigns, &c.

Now the whole effect of this system of reducing military operations under particular rules, learned in the closet, is to shackle the mind, and destroy the capacity of regulating all the particular arrangements, by the circumstances of the case.

This practice of particular rules and plans, to regulate all the operations of campaigns, has been the bane of the military system of Austria; whereas, the French, in their revolutionary struggle, by throwing off all these shackles and following the dictates of reason and common sense, laid the foundation of those splendid victories which followed the campaign of 1793.

The Austrian commander, field marshal Mack, could doubt less recite more rules than Napoleon; but what figure did he make when opposed to him at Ulm? This is all pedantry.

The truth is, Fellow-Citizens, that every part of education necessary to qualify a young man to discharge correctly and efficiently all the duties of an officer, in the line of the army, excepting the elementary instruction of the soldier, the company, the battalion, (in minature) and something of guard and police duty, (all of which can be learned by any person of ordinary activity and intelligence after he joins his company or battalion, in one month) can be just as well acquired at our colleges and ordinary academies, as at West Point. The more scientific duties of the military department belong to the corps of Engineers, for whom provision has been made.—There are also, at this time, in the United States, young men enough, who have received a regular scientific and practical military education (at their own expense) as an appendage to their civil education, to fill all the vacancies that would occur in the military establishment for twenty years; and there are also, at this time, four seminaries in the United States, (under competent instructors) where military science and duty constitute an appendage to the civil education of the pupils, and where youth can be as well, and I will venture to say, better qualified for military service than they are at the United States Military Academy; because the whole course of literary and scientific instruction is more extensive and more practical. Let it also be observed that these institutions are not supported from the public treasury. The plea, therefore, so often urged in favour of supporting this unconstitutional and aristocratical establishment, (by its advocates) that young men cannot be prepared to discharge their military duties without it, entirely falls to the ground.

I have said that the present system at West Point is *effeminate*, and calculated to unfit, rather than to prepare young men to encounter the fatigues and hardships of war. Such, I again assert, is the effect of it. Are they not there provided with the most comfortable quarters, and furnished with all the necessaries, and many of the luxuries of life? To what hardships are they exposed, more than to march a few limes each day over the beautiful lawn in front of their quarters, when the *weather* permits; to occupy tents pitched on the same beautiful lawn (and which are the most comfortable habitations they could select) during a few weeks of the warm season of the year? I answer none. Does this look like preparing youths to endure the hardships, fatigues, and privations of war? Did not the military committee of congress pass a practical commentary on the system, when they *gravely* asked an appropriation of 6000 dollars, at the last session, to erect a *comfortable* building, to shelter these *hardy* young soldiers from the weather, while they were discharging their ordinary duties? What would officers, thus educated, have done during your revolutionary struggle, when captains could scarcely provide themselves with shoes, and the march of your army over the frozen ground could be traced by the blood of their feet? Such a system will readily transform young men into military *dandies*, but never into soldiers.

But I had almost forgotten to tell you that these young men are to be *gentlemen* soldiers, and consequently it may not be deemed necessary, by the advocates of this institution, that they should be enabled to endure hardships, or that they should be exposed to the rude assaults of the weather, which might injure their *beauty*. Yes, Fellow-Citizens, they *are* int[en]ded for gentlemen soldiers—to sit high in *authority*, and exercise *command*, while you and your sons, who pay $200,000 annually for their education, must approach them cap in hand, and move at their nod. You and your sons are to march in the ranks; to carry the musket and knapsack; to be the *drudges*, yea, the mere *pack-horses* of military service. Are you prepared for this? If not, then prostrate this unconstitutional and aristocratical establishment, before the yoke is too firmly fixed on your necks to be shaken off.

But you are told it is a *national* institution, and its benefits *impartially* dispensed to all. Fellow citizens, be no longer deceived by this lullaby song; it is a gross imposition practiced on your honest and unsuspecting credulity. The truth is, that far the greater portion of the vacancies at West. Point, are filled by sons of members of congress; of persons holding offices of profit and emolument under the general government; of governors of states, and of persons holding offices of profit and emolument under the state governments, and of the most wealthy class of private citizens in the country.

I have, myself, known the sons of members of congress appointed cadets, whose fathers were in very affluent circumstances, out of the ordinary time of making such appointments; and without recommendations, unless by their fathers: and at the same time, the sons of honest, industrious, and useful citizens, whose pecuniary circumstances did not permit of their giving their sons a public education, put by, or entirely rejected; although their recommendations were of the most respectable and favorable character. I have also known persons holding offices of profit and emolument under the general government, who have had two sons at the military academy, (such was the case with the late head of one of the Executive Departments,) and in many instances the sons of particular individuals, some of whom were not of the proper age, have been appointed in advance, thereby engrossing the vacancies that might occur during the year, to the exclusion of others. But one of the most flagrant instances of this kind, I have ever known, is the following, which I will take the liberty to state more fully.

An individual, who, I believe, at this time, holds an office of profit and emolument under the general government, had both of his own sons educated at the military academy, the younger of whom, was appointed before he was of the legal age; and *his* eldest son at the time he was receiving from the public treasury not less, it is believed than $3,000 a year, had two of *his* own sons at the same institution; the younger one being appointed, as it is believed, a year in advance, and before he had attained the legal age of fourteen.

The foregoing, fellow citizens, will illustrate to you, in some measure, the *impartial* system that has been pursued in the appointment of cadets. Do not then allow yourselves to be longer imposed upon in this matter, for rest assured, the great body of you, and especially the industrious farmers, mechanics, &c., have no lot or share in this institution, except to pay for the greater portion of the expense for supporting it. The decided opposition of a portion of the members of congress, to any enquiry which would develope the system that has been pursued in the appointment of cadets, is strong evidence that they are aware all is not right.

A resolution was recently introduced, by an honourable member from South Carolina, calling upon the secretary of war. for information on this subject.

One of the clauses of this resolution, required him to report how many of those appointed cadets, since the commencement of the academy, were sons of members of congress, with their names. The whole resolution was well calculated, to develope such information, a[n]d such only, as ought to be before the representatives of the people, and the people themselves. This resolution was met, however, by a decided opposition from several members. One declared, that it had all been communicated some *ten* or *eleven* years ago—the *benevolent* feelings of others were much stirred up at the idea of *names* being exhibited to the public—it was *cruel* to do so. Sundry amendments were proposed, relative to striking out parts of the resolution, and especially those parts where *names* were to be embraced; but the *unfeeling* majority rejected all the amendments, excepting one or two, which went to make the enquiry more general, and were accepted by the mover. There was evidently a large majority in favour of the resolution, and the question was on the verge of being taken on its passage, when up rose an honourable and benevolent member from Massachusetts, and commenced a speech, secundum artem, which he concluded by a motion to refer the resolution to the military committee. The majority were apparently taken by surprise, and the reference to the military committee was carried.

It now becomes a subject of fair enquiry, what were the motives of this honourable member for moving the reference of this resolution to the military committee? Did he think the information could better be obtained in that manner, than by a direct call on the secretary of war? I for one do not believe he did; I believe his intention was to defeat the object of the resolution entirely. I have no doubt he perfectly well knew, that at least two, of the members of that committee, have had, or now have, sons at the military academy, and that a majority of it would be opposed to its passage. The result then of the reference would naturally be, either that the resolution would receive its quietus in the committee, or if reported at all. it would be so amended, as that the information obtained would be of no real importance. A few days however, will determine whether the will of a large majority of congress is to be thus thwarted, by the *management* of a small minority. It is an inauspicious omen, fellow citizens, when your representatives, to whom you have entrusted the guardianship of your rights, oppose a call for mere information on any subject, and especially on one, in which your interests are so immediately involved, as they are in that of the military academy.

You are often told, by the advocates of this institution, of Its great popularity. I ask, would not all the seminaries in the United States be popular, if they could furnish their pupils with all the means of obtaining an education gratis, and give them 28 dollars per month besides, with a promise that they should be supported at public expence, after they have graduated, during the rest of their lives. This, Fellow-Citizens, is the true secret of the *popularity* of the military academy—which *popularity* however, is limited to those who either have had, or expect to have their sons or relatives educated there. I repeat, it is the 28 dollars per month, and other emoluments and privileges therein enjoyed, amounting to about 500 dollars per annum for each cadet, together with the prospect of being supported in luxury and ease, at the public expence, that casts so many charms around this Institution; divest it of all these privileges and emoluments, and require that those who send their sons there, should pay for their education, as they do at other seminaries, and how many pupils under its present management would it number? I boldly assert not one. The whole system of discipline and government there, is totally abhorrent to the ingenuous and honourable feelings of the American youth. This is clearly proved by the general discontent, and frequent out breakings and even mutinies amongst the cadets; and also by the great numbers that resign, or otherwise leave the institution every year.

In order that you may understand, in some measure, how the cadets have been treated by the superintendent of the academy and some of his subordinate officers, I beg leave to call your attention to the statement of facts contained in the annexed document, marked (A.) You are there informed that a cadet, the son of one of the most respectable citizens of the state of Maryland, was treated with an indignity that would not have been tolerated in any well regulated military service, towards a private soldier, that his comrades, sensible of the injustice, as well as indignity of such treatment to one of their number, and considering also, (very properly) that such conduct on the part of their military commandant, was highly improper and unmilitary, appointed five of their number to make a respectful representation of his conduct to the superintendent, and ask a redress of grievances—that the superintendent treated this committee in a very haughty and supercilious manner, refused to hear their complaints, ordered them from his quarters, and then to leave the post in six hours—that they obeyed this order, and retired to a public house in the vicinity to wait the passing of the steam-boat—that they were ordered to leave this place in *one* hour—that in obeying this order they were exposed, almost destitute of necessary clothing, the greater, part of a day, to a severe storm, in the month of November, which caused the serious indisposition of one or two of their number, and is believed to have hastened, if not occasioned, the death of one of them. I challenge the admirers of this institution and of its superintendent, to produce from the military annals of any civilized nation, however arbitrary its form of government, an instance of more tyrannical and even barbarous treatment, than is here exhibited: and yet this treatment has been dignified with the name of *discipline* by many of the very *refined* and *benevolent* friends of the military academy.

As to captain B. I have nothing to say. Not being an officer of the corps of Engineers, he had no *legal* right to command the cadets at all, and they were not *legally* bound to obey any of his orders. But for the superintendent, there can be no justification or excuse; his conduct, by every principle of correct military discipline, was unjust and tyrannical, and he ought to have been dismissed the service, and would have been so, had those in authority over him, have done their duty. On what a proud eminence does the manly, dignified and independent course pursued by this committee, place them, in comparison with their tyrannical superindent[ent]? Yes, fellow citizens, the names of Loring, Ragland, Fairfax, Vining and Holmes, will long be respectfully remembered, and their memories affectionately cherished, after that of their relentless persecutor shall have sunk into oblivion, or if remembered, remembered only to be detested.

I now candidly ask you, fellow citizens, whether you will longer suffer 200,000 dollars to be drawn annually from your hard earnings, to support this unconstitutional, illegal and aristocratical establishment. Already have some of your number felt the effects of its tyranny, by being confined in the guard-house at West Point. If such things are done in the green tree, you may easily anticipate what will be done in the dry. Your only safety then is, in abolishing it; the evil can be effectually guarded against in no other way. If you ask me how you can accomplish this, I readily answer—tell your representatives in congress, in a language not to be misunderstood, that it *must* be abolished; that you will no longer support it: and if your mandate be not promptly obeyed, tell them, through the ballot box, that you have no further need of their services, and elect those who will obey your instructions.

If the money which is annually expended on this establishment, were appropriated to the instruction and disciplining of the militia, who are the only constitutional defence of the country, it would in the course of a few years, prepare them for the correct and efficient discharge of all the duties of the field. Would not this be a much more beneficial appropriation of it, than to apply it to building up a military aristocracy in the country? I have thus, fellow-citizens, candidly and frankly exhibited to you, some of the evils and abuses growing out of the military academy, under its present organization. If what I have said should induce you to apply the proper correction, I shall feel that I have rendered an important service to my country. If, however, such shall not be the result, I shall, at least, have the consolation of knowing that, I have honestly discharged my duty.

AMERICANUS.

NOTE. Americanus would beg leave to inform the cadets of the military academy, that whenever he has found it necessary to allude to them, in the preceeding pages, his remarks are to be considered as intended to be limited strictly to them, as connected with an institution, the principles, organization, and management of which, he highly disapproves; and that no reflections whatever, are intended to be cast on them personally, or in their individual capacities. He has the pleasure of a personal acquaintance with many of them, and entertains towards them, personally, the highest respect and esteem.

A.

[Memorandum!]

NEW YORK, Dec. 2d 1818.

On Monday the 23d day of Nov. 1818, the cadets of the military academy of the United States, in consideration of the indignities which they had received as a corps, from John Bliss, Captain in the 6th regiment of the United States infantry, and commandant of the corps of cadets, elected (without a deliberative assembly, or any deviation from the strict rules of discipline,) a committee, consisting of Charles R. Holmes, Wilson M. C. Fairfax, Nath. H. Loring, Charles R. Vining, and Thomas Ragland, to present to the superintendent of the military academy, major Sylvanus Thayer, of the corps of engineers, a statement of grievances, and to take such other measures as might appear necessary, to remove the growing evil—the committee in conformity to the sentiments of their fellow cadets, and adhering strictly to the principles of common right, framed a memorial, to be delivered to the super intendent, dictated by cool reflection, and expressed in the most respectful terms, noticing the most prominent acts of unjustifiable violence in the commandant, and shewing the effects on the minds of the cadets, at the same time assuring him, of their reliance on his disposition to administer justice, This letter was at first intended to have been delivered by the committee, (in person) to the superintendent, but Mr. Ragland a member of the committee was received (at the superintendent's request) as a *private individual*; in his conference, lie made known the object of the intended visit of the committee; the superintendent, while he approved the motives which actuated the cadets, requested Mr. Ragland to inform the committee, that, he could not receive from them their memorial—in extenuation of the commandants's conduct the superi[n]tendent observed, that although the instances of impropriety which the committee had noted, had occurred, there was no indication that the similar would hereafter take place.

286

The committee being apprized of these circumstances, and perceiving from the unaccountable conduct of the superintendent, that there remained but little hope of obtaining from that source the redress which they were in justice entitled to expect, resolved, rather than tamely submit to indignities as vile, as they were degrading, to enjoy those privileges which every citizen of a free country has a right to claim, the members of the committee, as individuals, waited in the evening on the superintendent, (by his permission) with a view of tendering their resignations as members of the military academy; these the superintendent peremptorily refused to accept, on any considerations, and not even hearing the motives, remarked, that had he known the business of the committee, he would have refused them an audience; at the same time giving them in substance, an order to leave his quarters; the committee obeyed, but resolved to frame charges predicated on the facts in their possession, against Captain John Bliss. The charge of ungentlemanlike and unofficerlike conduct, was prepared against him, each specification being founded on the surest and most undeniable evidence—cadets having certified on honor, to each of the following facts—

That J. Bliss, the commandant, did seize by the collar, shake and jerk in the most furious manner from the ranks, a cadet, who was stunned by the shock; the said cadet enquired if that was the manner in which he was to be treated? or words to that effect: the commandant answered—"Yes, God damn you, and I'll——" The remainder of the sentence was not distinctly remembered by the evidence.

That J. Bliss, the commandant, saw several cadets standing together, and without any provocation threw stones at them.

That J. Bliss, the commandant, passed a cadet who was sitting on the Barrack railings, and turning about, violently threw the said cadet from the railings, without the shadow of provocation.

That J. Bliss, the command[a]nt, actuated by the most ungovernable rage, with menacing gestures, and in a provoking manner, ordered a cadet from his quarters, and without allowing time for obedience to his order, violently seized and thrust the said cadet from his quarters.

As the superintendent had evinced his determination not to receive the committee in a public capacity, the charge was presented to him by Messrs. Fairfax and Ragland, in the character of private individuals; after a few cursory remarks on the discontent of the cadets, the superintendent thought proper to receive the charge; observing, that if it contained a statement of facts which were already known to him, the committee had reason to expect they would be properly noticed[.]

This charge was enclosed in a letter from the committee, assuring the superintendent that this measure, though it might be seemingly precipitate, proceeded from the coolest and most deliberate reflection; and proving by the fairest and most candid exposition of their sentiments, that no part of their conduct had been dictated by the effervescence of youthful passion.

This letter, in addition, expressed the intention of the committee (in case the superintendent deemed it inexpedient to give them the redress, they had a right to require,) to pursue the affair to the extreme of justice, and if necessary, sacrifice their individual interest in the cause, in which their duty as beings of reason and honor, had led them to embark. This letter the superintendent received, (enclosing the charge) from Messrs. Ragland and Fairfax as private individuals.

The subsequent morning (Friday, the 27th Nov.) the cadets of the M. A. were assembled on the parade ground, where the Post Adjutant published an order from the superintendent, in which that officer took occasion to remark, that the conduct of the cadets in the proceedings which are before stated in this paper, could in some measure be palliated, by supposing them impelled by the warmth of feeling natural to youth, but for the committee, no extenuation could be offered, they (it was said) acted by design, they knew the consequences; were aware of the impropriety of their proceedings, a[n]d it might be supposed were instrumental in seducing the cadets from their duty; the superintendent added, that the cadets would continue in the regular performance of their duties, but the committee would leave the post within six hours from the promulgation of the order, and repair to their respective homes to remain until further orders, while their conduct was reported to the secretary of war.

The committee do not wish to comment on an order, as singular as it was unexpected; but as it impeaches their conduct in so high a degree, it is deemed a duty, absolutely to deny the correctness of two assertions. To charge the committee with swaying their fellow cadets, is an imputation on the good sense of a large body of the first young men in the country, and from that consideration, independent of all others, cannot be confided in. But the committee are confident that such an allegation will be regarded by all who know them, as incompatible with their characters. The assertion, that the cadets were formed into a deliberative assembly, can be disproved by the honorable attestation of *two hundred* individuals.

The committee obeyed the order, by repairing within the appointed time, to a house (without the limits of West Point) occupied as an Inn; here they believed themselves authorized to remain until the next day (Saturday) at 1 o'clock at night, when the steam boat would pass for New York, as that was the first regular conveyance: but on the next morning, another order was served on them by the post adjutant, by which they were forced to leave the vicinity of West Point within *one* hour, being at the same time, to the knowledge of the superintendent, destitute of *necessary* clothing, which could, in part, be supplied by the evening. The barge was sent for the committee, and the whole of this extremely stormy and inclement day, was necessarily occupied in rowing about in the river and riding on the east side of the river to Peekskill, from whence they crossed to the west side, ten miles below West Point, in obedience to an order which has been productive of no other consequence than the serious indisposition of one or two of the committee, and the great inconvenience of the whole.

N. HALL LORING

CHARLES RUTLEDGE HOLMES

C. VINNING, *for themselves & in behalf of*

THOMAS RAGLAND, and

WILSON M. C. FAIRFAX.

Since this was written I have looked it over carefully, and can vouch for the general and particular correctness of the transactions, as herein above stated. Indeed they are of record in the War Office, contained in the proceeding of a court of enquiry, and also a court martial held at West Point, in 1819, relative to these transactions, where every possible light is thrown on the subject, from the attestation of witnesses, most of whom are living. I will add, that a member of the committee of cadets, Charles R. Vining, at the time of being ordered off from West Pont in the hasty manner above alluded to, was then in feeble health—that he died of pulmonary consumption the following winter, 1818-1819, and that it was the general opinion of all who knew the circumstances, that his death, if not occasioned by the harsh treatment of Colonel, (then Major Thayer,) was, at least, much hastened by the same.

WILSON M. C. FAIRFAX.

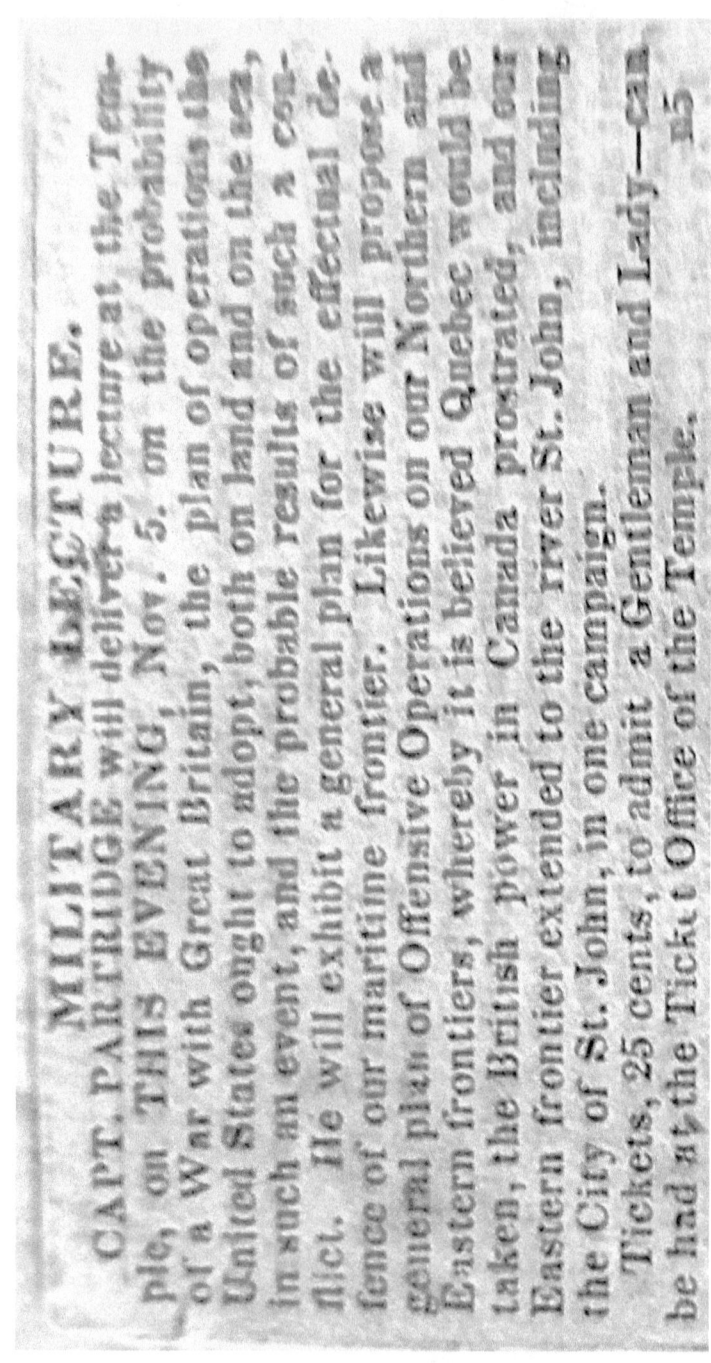

Figure 17—Advertisement for Military Lecture, 1841[660]

---

[660] Cutting from an unknown publication. Norwich University Archives and Special Collections.

Appendix H: West Point and Young Men Educated Elsewhere[661]

Memorial of Alden Partridge, relating to the military academy at West Point, and praying that young men educated at other military schools may have an equal chance for admission to the army as those young men have who are educated at West Point.

January 21, 1841. Referred to the Committee on Military Affairs.

To the honorable Congress of the United States:

The memorial of Alden Partridge, president of the Norwich University, at Norwich, State of Vermont,

Respectfully showeth:

That your memorialist holds it to be a cardinal principle of our republican institutions, that stations of honor, trust, and emolument should be equally open to all our citizens, to which all have an equal right to aspire, and from which none can constitutionally be excluded by any law, rule, or regulation whatever. Your memorialist has, however, witnessed, with deep regret, a direct violation of this vital principle of our constitution, by the rules and regulation adopted for the organization and government of the Military Academy at West Point. The cadets of that institution, all of whom are educated at the *public* expense, have, for many years, *monopolized* nearly, if not quite, all of the stations of honor, trust, and emolument, above that of non-commissioned officer, in the military establishment of the United States, to the utter exclusion of those who are equally well qualified, equally meritorious, and who are educated at their *own* expense. But, in order to place this subject more clearly before your honorable body, your memorialist would call your attention to the law of the 29th of April, 1812, entitled "An act making further provision for the corps of engineers." By the provisions of this act, no candidate can be admitted into the Military Academy who is under fourteen, or over twenty-one,

---

[661] Partridge, Memorial of Alden Partridge, relating to the Military Academy at West Point.

years of age. The effect of this provision is to exclude every young man in the United States who is above twenty-one years of age from the appointment of cadet, which the rules of the War department require that none except those educated at this academy can be commissioned in the army of the United States. The effect, then, of the law and regulation is to utterly exclude all the youth of our country, except such select few as the President may think proper to place in this "public charity school," from the military service of their country, who are above twenty-one years of age, unless they will enter in the humble capacity of *privates* or *non-commissioned officers*. And can such a system be in accordance with the principles of our constitution? Your memorialist believes not. On the contrary, he feels confident in the assertion that it is a most flagrant and palpable violation of them. The direct and certain effect of this institution is to extende *Executive patronage*; for the President has the entire selection of the *chosen two hundred* and *fifty* who are to be placed in the institution, and also to establish an *aristocracy* of the most dangerous kind, viz: a *military* aristocracy in the United States. What, your memorialist would ask, is an aristocracy? Is it not where any particular class in a State claims and exercises privileges of which the great body of the people are deprived? and if so, do they not constitute an aristocracy? Your memorialist believes that neither the fact nor the inference can be controverted. But your memorialist will go further, and aver that the regulations at West Point have not only constituted an aristocracy in the United States, but that this aristocracy has already become, in a great degree, *hereditary*. How many individuals, your memorialist would ask, who have held offices of honor, trust, or emolument, under the Government, for the last twenty-five years, have had their sons, brothers, nephews, or other relatives, educated at the public expense at West Point, to the entire exclusion of those who (to say the least) were equally meritorious, and equally capable of rendering service to the republic? And how many of thus educated have ever rendered any service whatever? A reference to the rolls of the institution will answer these inquiries. Your memorialist has *personal* knowledge of many instances. Your memorialist is well aware that it has been attempted, by the friends of this monstrous invasion of the rights of the people, to cast around it the mantle of Mr. Jefferson. Your memorialist is ready to grant that the institution was established during the early part of the first term of Mr. Jefferson's administration; but denies that any inference can be drawn from that circumstance to sustain the present system. The institution *then* consisted only of the corps of engineers, which was limited to sixteen officers and four cadets, without any of these exclusive privileges which have since been conferred upon it. On the 29th of April, 1812, (just previous to the declaration of war,) a law was, however, passed entitled "An act making further provision for the corps of engineers," by the provision of which, the whole number of cadets, whether infantry, artillery, or riflemen, was not to exceed two hundred and fifty; and the President was authorized, *at his discretion*, to attach them as students to the Military Academy. Now this provision of the law of 1812, authorizing the President to appoint a limited number of cadets, and conferring on him a *discretionary* power to attach them to them to the Military Academy, was evidently induced by the certainty of immediate war with Great Britain, and had a direct reference to a *war* establishment. Your memorialist would respectfully call the attention of your honorable body particularly to the provision of the law of 1812 just referred to; and, if he does not much mistake, it will satisfactorily appear that the President is not *required*, but simply authorized, to appoint a single cadet; and that it is left entirely discretionary with him, after they are appointed, to attach them to the Military Academy, or to attach them to their respective companies, agreeable to the provisions of other laws then in existence. And here your memorialist would observe that, in the *peace* establishment of the army previous to the late war, two cadets were allowed to each company of artillery, light infantry, and infantry, amounting, in the total, to a *larger* number than was authorized by the law in 1812. But neither President Jefferson nor President Madison considered that the law required of them to fill those vacancies so long as they considered their services were not required; and they consequently did not fill them. The largest number of cadets ever in service at the same time, previous to the late war, did not exceed forty and seldom exceeded twenty five. Do the necessities of the country require that any larger number should be retained in service now, than were deemed necessary by Presidents Jefferson and Madison during a time of peace? Your memorialist believes not. But it is urged, in favor of this academy, that it presents a most favorable opportunity for the education of meritorious young men who are poor, and, consequently, unable to educate themselves. Your memorialist, however, has yet to learn by what *constitutional* authority Congress is empowered to appropriate any portion of the public revenue for the support of a *national charity school* for the education of the poor. Besides, if this power did exist, (which your memorialist presumes no reasonable person will contend does,) *all* the poor in the United States have an equal right to the benefits to be derived from its exercise, and that, consequently, the institution at West Point is on quite too limited a plan for the accomplishment of the contemplated object. Either, then, the institution should be very much enlarged, or several other established in different parts of the United States, which would be far more convenient for the great body of the poor. If, however, the rolls of this institution for the last twenty years be examined, it will be found that many more of the *rich* and *influential* have been educated there, than of the *poor*.

Poverty, however meritorious the subject of it may be, is but a sorry recommendation for admission to this aristocratic establishment.

But it is further urged, that this institution is *necessary* for the education of the officers of the army; and that, were it abolished, the candidates for commissions would not be properly qualified for the discharge of their duties as officers. Before your memorialist proceeds to examine the truth of this position, he would inquire, at what institution, and at whose expense, Generals Washington, Greene, Knox, Putnam, Lincoln, Sullivan, Morgan, Wayne, Sumter, Pickens, Marion, and all the other officers of the revolutionary army, by whose valor, skill, and patriotic exertions, these United States now constitute a free and independent nation, received their education? The answer is ready: at the ordinary institutions of the country, and at their *own* expense; just as every *American citizen* should be educated. And have the *protégés* of the West Point Academy, on whose education so many millions of dollars of the people's money have been expended within the last twenty years, exhibited more skill, more valor, or more patriotism, than did the officers to the revolutionary army? Let the events of the Florida war, as compared with those of the Revolution, answer the question. The truth is, (and it cannot be much longer concealed from the view of the people, by the reports of *boards of visiters*,) that the whole system of education at West Point is well calculated to form *military pedants* and *military dandies*, but will never form *efficient soldiers*. Much more important to them is their attention to the *cut* of the *coat*, the placing of a *button*, and the *snowy whiteness of gloves and pantallons*, than those *physical* and *moral qualities* which are absolutely necessary to the correct and efficient discharge of the active duties of the field.

But your memorialist denies the truth of the position, that the West Point Academy is necessary for the education of young men of the army. There are other institutions where military science and instruction constitute a branch of education for the pupils. Of these institutions, however, your memorialist will particularize by one—and that is Norwich University, at Norwich, Vermont, over which he has the honor to preside. This institution was incorporated by the Legislature of Vermont, in November, 1834, with full power to confer diplomas, &c. By the act of incorporation, military science is made a part of the education of all the pupils. They are consequently correctly and thoroughly instructed in the theoretical part of the military science, and also in the *practical* duties of the soldier; and every one who graduates at this institution is well qualified to discharge the duties of a company officer (and even, if necessary, to command a battalion) in any corps of the army. In order further to prepare them to discharge the more hardy and active duties of the soldier, they occasionally perform military marches. In the month of July, 1840, they performed a march, under personal command of your memorialist, to the celebrated *military post* of Ticonderoga, carrying their arms, accoutrements, knapsacks, &c.; the whole length of which was 165 miles. Of this distance, 140 miles was on foot, and 25 miles on steamboat. The march on foot was performed in a little more than five days, crossing the Green Mountain range twice, and the ground, with the heavens for covering, constituted their only resting-place at night. The weather, during the whole march, was hot; and they were enveloped in a cloud of dust, occasioned by the severe drought, nearly the whole distance. They all returned in excellent health and spirits. They youngest member of the corps was thirteen years of age. The other branches of literature and science are attended to as extensively, and the latter much more practically, than at any other institution in the United States; and the students are consequently equally well qualified to discharge their duties in the *cabinet* and in the *field*. But notwithstanding the members of this institution are, to say the least, as well qualified for commissions of any grade, and in any corps of the army, as those of any other institution in the country, and have also obtained the necessary qualifications at their own expense, they are virtually excluded therefrom by the *arbitrary* and *monopolizing* regulations (established without the least sanction of law) of the Military Academy at West Point. In the month of September, 1840, a member of the Norwich University, the son of a highly respectable gentleman in the city of New York, well recommended, applied to the Secretary of War for a commission in the army, but was informed that there were *no vacancies*, and that the cadets from West Point were *more sufficient to fill all the vacancies*. On the 21st of December, 1840, your memorialist wrote to the Secretary of War, recommending three young gentlemen, members of Norwich University, for commissions in the army of the United States; and received an answer, dated War Department, December 29, 1840, from which the following is an extract: "I acknowledge the receipt of your letter of the 21st instant, recommending Messrs. Morris, Stevens, and Dorne, for appointments in the army; and I have here to inform you, in reply, that there being no vacancies at present, the application will be filed for consideration, when any occur, *to which they can be appointed*." Now your memorialist feel confident that the records of the War Department will show that a large number of cadets at West Point are commissioned every year; and he presumes that such will continue to be the case, unless a radical change is effected. But when young gentlemen of equal respectability and attainments, who have not been of the *favored few* whom *Executive favor* has admitted into this nursery of aristocracy, to be educated at the expense of the honest working men of the country, become applicants, their claims are entirely set aside. Against this *unconstitutional, unequal,* and *monopolizing* practice, your memorialist deems it his duty respectfully, but most decidedly, to protest; and to ask of your honorable body the establishment of some rule whereby the members of the Norwich University, at least, (to whom, in many respects, he stands in the relation of guardian,) may be restored to their *constitutional rights*; that when they become applicants for stations of honor, trust, or emolument, in the military service of their country, they shall stand on terms of equality with the cadets at West Point.

Your memorialist deems it proper here to remark, that in October, 1840, he addressed a communication to the President of the United States, on this subject, requesting to be informed whether, in the opinion of the President, he possessed the power to remedy the grievance of which your memorialist complains; and, if so, whether such power would be exercised for that purpose. To this communication no answer has been received. Your memorialist, availing himself of the privileges granted to every American citizen, by the first amendment of the constitution of the United States, would beg leave to call the attention of your honorable body to some subjects, which he considers grievances of a high order, and respectfully but earnestly solicits that they may be redressed, viz:

1st. Your memorialist considers the Military Academy at West Point a grievance. Under its present organization, it is unconstitutional, calculated to foster a military aristocracy in the county; calculated to depress the militia, (our only constitutional defence,) by engrossing all the patronage of Government; and is entirely unnecessary, as military science can be attained at other institutions, from which the necessary officers for the army can be supplied without any tax on the people. Your memorialist, therefore, asks that this institution may be abolished, and that the money that is annually appropriated for its support may be applied to aid in disciplining the militia, and disseminating military information amongst the people, who are its constitutional and safe depositaries.

2d. Your memorialist considers the Board of Visiters that annually assemble at West Point a grievance. This board never has any *existence whatever in law*, but was established by Executive usurpation; yet, to pay the expense of this illegal board, your memorialist believes that more than fifty thousand dollars has been drawn from the public Treasury. Your memorialist earnestly solicits that this appropriation, the making of which is a direct sanction of Executive *usurpation*, should be discontinued.

3d. Your memorialist considers the removal of the head-quarters of the corps of engineers from West Point to Washington a grievance, because it is a direct violation of the law of 16th of March, 1802, establishing that corps. That law requires the commandant of engineers to reside at West Point, unless ordered, by the Present of the United States, on duty at some other place in the line of his profession; and, when at West Point, the law makes him superintendent of the Military Academy; and when he is absent, the next in rank (who is then present) is made the *legal* superintendent. The appointment, therefore, of any particular officer as permanent superintendent, is evidently illegal, as the law was clearly specified who the superintendent shall be.

All of which is respectfully submitted.

A. PARTRIDGE,
President of Norwich University

January 13, 1841.

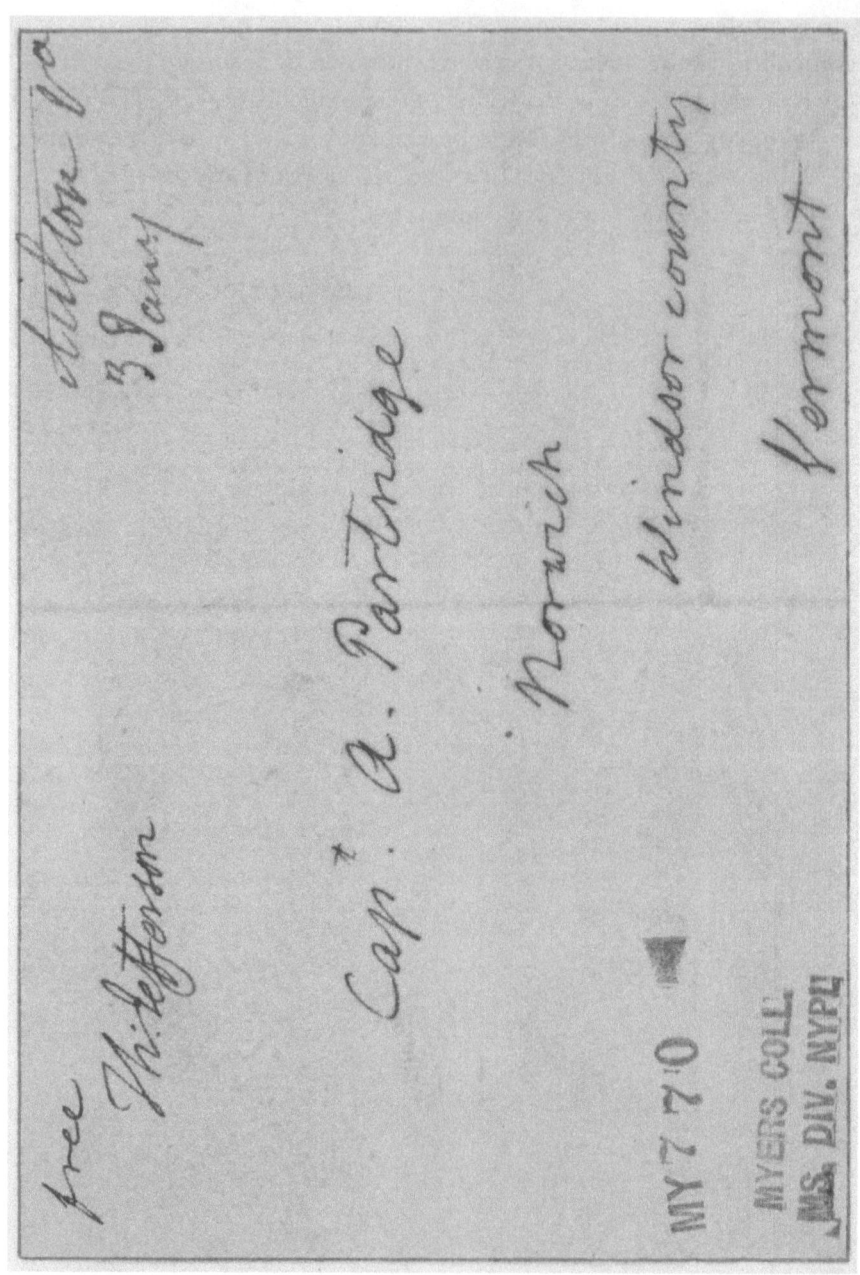

Figure 18—Envelope from Thomas Jefferson addressed to Alden Partridge[662]

[662] Manuscripts and Archives Division, The New York Public Library, "Jefferson, Thomas. Monticello. To Capt. Alden Partridge. Envelope," *The New York Public Library Digital Collections.* Jefferson would write to Partridge on January 2, 1816, to relay a few heights of mountain peaks and to discuss the methods of determining elevation through the use of a barometrical measurements.

Appendix I: Lewis Lochée's An Essay on Military Education[663]

By Lewis Lochée,
Master of the Military Academy, Little Chelsa.

----- Ego nec studium sine divite vena,
Nec rude quid prosit video ingenium: alterius sic
Altera poscit opem res et coniurat amice.[664] HOR.

I call A COMPLEAT AND GENEROUS EDUCATION, that which fits a man to perform justly, skilfully, and magnanimously, all the offices, both private and public, of peace and war.[665] MILTON.

The Second Edition.
London:
Printed for the Author;
And sold by T. Cadell, in the Strand.
MDCCLXXVI.[666]

To the Public
In this kingdom, through renowned for those great talents and generous virtues that are the source of the highest improvements, little has been written on the art of war. As the natives, therefore, for whatever reasons, have generally declined the illustration of military science, it is hoped that it will not be deemed impertinent in a foreigner impressed with strongest convictions

---

[663] Lochée, *Essay on Military Education*. The essay was edited to remove the use of medial S ("ſ") for the ease of modern readers. Variations in the spelling of the word "ancient" was also standardized to the modern spelling.

[664] Translation: "I see not what good can come from the study without a rich vein of genius untrained by art: thus one thing demands the aid of the other and both [science and art] unite in friendly assistance." Horace, *Ars Poetica*.

[665] Quote from John Milton, "Of Education."

[666] 1776.

of its vast importance, if he makes an humble offering of his best services to a People, for whom he has the highest esteem, and to whom he is under peculiar obligations.

Essay on Military Education

"The laws of education," says Montesquieu,[667] "are the first impression we receive; and as they prepare us for civil life, each particular family ought to be governed by the principles of the great family that comprehends them all." Hence it becomes necessary, that the education of youth, in every state, should be formed and adapted to the nature, end, and principle of its government.

Every kind of government has its nature, end, and principles: its nature is its particular constitution adapted to some end; its end is that to which its constitution is essentially disposed; and its principle is the active power of compassing that end. From this view it is evident, that the principle of the most essential part, the very soul of government, that puts it in motion, and gives it life and vigour. It follows, therefore, as a fundamental rule in education, that the principle by which the whole community is regulated and supported, cannot be too highly cultivated in the mind of every individual.

All governments may be reduced to three species, the republican, the monarchical, and the despotic; to each of which Montesquieu had assigned its proper principle: "to the republican, virtue; to the monarchial, honor; and the despotic, fear."

Virtue in a republic, is the love of country, and of the laws by which its welfare is secured and improved. This produces temperance, justice, fortitude; a suppression of all base and selfish passions; a continual sacrifice of private interest to the good of the community; and such universal benevolence of affection, and integrity of action, as give reciprocal strength to the noble principle from which they are derived.

Honour, the principle of monarchical state, according to Montesquieu, "estimates the actions of men, not as good but as splendid, not as just but as great, not as reasonable but as extraordinary." The spirit of monarchy endeavours to effect as much as it possibly can, without virtue. It subsists independently of the love of the country, of the thirst of true glory, of purity of manners, of the sacrifice of our dearest private interests, and of all those exalted virtues, which, in singular instances and during short intervals, we so much and so justly admire in the ancients. Honour is, indeed, inseparable from virtue; but no otherwise than as an effect from its cause: when it assumes the power of the leading principle, vice may be virtue, and virtue vice, as caprice

---

[667] French philosopher Charles Louis de Secondat, Baron de La Brède et de Montesquieu (1689–1755 A.D.)

and passion shall determine; for it is then wholly selfish, and is changeable in its objects and pursuits as interest predominates: in the vain-glory of pretension to sensibility of affection, and of possessing those arts and accomplishments that ensure conquest, it exalts the seduction of innocence, and the violations of conjugal fidelity, into acts of gallantry; and from a preposterous idea to greatness of soul, and true interests of the state, it descends even to intrigue and cunning, and deems the breaches of public faith refinements of policy.

But fear, the principle of despotism, is incapable of not only virtue, but even of those delusive appearances of it that honour may chance assume. It is the extinction of all intellectual light, the suppression of every generous affection: and as the tyrant and the slave are equally under its dominion, deep ignorance and multiplied wretchedness must be the lasting portion of both.

If we take a survey of all the associations, states, and kingdoms of the earth, we shall find, that they have risen to greatness or fallen to ruin, have been happy in themselves and beneficial to their neighbours, or rent with intestine commotions, or plunged in hopeless distress, of confederated for the violation of the common rights of mankind, according to the predominance of each of these three principles. But as the selfish spirit has always prevailed, even the republics of Greece, Italy, and Asia-Minor, the most renowned for urbanity and virtue, have made only transient appearance of true glory, like spots of azure in a cloudy sky, and have fallen at last a prey to that evil nature, which is the degradation misery and punishment of man. But human institutions must necessarily partake of the infirmities of human beings; and all states, like those that, compose them, have the principles of growth and dissolution in their own frame: like men, they are born and die, have their commencement and their period; they arise, like light, from the darkness of poverty, to temperance, industry, liberty, valour, power, conquest, glory, opulence, and that is their summit; from whence they decline to ease, sensuality, venality, corruption, cowardice, imbecility, infamy, slavery, and irremediable perdition. The only appearance of a permanent government, was the Theocracy of the Jews; till, though the extreme depravity of nature, the people became impatient and uneasy under the restraints of Infinite Wisdom of Goodness, and would no rest till the constitution was changed into the regal dominion of ignorant, selfish, and arbitrary man; when declining from one degree of evil to a greater, they were at length scattered over the face of the whole earth; in which unnaturalized condition they still remain, individual monuments of the punishment due to revolters from the eternal laws of righteousness and beneficence.

From despotic government, the understanding and the heart of man recoil as from darkness and death. It is essentially evil in its nature, and wholly and unchangeable evil in all its effects: and, therefore, in a political and moral view, it can be considered in no other light, than as an enormous whirlpool situated between monarchy and a republic, to which both have a disposition to tend, and in which both may be swallowed up and lost for ever. But in a review of all human governments, where shall we find such a harmonious union and just counterpoise of regal aristocratical and democratical powers, as that by preventing all preponderation each derives reciprocal strength from the other two, and the freedom and peace of every individual is essentially included in the liberty and safety of the whole; but in the Constitution of Great-Britain? To the youth, therefore, of Great-Britain, of every class and condition raised above the labouring peasant, who may possibly be contented with the protection of his property, liberty, and life, without inquiring into the source from whence that protection flows; the principles of the wonderful Constitution, and the high duties it requires, cannot be too clearly explained. On those who enter into the profession of arms, the inculcation of this knowledge is indispensably necessary, that they may not only learn what a treasure of blessings they have to defend, but why and how it ought to be defended; and that their dearest blood will always be nobly shed, in the protection of this illustrious offspring of the most improved human wisdom and the most exalted human virtue. As I have already had the honour of being entrusted with the care of some, who by their birth and fortune may be entitled to a share in the legislature of this country, the highest trust that can be confided by man to man; and as from a consciousness of my own principles and views, and the growing favour of a generous people, I have reason to hope for repeated instances of the same distinction; it has been though expedient, as a necessary part of the system of education that I adopt, to close the following observations with a short sketch of the nature, privileges, and principle of the British Constitution, drawn according to the best of my judgment from the ablest writers on the subject.

In the perusal of history, curiosity urges us impatiently on with the rapidity of events, and we are wholly engrossed by the emotions of pleasure and disgust, of joy and sorrow, excited by the revolutions of great states, and the characters and fortunes of the principal agents engaged in effecting or opposing them. But in a review of the knowledge thus acquired, either for the occasional or general application of it to the purposes of life and duty, when the mind is at leisure to trace back effects to their causes, to follow the progress of virtue and vice, and mark their gradual influence on individuals, and from individuals to the community; it then discovers the common source of that almost boundless variety of design, action, and character and event, which history has displayed; and finds, that not only the various customs and institutions civil and religious, and the different systems of political and moral of different states, but their rise, establishment, grandeur, declension, and ruin, are to be ascribed to the nature and power of education. This raised the petty state of Athens to its amazing heigh of power and glory; this preserved in vigour the Spartan commonwealth for more than seven hundred years, when Philopæmen subverted the institutions of Lycurgus, and abolished the ancient laws for the instruction and formation of youth; this made Rome the wonder of the world: and to the neglect or perversion of this, thro' the depravity of nature, it has been owing, that the wisdom, virtue, valour, and power, of Athens, Sparta, and Rome, line now only in idea.

In all well regulated states, that first step in the education of youth is to make them good men, just and benevolent members of the universal society of mankind; and the next, to qualify them for the highest usefulness to their own country; to inform their understandings with the kind of knowledge, and to apply their talents to those objects and pursuits, that will render them most serviceable in support of the government under which they are born, and on the strength of prosperity of which their own welfare as individuals necessarily depends.

But great as the object of education is as the source from when all happiness or misery of the rising generation must flow, it seems not, in modern policy, to engage that care and attention which the vast importance of it requires. Though the principles and forms of education in every state may have a general correspondence with the principles of government, yet if some discrimination is not observed, and particular genius and particular disposition directed and applied to those employments to which they have a prevailing bent, and for which they are peculiarly fitted, disorder and weakness will be consequence: the individual must suffer, from the consciousness of inability to discharge the duties to which he has been called; and when, from a department of importance, a qualified member is excluded by the appointment of the one unqualified, the state must also suffer, not only by the failure in the duties of that department, but in the loss of two members whose talents have not been suitably applied.

The fame from of what is called a learned and liberal education in this country, is adopted for the youth of all ranks and conditions, let their disposition, and their future prospects in life be ever so various. A boy totally ignorant of the rudiments of his own language, and the scarcely able to read it with propriety, is sent to school to learn Latin and Greek, where seven years at least are spent in acquiring only a moderate degree of skill in those languages: after this he is transmitted to one of the universities, where he passes four years more in procuring a more competent knowledge of Latin and Greek, and in learning the rudiments of logic, natural philosophy, astronomy, metaphysics, and the heathen mythology and morality: at the age of twenty or a little more, he perhaps takes a degree in arts, and then education is finished.

When education is said to be finished, it is natural to expect, that a young man is compleately qualified to fill and sustain some useful character on the public stage: and yet it will be difficult to say, what single important duty of society he is able to discharge, what single office as a citizen he is qualified to execute. It may, indeed, be asked with Seneca, where are the promised fruits of this learned and liberal education, that has swallowed up so many of the important years that give to future life its form and tincture? whose errors will it diminish, whose passions will it restrain, whom will it make more brave, more just, more liberal? Excepting the lawyer, the physician, and the divine, who may convert a small part of these rudiments of science to their own and the public advantage in their different professions, the youth thus educated will be excluded from service, and obtain no establishment for future life; and he who, after so many years of laborious application, finds the knowledge he has acquired of no immediate use, is too strongly tempted to forget it all.

But this mode of education has been so long and so generally adopted for the gentleman, that it is even stiled the gentlemen's education. The character of the gentlemen is, indeed, the most respectable and important in this country, and therefore demands the greatest care in the formation of it. In public life, gentlemen are born to assist either in composing the councils of the nation, or in conducting her fleets and her armies; to be the bulwarks of the constitution; to sustain parts that require continual exertion of wisdom, fortitude, and the most highly improved talents: and, in private life, to contribute by study to intellectual and moral improvement, to be depositaries of upright principles and pure manners, illustrious examples of temperance, justice, benevolence and piety, disusing order and happiness all around them. But what benefit does this education produce to the gentlemen, if, so far from being thought compleat, it is not considered merely as the ground-work of a more enlarged and more active cultivation? After the restraint of so many years in the school or the college, the mind is too apt to exult in a fancied deliverance from discipline and duty; and where amplitude of fortune and the false indulgence of parents unhappily concur, it seldom fails either to hide itself in idleness and amusement, or to seek distinction only from the indulgence of brutal appetite, and the exploits of boisterous passions. If then the gentlemen, when he leaves the school or the college, does not truly consider the education he has received merely as the conductress to much higher attainments, and firmly resolve to devote himself to the acquisition of more enlarged knowledge and the rules of stricter discipline, either in the study of laws or government in the most extensive and liberal form, or, if his spirit and genesis lead him to it, in the profession of arms, till he is called forth to public service; his education, so far from being either honourable or useful, will only help to render him more base and wretched in himself, and more disgraceful and injurious to his country.

I have here connected, not only the character, but the most enlarged education of a gentleman, with the profession of arms; and it will be found to require all the intellectual improvements implied in the character, not less than all its graces and virtues.

Many groundless prejudices and weak and unworthy practices have concurred, to render that profession in some degree odious. In all modern free states, like this, the common censure describes it as a necessary evil. But in what light did it appear at Athens, Sparta, and Rome? in the zenith of their glory, did they derive less honour from their valour in the field, than from their wisdom and virtue in the senate? But, in their ideal, valour, wisdom, and virtue, were the three inseparable constituent qualities of true greatness: and the retreat of Xenophon,[668] the defence of Thermopylæ by Leonidas,[669] and the generous and seasonable interposition of Camillus,[670] have in all succeeding times received the tribute of praise which they demanded, not only as instances of genuine valour, but of the highest skill and address in the art of war. How unsafe would stand the noble fabric of the British Constitution, though so deeply founded and so many ages in erecting, if the aid of equal valour, and equal skill in the art of war, was wanting for its protection and support!

The censure on the profession of arms as a necessary evil, admitting it to be just, only adds strength to the many reasons arising from the profession itself, for enforcing the strictest discipline on the student, for training him to the habitual love and practice of order, decency, and virtue. But it is known, that a soldier ever dared to plead his profession as an exemption from the obligations of duty in civil and social life? Are the intemperance, injustice, cruelty, profaneness, and impiety, less cognizable by the laws, less culpable, and less disgraceful, in him than in another man? If the private men should happen at any time to be guilty of tumult and outrage, it could arise only from ignorance presumption, and would be immediately checked and severely punished: if officers are habitually addicted to gaming, drunkenness, and lewdness, they dishonour their character not less than by cowardice.

But the profession of arms is also censured as an idle profession, that requires little intellectual ability, and less application. Impressed with this

---

[668] Xenophon of Athens (ca. 430–354 B.C.)
[669] King Leonidas I of Sparta (ca. 530–480 B.C.)
[670] Marcus Furius Camillus (446–365 B.C.)

character, parents, after many fruitless trails of the uncultivated talents of their sons, are apt to make choice of the army as a sure retreat for ignorance and indolence; and if the lads have but the kind of spirit and address, which, in its rude state, must approach much nearer impudence than intelligent and manly intrepidity, they suppose them possessed of the essential requisite in a soldier, and that a suit of regimentals will hide all little defects. The young men themselves help forward this gross and most pernicious delusion: unaccustomed to look beyond the fancied importance of external qualities, and the glitter of external appearance, they are captivated with the habiliments of an officer, with the respect that is shewn him in all companies, and with the unreserved freedom of his own behaviour, which, if not governed by that politeness which is the pure effect of sensibility and knowledge, must be worse than the rudeness of a savage; and they also captivated with an idea of the authority they expect to be invested with over the private men, not reflecting, that they themselves must be subject to authority, and that the degree which is to devolve upon them, if not maintained upon right principles, will be in a continual danger of being abused.

But are these notions and practices calculated to change that which is censured as positive evil, even into negative good? Or rather, have not these and other similar notions and practices continually operating upon and giving strength to each other, contributed to increase the disgrace that has fallen upon a profession, in which excellence cannot be attained, without the concurrence of the most highly cultivated understanding, with the noblest principles and the strictest manners?

In all ages, and all nations, the profession of arms has at least been deemed the profession of honour: in some nations, indeed, whose form of government was monarchical, it has not been capable of a higher principle; and that has served to raise it to great glory: but, in the British nation, it is the profession of honour, only as it is animated by the most generous public virtue; and whether the most generous public virtue can be attained, without the enlarged cultivation, and those principles, habits, and acts, that constitute the intelligent and generous virtue of a private individual, let the history of mankind determine. By this subordination, however, the peculiar rights of honour are not in the least degree violated: though she is only the attendant of virtue, on whom her very life depends, yet she may boast of ornamental service, and that with her own native delicacy and grace she can soften the austerity and even adorn the dignity of her noble mistress. As much, therefore as the British government is superior to all other human governments that ever had a being, so much ought a British soldier to rise above the character of every other soldier, in the acquisition of all useful knowledge, in discipline, in military skill, in self-denial, in the patient endurance of toil and hardship, in valour, in generosity, in love of country, and every other intellectual and moral accomplishment.

The degrading notions and practices I have been speaking of, would, however, scarcely have had a being, but for a principle of a much higher original, and of a more authoritative and extensive influence, than the misconceptions of parents and the effeminate ambition of ignorant young men. In forming the British army, it has been too much the prevailing maxim, that practice alone is sufficient for of a soldier; that he will learn the business of his various stations, as he advances from one rank of service to another; and that he will be qualified for the discharge of his numerous and important duties, by being called to the exercise of them. Hence have arisen all the false notions that ever prevailed of the sufficiency of bodily strength and bodily exercise, to the neglect of preparatory education, and the exclusion of intellectual and moral qualifications.

In early times, indeed, when this nation had few foreign connections, and little foreign territory; when she depended for protection on her own situation, and the superior strength of her fleets; when her people where soldiers only by accident, and for a short duration; and an extensive military institution would have appeared rather useless, and in some respects dangerous; such a maxim might have admitted of some apology. But even then, so far as the people were required to be soldiers, so far was a previous knowledge of the art of war indispensably necessary; their innate bravery could effect much, but bravery without skill will not always insure success: and the nation would have though herself unable to maintain the dominion of the seas, and unsafe in the protection of her coasts, if her numerous naval officers had not been compleatly qualified, for the service which had rendered them superlatively eminent, by an early, well cultivated, and profound knowledge of the theory of navigation in its utmost comprehension.

But new and different states and situations, grown in establishment, necessarily require new and different maxims and pursuits; and as Great-Britian, by her multiplied connections and enlarged empire, is brought near the condition of a continental power, why the officers of her army should not emulate the officers of the continent in that military skill which has rendered them so illustrious and so formidable, the warmest advocate for practice to the exclusion of theory will find it difficult to align a reason. Is there any thing in nature from which the boasted superiority of French officers over British can arise, but this rejection of preparatory theory from British military education? A Briton is as brave, and as intelligent, as a Frenchman: he is less volatile and desultory, indeed, and less quick at invention; but he is more patient, and more persevering in the improvement of suggested knowledge: he has a more generous sense of the obligations of duty in civil and political life; and a much higher principle, to carry him through all the difficulties and obstructions of the greatest enterprizes. That he should fall short, therefore, only in one attainment of such great importance to his honour and welfare, must be wholly ascribed to a want of resolution break the chains that blind and obstinate custom has imposed upon him.

The minds of youth, if suffered to rove at large without employment, and not habitually tuned, by repeated exercises, to the pursuit of good, will necessarily decline into evil: and shall a candidate for the army be more ripe for any mischief that ignorance and ungoverned passion can suggest, because he has had less culture than another? And what culture can be more suitable to his views and duties, than of that science, and of those principles and habits, on which all his future excellence and usefulness depends? One man many be honest, though destitute of dexterity of learning; another may be learned, without corporal or mental strength; and another may succeed in various employments, without either courage, wisdom, or virtue: but a soldier requires the united force of all these qualities, and the want of any one of them must render his character defective.

In most manual arts, and in some sciences, a stated course of several years is allotted for the instruction of students, when theory and practice mutually contribute to their improvement; and he will be found the greatest proficient, who has established his practice on the most accurate and best digested theory: and though, in the effusion of that esteem which is always felt for eminent talents, it may be said, that a man is born an orator, a poet, a painter, and even a soldier; yet it is certain, that correctness and great mastery are derived, not from nature but art, not from genius but study.

It is the observation of Folard,[671] that "war is a science for the ingenious, and a calling only for the ignorant:" but as a science, it cannot possibly be acquired, without a continual study and application of the best formed rules and precepts, a combination of all possible events in their minutest complications, and conclusions deduced therefrom and treasured up in the mind ready for service: whereas in practice along, though enlarged by the experience of many campaigns, the mind must remain blind to the event of numberless operations, and always liable to be disconcerted by the capricious turn of fortune.

But, between theory and practice, when placed in competition, let history determine which has the superior excellence; and history has given

---

[671] Jean Charles, Chevalier de Folard (1669–1752 A.D.)

many examples of great generals suddenly produced from the fruitful source of anteriour study, but none who left their knowledge to be acquired solely by practice. Alexander, Cyrus, Pyrrhus, Scipio, and many more, stepped at once from theory to the chief command of great armies, and almost as soon as they appeared were known in the characters of commanders and victors. That the greatest success, and the most useful instruction, may arise solely from the operations of the mind on a rational and well digested knowledge acquired by study, the examples of Zisca,[672] general of the Hussites, and of Count Pagan, are incontrovertible proofs: both, in the prime of life, furnished the most important lesson in the art of war, the one by his victories, and the other by his system of fortifications; both were blind, and therefore, had theory chiefly for their guide. Instances like these, even in barbarous times, must have evinced the superiority of theory over practice, when placed in comparative view.

The ancients, not less eminent for a profound knowledge of the mysteries of war than the moderns, were more convinced of the vast importance of preparatory theory, of the absolute defect of practice singly, and of the indispensable necessity of both in continual union. They established public school for instructing their youth in this science by rule and precept illustrated by practice: and those republics that could not bear that expence of a military institution in its fullest extent, encouraged masters by honorary rewards to read stated lectures on the subject: and this, in the leisure of peace, and by the extensive theory, the Greeks and Romans prepared themselves for war. Xanthippus and Gylippus, two subaltern officers of Sparta, illustrated their training in this form and brought lasting honour to themselves and their country, the one by gaining the battle of Tunis, in Africa, against Regulus, and the other by saving Syracuse.

But convinced as I am of the necessity of the most highly improved theory, it is far from my idea that it should ever be separated from practice: I have been arguing only against those, who are best upon the disunion, upon repudiating theory with indelible disgrace. I consider theory only as anticipated practice, and practice as the natural illustration of the proof of theory; and the warmest advocates for theory have never considered it in any other light: Xenophon says, in many passages of his works, that "every thing must be referred to action." They are like light and substance, irradiating and giving strength and beauty to each other; like the soul and body of the art of war, whose dissolution must be succeeded by darkness and death.

I will even venture to assert, that the continual application of theory to practice, is more necessary in military science, than in any other. This forms the

---

[672] Count Jan Žižka (1376–1424 A.D.)

criterion, by which along the different attainment of young students can be justly estimated; and most honour is due to him, who has secured his advances step by step, confirmed his theory by an accurate practice, and acquired a facility of applying his knowledge to real occasions, by continual proving and illustrating the truth of it by those that are feigned. The ultimate end of war is singularly great and important; and the attainment of it depends upon means various in their nature, and almost endless in their combinations: and from the opposition of art to art, and of strength to strength, added to the intervention of uncontrollable natural causes, such new, sudden, and critical revolutions must frequently arise, as required the immediate exercise and application, not only of all the fortitude, but also of all the knowledge of which the human mind is capable.

In the course of the preceding observations, it is hoped that the principal points I have endeavoured to illustrate are evinced, and it is sufficiently evident,

That as mankind, with respect to the natural powers of the understanding, and the great tendencies of the will, are nearly upon a level, the diversified distinctions of intellectual compass, and of vicious and virtuous manners, that have appeared among them either as individuals or as formed into states and kingdoms, are to be ascribed to the nature and power of education:

That no state can long subsist, let its form of government be what it will, if the education of its youth is not peculiarly adapted to the nature, end, and principle of that government:

That, in the British state, education implies that attainment of all that enlarged knowledge and generous virtue, by which its constitution of government has been formed, and by which alone it can be supported; and that, therefore, a British soldier, whose profession is not valuable for its own sake but for its subserviency to the welfare of the state, is in a more peculiar manner bound to make such attainment, lest that which is intended as a benefit should be perverted into an injury:

That a British soldier, to answer the ends of his institution, requires the superaddition of the highest excellence in his own profession considered as a science:

And that excellence in military science cannot possibly be attained, without the knowledge of the most extensive theory, illustrated and confirmed by the continual application of it to practice.

Hence, though military education will naturally divide itself into the two branches of the exercise of the body, and the operations of the mind; yet it is to be understood, that such a degree of corrective intelligence and virtue must always accompany even mere bodily exercise, as will keep them within due limits, and improve their subserviency to the great ends of a soldier's duty. Thus a robust constitution that is fitted by nature to bear hardship and fatigue, must be animated by the love of hardship, that despises ease and indulgence as unworthy of a man; and indured chiefly to the kind of hardship to which military service is most exposed, such as long marches, extreams of heat and cold, course food, hard lodging, and short and interrupted sleep.

A young man brough up in the prevailing delicacy of the times, will never make a soldier, without such an effort of virtuous resolution as the spirit of the times does not greatly encourage us to expect. Much, however, may be done by gradual advances: he must relinquish the amusement of the town; decline the pleasure of the table for the sake of plain food; abandon the refinements and fopperies of dress; love cleanliness only for the sake of decency and health, and the distinctions of apparel only for the respect due to his profession; and thus, by repeated acts of self-denial in personal accommodations, he may acquire counter-habits, and give life to that manly virtue which an effeminate education had suppressed.

The principal article of bodily exercise, are dancing, fencing, , riding.

Dancing gives grace to all motions of the body, ease and elegance manner and address, and that manly confidence which even the best characters require for their deportment in public life. But here, in conformity to the pervading principle, dancing must have its limits: if it is practiced merely as a diversion, if it is loved only as an accomplishment for attracting female notice, if too much time is devoted to it, and if it gives birth to affection and grimace, it had better not be learnt. Much, however, both of its true use, and its perversion, depends upon the choice of the master.

But fencing, for its manifold advantages, can admit only of one limitation. As an exercise, it produces great muscular strength, and is peculiarly adapted to improve health and invigorate the constitution; as an art, it gives quickness of fight, agility of motion, a martial air, a firm and graceful attitude, and such a knowledge of the structure of the human body as will direct the practitioner in the nicest application of its strength. But, notwithstanding there advantages, it is objected to by some, as inspiring too much confidence, and consequently producing competition and quarrels, which, from a false notion of honour, end in the pernicious practice of duelling. There is, indeed, some apparent reason for charging fencing with this evil; which, however, upon examination, will be found to arise solely from the perverted notions of honour, gallantry, and courage, inculcated in some schools; and must, therefore, not be imputed to the art itself, but to the abuse of it. But here the restraining principle will interpose for the safety of the pupil, for keeping his skill in fencing within due limits, and directing it to its proper ends. He must be taught, that true valour is cool and patient, not apt to take offence, willing to hear the pleas of reason, and most ready to forgive upon the least acknowledgement; that if, after repeated provocations, insolence is not to be repressed, and the quarrel must be decided by the barbarous and superstitious custom of single combat, the decision cannot be made by an weapons fitted for the destruction of life, without infringing the laws both of God and man, which have been always united for its protection; that though, as a soldier, it is proper he should be expert in the use of the sword, yet, as a British soldier, his own sword is hallowed in the cause of virtue and the public; and that neither his sword nor his life are at any time at his own disposal. His mind must be strongly impressed with the just and noble motives that influence the conduct of the Roman, who, upon being urged to single combat, made this answer to the challenge: "To-morrow we are both to face the enemy; let it then be decided, which of use has most courage, and is the best citizen: for remember, friend, that our lives are not our own, but our country's."

If we look no further than trifling punctilios on which the practice of duelling is founded, and the solemn ceremonial with which it is conducted, it is most Quixotic absurdity that can disgrace the understanding of man; and, in this view, as those devoted to the practice of it are less insensible to the poignancy of ridicule, than the rebukes of conscience, it is to be wished, that the wit of mankind was united to laugh it out of the world. But it is born of such passions, and produces such horrible effects, as to demand the utmost exertion of legal authority for its total extirpation. For an imaginary affront given to imaginary dignity, substantial and lasting disgrace and misery are incurred; families are plunged into irremediable distress, and the authority of government is daringly infringed, and the unalienable prerogative of the Great Author of Existence impiously invaded. The poison of this infernal spirit has, indeed, no limits: it can so transform the understanding and the hear of man, that the wretch who has corrupted his friend's wife, can think cooly and deliberately of killing the husband; and hope, by fighting a duel, to justify his abused character, and render both adultery and murder honourable. If there is an extravagance of folly to equal this extravagance of wickedness, it is that of the much injured sufferer, in putting his own life upon a level with the life of the abandoned aggressor.

Though it is always desirable to secure obedience by the pure influence of truth benevolence and piety, rather than by restraint of positive law; yet young minds, in the ardour of high courage, and controlled by those dazzling but deceitful notions of honour which folly has introduced and custom only has established, may stand in need of some external check: military students, therefore, must be informed that so far is the violation of the laws against duelling necessary to prevent an imputation of cowardice, that, as soldiers, in whom cowardice in the time of action is punished with death, the laws against duelling are strictly enforced upon them by the following clause in the nineteenth article of war:

"Nor shall any officer or soldier upbraid another for refusing a challenge; since, according to these our orders, they do but the duty of soldiers, who ought to subject themselves to discipline. And we do acquit and discharge all men who have quarrels offered, or challenges sent to them, of all disgrace or opinion of disadvantage in their obedience hereunto: and whosoever shall upbraid them, or offend in this case, shall be punished as a challenger."

Swimming, besides its natural power of conducing to the refreshment and health of the body, is so necessary to a soldier, who must often pass the seas, engage near great rivers, and sometimes cross them when bridges are not ready or cannot be used, that his common preservation not less than discharge of his duty, requires an expertness in it which frequently practice only can give. The Romans considered it as such an indispensable article in an improved education, as to rank it with literature; and it was the common censure on one poorly educated and good for nothing, "that he ha neither learnt to read nor to swim." It was no wonder, therefore, that, with so many additional inducements, the soldiers at Rome, after business of the day, should make it their constant practice to bathe in the Tiber.

The great dramatic poet of this nation, who knew in what high estimation the art of swimming was held by the Romans, and that their greatest men, especially if they were warriors, were the greatest proficient in it, has described with his peculiar force and beauty a contest in swimming between Caius Cassius and Julius Cæsar; and to intimate that it was a practice as customary among them as contests in other athletic exercises, he has most judiciously divested it of all parade and ceremony, and represented it as the sudden effect of a sudden challenge given and accepted in sport. Cassius himself relates it, among other instances, to Brutus, to heighten the contrast he was drawing between the inferiority of Cæsar's personal talents, and the ambitious and dangerous designs he was forming against his country.

I was born free as Cæsar, so where you;
We both have fed as well; and we can both
Endure the winter's cold, as well as he.
For once upon a raw and gusty day,
The troubled Tiber chasing with his shores,
Cæsar says to me, "Dar'st thou, Cassius, now
Leap in with me into this angry flood,

And swim to yonder point?" Upon the word,
Accoutred as I was, I plunged in,
And bid him follow; so, indeed, he did.
The torrent roar'd, and we did buffet if
With lusty sinews; throwing it aside,
And stemming it with hearts of controversy.
But ere we could arrive the point propos'd,
Cæsar cried, "Help me, Cassius, or I sink!"
I as Æneas, our great ancestor,
Did from the flames of Troy upon his shoulders
The old Anchises bear; so, from the waves of the Tiber
Did I the tried Cæsar.---

Upon this passage Mr. Theobald[673] has inserted the following note, which he ascribes to Mr. Warburton[674]: "This may, perhaps, appear a very odd amusement for two of the greatest men in Rome. But that this was an usual exercise for the nobility, that delighted in the hardy use of arms, and were not enervated, appears from the passage of Horace, book I. ode 8.

Cur timet flavum Tiberim tangere?

Upon which Hermmanus Figulus make this comment: Natare. Name Romæ primæ adolescentiæ juvenes, præter cæteras gymnasticas disciplinas, etiam NATURE discebant, ut ad BELLI MUNERA FIRMIORES APTIORESQUE ESSENT. And he puts us in mind from Suetonius, how expert a swimmer Julius Cæsar was." See Thoebald's Shakespeare, edit. 1767, vol. vii. p. 10,11.

The benefits of riding for the preservation and improvement of health, is universally acknowledged. The particular form of this exercise more immediately necessary for a soldier, is the expert management of the great horse; a proper training in which will qualify him to act with greater usefulness, and more personal safety, in various emergencies. The military service of riding, however, seems rather limited by its encumbrances: the success of a sudden enterprise may require the instant dispatch of new orders; and it is grievous to think it must fail, because a horse is not ready with all his accoutrements, and there is no rider to back him without them: if, therefore, military students, like the Numidians of former times, and the Tartars and Morattoes of the present, were to accustom themselves to mount, and leap, and ride at full speed, without a saddle, they would at least have the satisfaction of knowing, that they are qualified for any service that depends on celerity and instant execution.

---

[673] Lewis Theobald (1688–1744 A.D.)
[674] William Warburton (1688–1779 A.D.)

The use of this exercise in hunting, is also a singular advantage; but the regulating principle must confine it to the great purpose of military service. It forms insensibly a most necessary quality in a soldier, the coup d'oeil, which consists in judging so precisely of a country, as to tell, at first view, the number of troops which the whole or any assigned part of it will contain, the best situation for an encampment, and the best possible dispositions for the order of battle.

It is from hunting that the Indians in America have derived their chief knowledge of the art of war; and it has made them compleat masters of the coup d'oeil, and enabled them to contrive uncommon ambuscades, safe for themselves and dangerous to their enemies.

The great Cyrus, remarkable for the frequent use of this exercise, pursued it more for his instruction than his pleasure. In his expedition against the Armenians, he sent Chrysontas to possess the avenues of the country; and in his orders for that service, he alludes to the stratagems used in hunting, and supposes in his officer the ability to judge of the county which much hunting had given him:

> "Consider," says he, "as in hunting, that we are to be the finders, and that you stand at the nests: remember, therefore, that the passages must be stopped, before the best is roused: and that they who are appointed to the station, ought to be concealed, if they would not turn off every thing that takes its course towards them. Do not march through the woods and such difficult places; but order your guides to lead you the easiest way, unless there be one that is abundantly shorter; for to an army, the easiest way is the quickest."

Where hunting is not often practicable, riding, and even walking, for the sole purpose of acquiring the coup d'oeil, must be frequently used. In cultivated countries there is a general conformity of disposition and situation; and he that has been used to judge of small spaces, will easily be led to judge of greater; so that, with a little aid from mathematics, he will soon be qualified, upon the first sight of a new country, to determine, with precision, the extent of a plain the course of a river, the altitude of a hill, and even of two hundred different situations for an army afforded in a space of near two leagues square, to point out which is the best. Instances are to be found in history, of many armies saved by this faculty of nice discernment, and of many lost by the want of it. In the war with the Samnites, Cornelius the Roman Consul was repulsed in a valley, and in the greatest danger of being surrounded and totally cut off; when Decius the tribune, upon an accurate observation of the country, discovered an eminence and the avenues that led to it, and by securing a retreat saved the army from perdition.

Many other exercises for the improvement of agility and strength may be introduced, and tennis, cricket, foot-ball, wrestling, and running, occasionally take their turns: but those are to be used the least, and for the shortest space, that are only mere diversions; that have the least tendency to promote that kind of activity and strength which a soldier most requires, and the least connexion with his advancement in military science. Whatever is done for the sake of inspiring true courage; particularly, what is done from a love of hardship and fatigue, such as traversing rugged surfaces, penetrating thick woods, and climbing steep and difficult ascents, especially if the mind is all the while usefully employed, cannot fail to produce lasting benefits. The views of the ancients in the institution of their Olympic and Pythic games, extended as much to the soldier as the man; and the many noble examples which they contributed to form, should have engaged the moderns in some public institutions of the same kind, that while the bodies of their youth were exercised and strengthened, their minds might have been delivered from effeminacy and sloth, and fitted for the highest attainments in wisdom and virtue.

We proceed now to the operations of the mind, which, indeed, have an unlimited scope: but as some attainments are more adapted to the views and duties of soldiers than of other men; as some subjects of study require a greater degree of his peculiar application, and others less; it is necessary to direct his progress with a judicious discrimination, that every part of the time allotted for preparatory education, which grows more valuable in proportion as it grows more short, may be advantageously employed.

The first and most important object in the cultivation of his mind, is to establish those principles of moral truth and duty, and form those habits of severe virtue, that will support the dignity of his character in all situations, and render his actions not less honourable and useful in the stillness of peace, than in the activity and bustle of war. This is to be effected only by gradual advances; and therefore must never be out of sight, but constantly accompany his progress in science, as the only ground upon which the advantages of sciences can be permanently secured. With this, where brilliancy and vigour of natural parts are wanting, an inferior degree of skill in his profession will preserve the honour and usefulness of his character; but, without it, the higher his attainments in science are, he will only be rendered more insolent, more presumptuous, and more mischievous.

But as military students are designed, not for learned casuists, but for accomplished offers and useful men, this great business is to be pursued more by practice than by study, more by the vigour of the heart than the speculation of the head. His common course of reading, of which some notice is hereafter to be taken, will furnish him with many instances of the lasting influence both of good and evil principles and habits; and a little superintendence will assist him in the just and discriminate application of them to himself: but the great foundation of wisdom and virtue is to be laid in his own heart, and it consists in self-denial. Mental not less than corporal vigour depends on the ability to bear hardship; the love and desire of this ability will promote the attainment of it; and when by repeated efforts controlling the strong tendency of the will to ease, indulgence and pleasure for the sake of a strict and honourable conformity to discipline and duty, a vigorous habit of self-denial is once established the mind will be fitted for all the various scenes and offices of duty in the higher walks of life, and be always able to repel the influence of the three great corrupters of the intellectual and moral excellence, gluttony, wine, and wanton women. Reasoning minds have not found it difficult to trace back all the virtues that can adorn the soldier and the man, to self-denial as their common source; the very ideas of temperance, patience, justice, and even liberality, in their utmost latitude, being apparently involved in it. And, indeed, he that has obtained the mastery of his appetites and passions by continual resistance, is in no danger of being tempted into excess and riot: he that has cheerfully submitted to voluntary hardship for the love of order and due obedience, will not be apt to murmur at severe service in the order of military duty, nor at disappointment and distress in the order of providence: he that can easily surrender his own alienable rights for the common good, is not likely to appropriate the alienable rights of others; and knowing life and liberty to be unalienable rights in himself, he will hold them sacred and inviolate in all mankind: he that can despite ease, indolence, and pleasure, will set no value on the means of gratifying them; and therefore, will look on the possession of wealth, not as a gift for his own use, but as a deposit for the reward of indigent merit and the alleviation of helpless wretchedness.

Fortitude, however, that distinguishing characteristic of a good soldier, is the genuine offspring of self-denial; and if it does not include in it all other virtues, is, at least, best protection and support. Insensibility of danger cannot constitute this character; it knows all that it has to suffer, and fears nothing but the abatement of it own vigour: it is a calm undisturbed unchangeable resolution of persevering in duty, whatever danger lies in the way, or whatever evil may be the consequence; and is, therefore, the irreconcilable and determined enemy of vice, in every form both of violence and allurement. No people, however renowned for courage, can long preserve their character when once corruption has broken down the boundaries of duty, dissolved the restraint of discipline and given humour appetite and passion their own scope.

In this kingdom, where moral science has been more cultivated than in any other, the student, out of the vast variety before him, may be assisted in the selection of such truths, as will illustrate the principle of self-denial, and by convincing his judgment confirm him in the practice of it; and if, as indispensably necessary to compleat his system of duty, his attention is turned to the spirit and precepts of Christianity, he will find, that the same self-denial which he has been practicing as the sure ground of wisdom and virtue, is made the sole foundation of every exalted attainment that Christianity requires.

It may, indeed, be objected by some, that it is rather an impropriety to recommend Christianity to the notice of a soldier, when Christianity is supposed to discountenance war as repugnant to its heavenly spirit. The frame objection, however, from the precepts of Christianity, lies equally as strong against the use of oaths, and, by implications in the nature of spirit, against a boundless commerce, and against litigations in courts of law, as against war. But, in the present imperfected state of things, the people of Christian nations practice oaths, devote themselves to the extension of commerce, and harass and distress one another to obstinate and expensive lawsuits, even more than they engage in war; and therefore, a soldier who is trained up only for the occasional service of his country, is not more exempted from the obligations of Christianity on account of his profession, than he who practices oaths, or he who devotes himself to trade, or he that goes to law. It is not necessary, indeed, that he should perplex his mind with the various speculative opinions, that divide, inflame, and nearly disqualify its various professors: let him consider it only as a dispensation of Mercy from Supreme Wisdom and Goodness, for the restoration of all mankind to the original knowledge benevolence and dignity of their nature; let him make it percepts, as much as possible, the rule of his practice, and its promises the foundation of his hope; and as a soldier, who ought to value obedience more than life, he will always be the wiser, the better, and the happier, for his religion.

Upon this foundation, and with the progress of the noble structure to be raised upon it, the student may securely advance in the attainment of all necessary science: and of languages, which ought to be the earliest object of his notice, the French and German are the chief. The French language, though inferior in strength and beauty to most other modern languages, yet by the insuppressible vanity of insinuating policy of the people is become the general medium of communication throughout the civilized world; and as Germany has long been the common theatre of war, the language of that country has the next claim to the attention of a British Soldier, who, as he will stand in continual need of both, for occasional intercourse with prisoners spies and peasants, and for general consultation and converse with foreign officers, must learn to speak them colloquially as well as with force and elegance: and if, to these, he can add the Spanish and Italian, his personal advantages will be improved, and his military usefulness much enlarged. These, as living languages, will be more easily and more effectually attained, by constant conversation and the frequent perusal of good writers, than by the rules of grammar. The knowledge of grammar, however, that he has brought with him from school, it would be folly to discard, if, without laborious study that will rob him of his time, it can be advantageously applied to the languages he is learning, and chiefly to his own, which merits his highest respect and most assiduous cultivation. So also, if, by a singular proficiency in early education, a young gentleman is able to read the Latin and Greek historians in the original, it is a qualification too commendable not to be preserved: he must only beware of making it the ground of self-admiration, of a neglect of better knowledge, and of a supercilious contempt of his comrades; for as translation has long been industriously and wisely employed, and as a Soldier is in quest of the knowledge of things and not the words, all the knowledge he requires he will find abundantly supplied, either in English, French, or German.

This attainment of language, like that of generous virtue, will also be progressive, and accompany the student through the whole course of his education. The articles of science that require a more distinct and separate study, may be ranged under the general heads of mathematics the natural and civil rights of mankind, ancient and modern history, and the constitution of his own country.

Mathematics, which contemplates whatever is capable of being numbered or measured, is as necessary to the soldier, as it ever was to the astronomer; it contributed no less to form a Turenne[675] and a Vauban,[676] than an Archimedes[677] and a Newton.[678] Mathematical knowledge delivers from credulity and fortifies against scepticism, restrains from rash presumption and disposes to a rational assent; and while the mind is engaged in contemplating distinct ideas, and arranging beautiful forms, and adjusting the harmony of proportions, the affections are composed, the fancy restrained, the understanding fitted for still higher exercises, and the manners corrected and improved.

But though mathematical investigations have these general advantages, the military student, who is designed not for a contemplative but an active and

---

[675] Henri de La Tour d'Auvergne, Viscount of Turene (1611–1675 A.D.)
[676] Sébastien Le Prestre de Vauban, Seigneur (later Marquis) de Vauban (1633–1707 A.D.)
[677] Archimedes of Syracuse (287–212/211 B.C.)
[678] Sir Isaac Newton (1642–1726/1727 A. D.)

busy life, must generally decline all the serve only for meer speculation and amusement, and confine himself chiefly to those that are immediately applicable to his various duties and employments. The occasions for the practical use of the science of numbers in military duties, are too minute and various to be recited here; but in a System of Mathematics[679] which I have been forming for the sole purpose of military service, and is just published, I have adapted it to all the possible situations offices and employments of an army; by which, I trust, military students will be convinced, that excellence in their profession indispensably requires a most comprehensive knowledge of numbers, and the greatest expertness in the application of it. The same may be said of geometry, or the science of extension, the practical uses of which I have also illustrated in a variety of important instances. Upon this ground alone the student can be led to the knowledge of the nature laws and effects of motion, and moving powers; and to the more relative sciences of fortification, tacticks, and artillery.

The science of fortification is intimately connected with that of tacticks, and when rightly attained is the utmost importance to every military man. That the great advantages to be derived from it are not obtained, must be ascribed to the general and superficial manner in which it is taught; for though the theory is very simple, yet the modern treatises, some of which have been written even by ecclesiastics, are not calculated for its elucidation and improvement. It is customary, I know, to adopt either the system of Vauban[680] or Coehorn,[681] as prejudice and indolence may determine; and to confine the student to the chosen one, without pointing out the ground or preference, or assigning the reasons which determined each of those engineers to forsake the accustomed path, and strike out a new one on his own. To say, as many do, that the difference in these systems consists in the greater or less covering of the flanks, may, indeed, characterize the system, but will not instruct the student; who, if he can produce a plan regularly drawn and well washed, and say that it is conformable to this or that system, is, together with his friends, deluded into an opinion of possessing a knowledge that he is far from having attained. It is, therefore, much safer for him, to go a little out of the pedantic and systematical road of what is called great fortification, and applying himself to field fortification to acquire an experimental knowledge of all the different works, and in what circumstances, and with that limitation or extension, they are to be erected and applied. The former, though comprehending the

---

[679] Lewis Lochée, *A System of Military Mathematics*, 2 vols. (London: T. Cadell, 1776).
[680] Sébastien Le Prestre, seigneur de Vauban, (1633–1707 A.D.)
[681] Baron Menno van Coehoorn, (1641–1705 A.D.)

foundation, is generally learnt trouble, is sometimes recommended to students as the utmost extent of the art: but of the latter the mastery cannot be obtained, without a diligent and rational exercise and application of great principles of arithmetic and geometry.

The knowledge of artillery also cannot possibly be acquired on any other ground: and yet the knowledge of artillery is not only as proper for a soldier, as the knowledge of structure power and use of the tool he works with, is to every manual artist; but is also indispensable necessary in the attack and defense of places, which cannot be understood without it.

An expertness in drawing acquired on the principles of perspective, must give the finishing to all these studies. Without this, the student will be retarded in his own progress, and unable to impart his knowledge on others; but with it, he will not only find the illustrations of science easy and delightful to himself, but will also be capable of rendering important military service on a variety of subjects and occasions. Skill in drawing extended beyond the linear part, to plans, elevations, finished edifices, figures, landskips[682] and composition, qualifies the artist to express with precision those ideas in a few minutes, which he could not so clearly communicate by writing or speaking in as many hours.

The science hitherto proposed, with all that are dependent upon and connected with them, where there is an equality of natural talents and of vigorous application, may be equally attained by the contracted and the liberal, the selfish and the benevolent, the obstinately evil and the generously good; and, therefore, to prevent misapplication and abuse by some minds, and render them of substantial and lasting benefit to all the student must be led to the completion of that structure of intellectual and moral excellence, which is supposed to have been gradually advancing on the foundation of self-denial. Indeed, where an unremitted attention has been paid to the cultivation of that great principle, and the habit is nearly confirmed, the predominance of natural temper and disposition must be already changed, the obliquities of baseness restrained, and the aspiring of generosity animated: and when once the power of meer self-will is destroyed, whatever be the prevailing complexion and bent of nature, the path of virtue will be made easy and few obstacles can arise that will require much labour to remove; the tyranny of appetite and passion, that darkens the understanding and contracts the heart, will be shaken off, and the mind left at full liberty to pursue the public good wherever vigilant experience and unbiassed judgement point it out and to exemplify on all occasions that universal righteousness and beneficence which Christianity inspires. On this

---

[682] Archaic spelling of the modern term *landscapes*.

ground, nothing more seems to be necessary for persevering in the path of rectitude, but to be acquainted with the great lines of duty in real life, and the manifold deviations to which they are exposed in the present complicated and imperfect state of human nature. The student, therefore, will regularly proceed to the study of civil law, separated from the refinements and sophistry of private contests, and confined to the great and liberal parts respecting the natural rights of mankind, the origin and foundation of society and government, and the civilized nations: and on this interesting subject, Grotius,[683] Puffendorff,[684] Heineccius,[685] and Montesquieu, will give him ample information.

But, to obtain a true knowledge of human life and manners; to penetrate into the recesses of the hear; to follow the progress of virtue and vice in their minutest gradations; to estimate the power of habits; to make the manifold diversification to temper and character; to distinguish between essential and adventitious good; to behold the rise and fall of great states, their gradual advance to virtue and glory, and their rapid declension to vice and ruin; to discover the excellencies and defects of government; to perceive what and how much is to be sacrificed to fulfill the obligations of social civil and political duty, and what chiefly disposes to the violation of them; to form maims of true wisdom and generous virtue for the conduct of private and public life; and, indeed, to bring to the test of example and experiment all the knowledge he can acquire from observation, from a view of others or from an intimate inspection into himself; he must study history, a science that cannot be begun too soon or continued too long.

But as history is a recital of past events that have taken place in different countries, and in a series of succeeding ages, it is proper to obtain some preparatory knowledge of chronology. The revolutions of one age often give rise to and are intimately connected with those of another; and, therefore, to prevent confusion, the student should have such a general comprehension of the whole current of time, as will enable him to trace out distinctly the connection and dependence of great events, to arrange them into distinct periods, and by adjusting the whole to form general period as the standard, to preserve it in one regular and uninterrupted series.

In like manner, as the situation of different kingdoms with respect to each other, and their interfering views and interests, have been the fruitful

---

[683] Dutch humanist Hugo Grotius (alternately Huig de Groot and Hugo de Groot) (1583–1645 A.D.)

[684] German philosopher Samuel Freiherr von Pufendorf (1632–1694 A.D.)

[685] German jurist Johann Gottlieb Heineccius (1681–1741 A.D.)

source of invasions, and of long and obstinate wars that have ended in great revolutions; it is necessary to have some knowledge of the various distributions of the earth, the extent of states and kingdoms, and their subdivisions and dependences: and this makes geography another part of preparatory science for the study of history. But though a general acquaintance with the terraqueous globe, and the great changes that have happened on it from natural causes as well as from the agency of men, may be sufficient for the purpose, yet there is a dependent branch of this science in which a military student cannot be too expert, and that is an exact and even a minute knowledge of particular provinces and kingdoms. It is by this skill topography that he will be qualified to chuse situations for encampments, direct the march of an army, dispose the attack and defence of posts and places, and order other military service, in which a rivulet, a morass, a flight eminence, a hallow way, and even the minutest circumstance of variation, is often of great importance. It is the knowledge of the difference in the soil and external formation of countries, that determines the different mode of carrying on war, in Flanders, Germany, France, Italy, and Spain; and it was a singular attainment of this kind, that the great Turenne chiefly owed the successes of his tow last campaigns.

But though the enlargement of the mind by an intimate and practical knowledge of human nature, life, and duty, is the primary advantage of studying history, yet the military pupil is also to make a subordinate use of it in an immediate reference to his own profession; and, in this view, to study history, is to study war. Lucius Lucullus,[686] who supported the triumphant progress of the Roman arms in the east, and defeated the great Mithridates in every encounter, derived his knowledge chiefly from reading the history of former wars, to which, it is said, he applied himself only in the passage from Rome to Asia; but this application would have produced little benefit, if he had not scrutinized and pondered every circumstance that tended either to the miscarriage or success of the great generals whose actions he was reviewing, and deduced maxims from the whole for the direction of his own conduct. It is presumed, that there are few officers who have not read Xenophon, Quintus Curtius,[687] Cæsar, Vegetius;[688] but unless the various maneuvres described by those writers are carefully analyzed and compared with those of modern times, and their degrees of usefulness and importance proved by the test of application in similar instances, such reading ill deserves the name of study for the acquisition of military knowledge. Indeed throughout the whole extend of

---

686 Roman general Lucius Licinius Lucullus (118–57/58 B.C.)
687 Roman senator Quintus Curtius Rufus (died 53 A.D.)
688 Publius Flavius Vegetius Renatus (died after 383 A.D.)

historical learning, whether from general purpose of life and duty, or the particular improvement of his own profession, the ardour of mere curiosity must be restrained, and the student habituate to ponder, with diligent inquisition and severe reflection, the characters actions designs and events that rise before him; till he derives the same advantages from this more enlarged knowledge by the continual application of it to his business and himself, as he had been used to derive from mere military science by the continual application of it to military service. It is this friendly harmonious indissoluble union of genius and learning, of nature and art, of theory and practice, throughout his whole course of studies, that must crown all his labours, and make him an accomplished soldier, as well as a wise virtuous and useful man.

Of the books necessary for all these studies there ought to be a scrupulous and discriminate selection: if the best are chosen, a few will be sufficient. It is, indeed, to be lamented, that in every garrison of this great and respectable kingdom, there is not a public library appropriate to the use of the army, to which young officers may be privileged but obligated to resort, for the preservation and advancement of that science which they have acquired in their pupilage at an academy: and till this could be accomplished in its most liberal extent, it is to be wished, as a partial substitute, that every regiment was furnished with a small collection of the best books as part of its common baggage. My zeal for the honour and perfection of the service, emboldens me to mention this; and if I should ever fee the establishment of such an extensive fund of military science, it would give me inexpressible satisfaction to reflect, that I once took the liberty of publickly suggesting it. It would at least take away all pretensions for waiting the precious leisure of peace in idleness, that paralysis of the mind, which not only disqualifies it for the future acquisitions, but renders it incapable of retaining those that are past.

Though many private academies for military education are encouraged in foreign states, in which the science and discipline proper for a soldier are most assiduously and rigorously inculcated, yet so active is the vigilance against the encroachments of idleness, that there are also academies erected in the most capital garrisons, to which, in the winter, young officers are sent in rotation, to improve the knowledge that had gained in their private seminaries. This has raised such a general spirit of emulation, that even the private men of several companies in a corps of artillery have voluntarily formed themselves into a society, for a daily review of their progress in military knowledge; and the few who from indolence or the pursuit of pleasure neglected to attend, incur great disgrace, and hear from their wiser comrades the same reproof as some of the Lacedæmonian soldiers once did from the Ephori, who told them, that sauntering about town in quest of amusement was unworthy of a Spartan, whose whole time ought to be devoted to the attainment of virtue and the welfare of his country. For the same wise purpose of suppressing idleness, it is now a custom in some parts of Germany, for the subaltern officers of every regiment to repair to the major's house at six in the evening, where they are strictly examined in every branch of military science and duty: this mode of instruction is managed by a series of important questions, containing general maxims that lead to particular illustrations, and is common to every corps. I have myself so sensibly experienced the benefit of such a rigorous and unremitted interrogation, as to be induced to introduce something similar into my own plan, and to adapt it as well to officer of hose, the engineer, and the artillerist, as the officer of foot.

It is usual to close all systems of education, with recommending travel, as that which is to give the high-wrought form and polished lustre to the whole body of preceeding discipline. But however great the advantages of early travelling may be supposed, they are remote and difficult of attainment; while the disadvantages, if not much greater, are immediate and easily incurred. A foreign country is certainly the most improper stage for a British youth's exhibition of the untried efforts of practical wisdom, the first application of his attainments to the realities of life and manners. An intimate acquaintance with his own country, in every useful sense in which it can be understood, is a previous qualification for travelling indispensably necessary; and therefore, instead of being the concluding part of a plan of education framed and imposed by others, it should be left to the voluntary choice of mature years, when experienced judgment, and a generous thirst after new knowledge concur to produce it. But if a youth had not enjoyed a free and enlarged communication with men of genius, learning, wisdom, and virtue, at home; can he be fit for such an intercourse abroad? and will he not then be left to the company of fools, fops, and libertines? If he had no idea of what, as truly valuable, would attract the notice of a foreigner here, can he form any intelligent rule for the government of his own curiosity elsewhere? If he has not attained that dignity of sentiment, and that simplicity of life and manners, to which the genius of his own country, after much discipline, would lead him, will he not become a motley composition of the prevailing fashions, follies, and luxuries, of foreign courts? And if he does not perfectly understand and deeply reverence the liberal Constitution of government in Great Britian, can he remain wholly uncorrupted by the slavish principles of foreign states? Let not then the military student repine, because the time and application requisite to prepare him for the great duties and offices of his profession, will not suffer him to put a period to the series of his studies by this customary desertion of his native land: he is now in the path of wisdom and virtue, leading to extensive usefulness; and a due perseverance in it, by diligent exercise and application, will in time qualify him to set his foot on any part of the globe, with safety and advantage to himself, and with honour to his country.

There is, however, a mode of travelling highly useful to the army, that I wish to see adopted; but it is limited to the mere attainment of topographical knowledge, and the visitation of fortified places, fields of battle, and the vestiges of camps. Besides the imperfect and sometimes false descriptions given in books, time that changes the views and interests of men, superinduce a change also in the places they inhabit; and, therefore, they cannot be accurately known, without a frequent visitation accurately made: Flanders, which was once an open country, by a spirit of cultivation and a care to fix the boundaries of property, is now become much inclosed. If, for there purposes, a few subalterns in every corps were sent out under the conduct of an officer of tried judgement and integrity, every part of the army would soon be furnished with a guide intimately acquainted with the different qualities and appearances of the different countries in Europe, and every other circumstance proper for the notice of a soldier. But in all these expeditions, regular views and plans must be taken as often as it is practicable, and an accurate and circumstantial journal kept of every day's progress.

Upon the liberal principles and with the great and extensive views suggested in this essay, I have at my own expence, and in struggle with many difficuluties and obstructions, established A MILITARY ACADEMY, of the regulations in which the following is a short abstract.

The members of this academy constitute a military republic; the laws of which are so framed, that the liberty and accommodation, as well as the discipline and improvement of each individual, are inseparable connected with the order and good government of the whole. Of these laws there is a written code, which is regularly read over to the whole society every week.

As the degrees of offence and punishment are adjusted with an equal and impartial discrimination, so also are the degrees of merit and reward. The rewards consist of silver and gold medals, and superiority of station and honour lasting as the superiority of merit that gained it. The punishment are either pecuniary (Guards extraordinary, arrest, and banishment to Coventry), corporal, or degrading; and to impress delinquents with a deeper sense of their offence, and also to deter others, they are immediately inflicted.

If any offence is committed of which the laws already enacted do not take cognizance, the offender is tried by a court-martial, and if found guilty the sentence passed upon him is instantly executed.

A faithful respect on past conduct restrains from the repetition of accustomed evil, and animates to the pursuit of higher good; and therefore a circumstantial journal is kept, as well of all offences and the punishments inflicted for them, as of all works of merit and the rewards by which they were distinguished; and this lies always open for public inspection.

These are general principles of the constitution and government of the society: the more particular rules of it with respect to business, are as follow.

Every candidate for admission has a week's probation, to acquaint himself with the laws of the society, and obtain the approbation of the members. After admission, he is to provide himself with the uniform, and other accoutrements.

All the members are to attend prayers at a stated hour every morning and evening; and also to attend the public service of the church every Sunday: and they are bound to the observance of this rule, under penalty of a fine for every violation of it.

The members are distinguished into three separate corps, each of which in rotation daily mount guard, and perform other military duties.

Each member, in rotation, is invested with the character of Officer of the Week, and is then distinguished by some external mark of dignity: but as it is an established principle, that those are best qualified to command who have most learned to obey, none is admitted to the privilege of commanding, till full probation has been made of his due submission to military discipline.

The week is divided into theoretical days, and practical days.

The theoretical days are Monday Wednesdays and Fridays, on which the study of the sciences and the attainment of all useful knowledge is pursued, such as languages, drawing, arithmetic, algebra, geometry, mechanics, fortifications, artillery, chronology, geography, civil law and history.

The practical days are devoted to the exercise of the body in dancing, fencing, riding, and the manual exercise, during the day; and in the evening, in sports for the improvement of agility and strength.

With respect to military exercise, the members are instructed in all the various maneuvres practiced in the field.

In the summers of 1773 and 1774, they formed an encampment on Banstead-downs; and the two succeeding summers, though the kind and condescending offices of Lord Barrington,[689] permission was granted them by Lord Spencer,[690] to form an encampment on Wimbledon Common, when mines have been sprung, shells thrown, redoubts raised, intrenchments forced, and every thing performed that can occur in a day of battle. In the manner

---

[689] William Wildman Shute Barrington, 2nd Viscount Barrington (1717–1793 A.D.)
[690] Lord Charles Spencer (1740–1820 A.D.)

Lollius[691] exercised the Roman youth, instructing them in all the operations of war, massacre excepted.

The hours employed on theoretical days are from six till eight, and from nine till twelve, in the morning; and from two till five, from six till eight, and from nine till ten, in the afternoon and evening: on practical days, the hours are from six till eight, and from nine till one, in the forenoon; and in the afternoon and evening, from three till five and from six till eight.

The French language is spoken throughout the day, under penalty of pecuniary fine.

Lectures are read twice a week on tactics, artillery, fortifications, geography &c. Each science is traced from its origin through all its variations and improvements in ancient and modern times; and to impress as well as to facilitate this progressive illustration, every part of it is demonstrated by models. In fortification, besides models in wood, various works have been raised on Wimbledon Common, by the students themselves.

In imitation of what is practised in the German corps, every member is to learn, by heart, the principal maxims in the different branches of the profession, as the ground of various questions to be proposed to him in a regular and stated course of examination.

Saturday is constantly employed in collecting the pecuniary fines, and in reviewing the progress that has been made the preceeding days of the week; and every one neglecting this, is punished by confinement to his room, till the neglect is repaired. A plan of fortification, &c. is also producted by every member; the whole of which are exhibited to public view, and, with a design to inspire emulation, are ranged according to their different degrees of merit.

The money accumulated by the fines, is appropriated to the purchase of medals, one of which is always adjudged to that student, who, on a public examination, excels his competitors.

This course of exercise, discipline, and study, is pursued, without intermission, throughout the year; and, in every scientific part, it is my constant endeavour to inculcate upon all members, the indispensable necessity of ascertaining their progress in theory, by a ready and accurate illustration of it in practice. It is in this way alone, that genius and industry will always be sure of meeting their just reward; and that no one will be in danger of deceiving himself and his friends, by pretension to a knowledge that he is not able to make use of.

And not let the impartial reader determine, whether, it this plan of education is properly pursued as the essential ground of attainments that

---

[691] Marcus Lollius (ca. 55–2 B.C.)

require unremitted cultivation, the profession of arms can possibly be an idle profession: and let him also determine, whether, to preserve its subserviency to the welfare of the state whose Constitution of government is founded in wisdom and virtue, it does not require such acquisitions of wisdom and virtue, as will not fail to render it ornamental in peace, and great and glorious in war.

Temperance and patriotism were the basis, on which the ancient republics rose to their highest glory; and those individuals who served their country in war, were only distinguished from others by the severer and more resplendent exercises of those two virtues. But the soldiers of Britain have much higher motives to the practice of the most exemplary self-denial, and the most generous public virtue: they have a better form of government, for the security of their dearest rights; and they have a divine religion, inspiring universal temperance, righteousness, and benevolence. "Citizens of this profession," says Montesquieu, "being infinitely enlightened with respect to the various duties of life, and having the warmest zeal to fulfill them, must be perfectly sensible of the rights of natural defence: the more they believe themselves indebted to religion, the more they will think is due to their country. The principles of Christianity deeply engraved upon the heart, will be infinitely more powerful, than the false honour of monarchies, the human virtue of republics, or the servile fear of despotic states: and that religion which ordains that men should love one another, would doubtless have every nation blest with the best civil and best political laws; because these, next to such a religion, are the greatest good that man can receive or impart."
When gentlemen bred for the army, are early impressed with the idea of a good citizen under the best form of government united with the profession of Christianity, as their principal and distinguishing character; and are continually reminded that their military character arises out of it, is subordinate to it, and dependent upon it; there will be little danger of their looking upon the profession of arms in any other light, than bound to the attainment of such knowledge and virtue, as will be always instrumental to the public good, whenever the public good requires their exertion. "The righteousness and mercy," say a very elegant and judicious writer, "which is due to all men, is the law of nature, the command of religion, and ought to be the first and leading maxim of civil polity. Not many ages ago, it was customary to engage in war without a reasonable cause of provocation, and to carry it on without humanity or mercy: since then it is happily become necessary, for states to explain their motives, and justify their conduct, before they begin to destroy their fellow creatures. And blessed be his memory, who first taught the soldier to spare the useful husbandman, and to feel a horror at shedding innocent blood!"

The Theban general Epaminondas,[692] though so highly renowned for his military talents, yet valued not the character of a warrior merely for itself. Distinguished as he was for his abilities and virtues in the field and the senate, he was still more eminent for his wisdom as a philosopher, and for his amiable qualities in private life; so that historians unanimously represent him, as a pattern of all that is great and excellent in human nature. Let military students be prompted to emulate this exalted character; and as Britons professing Christianity, let them crowd as much knowledge and as many virtues and graces into it, as the character is capable of receiving. The wiser and better they are, as men and citizens; the more useful and illustrious they will be as SOLDIERS.

THE END.[693]

---

[692] Epaminondas (419/411–362 B.C.)
[693] Lochée followed his essay on education with a sketch of the British Constitution in the original publication.

Appendix J: Partridge's Proposal for Primary / Secondary Education[694]

GENTLEMEN: The accompanying communication contains a synopsis of my views, hastily draw up, on a subject which I consider of great importance. Of the utility of such seminaries as are there proposed, in the vicinity of our larger cities, for the education of the youth of those cities. I have no doubt; and am well convinced that, if the experiment were fairly made by the establishment and organization of an Institution in the vicinity of one of our cities on the plan there in proposed, it would serve as a model which would soon be imitated by other cities. Boston, New York, Philadelphia, and Baltimore, all possess the requisite advantages for such an experiment. Of these, I should think New York, in consequence of its central situation, its rapidly increasing wealth and population, combined with the peculiarly liberal policy which has been adopted there on the subject of education, would be the most favorable. There are many situations on York Island, in the vicinity of the city, which are well calculated for the location of such an institution. The vicinity of Brooklyn, Long Island, also contains several sites which appear to me peculiarly well adapted to this purpose. I know it may be said, it is very easy to project plans of education which will appear well on paper, but which will be found difficult in the execution. On this subject I would observe, that I never project a plan on this or any other subject, which I would not-the means being placed in my power-undertake to carry into execution. I therefore, without hesitation, declare that I will undertake to organize and put into operations a seminary on the plan I have proposed, in the vicinity of either of the cities

---

[694] The Norwich University Archives and Special Collections has an article fragment of Partridge's design for primary and secondary education published in the *National Intelligencer*. This fragment terminates abruptly mid-sentence. Also held within the University Archives is a handwritten draft of this article. It appears to have originally been written on October 12, 1826, with the date later updated to the 14th, the same date as noted in the letter to the *National Intelligencer*. This possible original draft has some notable differences, namely there is an absence of a suggestion that these academies could be used to suppress slave revolts. The shift from the *National Intelligencer* fragment to the letter is indicated through a footnote.

before mentioned, provided a suitable site shall be prepared, for any number of pupils from 250 to 500. The latter number could be perfectly well instructed and governed, in one seminary, on the plan proposed. I will further hold myself in readiness to furnish plans of buildings, &c. for such an Institution, and to propose a method for constructing them, whereby those may furnish the capital for that purpose, would receive a liberal interest for the money invested. In addition to what I have stated, I would observe that the expenses of education a youth in such an Institution, would probably be less than what is generally incurred at the boarding schools in the vicinity of these cities. There is one further consideration connected with this system, which appears to me, would bear with peculiar force upon our Southern cities, as Charleston, Savannah, New Orleans, &c. It is, I say presume, a fact well known, that the inhabitants of insurrections, &c. amongst the colored population; and that to guard against this change a city guard is maintained in one, if not all of them, at very considerable expense, and, after all, an insecure practice is afforded. Now suppose that these was located, in the vicinity of each of these cities, a seminary on the plan I have proposed, were there were from 250 to 500 pupils, completely equipped, and in a state of perfect military organization; I ask, would not such a battalion which could go into action, at a moment's warning, constitute a most efficient and safe defense in such an emergency? It may, perhaps, be said they are boys, and consequently could render no efficient service: I would, however, remind those who doubt the prowess of boys, when properly disciplined, of the formidable resistance made to the allied armies by the pupils of the Polytechnic School at the battle of [illegible, assumed to be the Battle of Paris in 1814].[695]

By giving the foregoing a place in your valuable paper, together with the accompanying communication, you will much oblige.

Your obedient servant,
A. PARTRIDGE.

Middletown, October 19th, 1826.

Middletown, Oct. 14, 1826.

GENTLEMEN: In a lecture which I delivered during the recent examination of Cadets at this place, I gave a description of the organization of a Seminary proposed to be located in the vicinity of each of our larger cities,

---

[695] The clipping from Alden Partridge's personal collection, housed in the Norwich University Archives, had the highlighted area crossed out. The justification of establishing Partridge's school for the purpose of suppressing slave revolts is not listed in any other known source.

and which, it appears to me, would be much superior for the education of the great body of youth of such cities, than those established on the present plan. As the subject of education at the present time excites much of the public attention, and believing that some good may result from the plan I proposed I take the liberty to transmit you a synopsis of my views on this subject, as then delivered, with a request you will insert it in the National Intelligencer. The difficulties attending the bringing up and educating a family of sons in our larger cities, are best known to those on whom the care and duty devolves, but must be evident to every reflecting mind. From the ages of eight to fourteen or fifteen years, their parents hardly know how to dispose of them, for, until they attain the latter ages, they are too young to enter the counting house, or engage in other active duties. At this early age, it is dangerous to trust them much in the streets, where they are liable to personal injury, and where their morals are constantly exposed to be contaminated, from frequent exhibitions of vice in its grosser as well as more fascinating forms. Under these circumstances they are deprived of other manly and healthful exercises so absolutely necessary for developing and maturing the physical and more energies of youth, and without which, they grow up puny and debilitated, incapable of either mental or physical exertion. If they are sent to the boarding school in the vicinity, there are still many evils attending it. During the six to seven hours a day that they are engaged in school, the mind may in a degree be improved, but during the remaining portion, having no regular employment, they generally go at large, and without restraint. It is during this portion of their time, that they imbibe many vicious and even ruinous habits and practices. The absurd and foolish idea that boys must be furnished when at school, with a regular allowance of pocket money, still pervades the minds of the great body of parents, especially in cities. They are little aware, however, of the ruinous consequences attending it. Let us suppose a youth eight or ten years of age, furnished with his regular weekly allowance of pocket money, 12, 25, or 50 cents, as the case may be, the fact of his having money in his pocket draws him to same public place, confectioner's shop, grocery, or tavern, for the purpose of spending it. He probably at this early age, will commence with purchasing sugar plums, candy, &c. but soon [discover] that boys a little order than himself, drink lemonade, and he readily imitates their example. Soon lemonade becomes insipid, and punch is substituted in its place. As he grows older, he deems it manly to smoke, and cigars are consequently put in requisition. In the train of these practices, naturally and almost certainly follows profanity, gaming, and retinue of other evils, which ultimately involve the unhappy subject in disgrace and ruin. I am well aware, that to many this picture will be considered too highly colored. I know how difficult it is for the fond mother and the doting father to be made to believe that their darling son has any faults, or that he is not almost a model of perfection. Still I feel perfectly convinced, that there is no exaggeration, no fallacy of deviation, and confidently appeal to the habits and practices of hundreds of the youths of our larger cities, and these, in many instances, of the most wealthy and respectable families, as a melancholy proof of the truth of my assertion. Now all these evils, in my estimation, may in great degree be remedied, by a proper system of elementary education, based on non-allowance of pocket money, except on particular occasions; constant useful employment, and a correct but energetic system of discipline. For the accomplishment of this important object, I would propose the establishment of one or more Seminaries in the vicinity of each of our large cities, on the following general plan, viz:

1st. Let each Seminary be placed under a military organization; the Government should be strictly military in principle, and in the practice adapted to the various ages of the pupils; this would enable us to enforce a more strict discipline than could otherwise be done.

2d. Let each Seminary consist of three distinct schools, or departments, viz: The Primary, Junior, and Senior.

3d. Let the branches of instruction in each department be adapted to the ages of the candidates for admission into them.

4th. Let youths be admitted into the primary school, at as early an age as eight years—into the junior department at ten years—and into the senior department at thirteen years of age, provided they are found to possess the requisite qualifications for the two latter.

5th. In the primary department, the pupils should be correctly and critically instructed in the elementary branches of an English education, viz: Reading, writing, spelling, the elements of arithmetic, &c. &c. In the junior department might be added, the French and Spanish languages, also, the commencement of the Latin and Greek, for such as are to attend to the classics. A further prosecution of arithmetic, and perhaps the commencement of the higher branches of mathematic, the elements of geography, astronomy, &c. In the senior department, the fore going branches should be continued to which should be added such others as would fit a youth when he should attain the proper age, say fifteen years, to enter upon the active duties of the counting house, or other active pursuits for which his age would qualify him, or for admission into a higher Seminary, provided, it is intended he should obtain a finished classical and scientific education. Those entering the primary and junior departments would of course be transferred to the next higher, when they attain the proper ages, and have completed the course in that into which they entered.

6th. The pupils, especially those of the primary and junior departments, should attend to their studies and recitations in their recitation rooms, under the immediate inspection of the instructors, from eight to ten hours each day, allowing an intermission of thirty minutes every two house for exercise and amusement.

7th. The pupils should be regularly organized into a military corps—should wear an uniform dress—be completely equipped with arms, accoutrements, knapsacks, &c. and systematically and correctly instructed in all duties of the soldier, the non-commissioned officer, and the officer; should frequently be taken on military marches, and pedestrian excursions into the country, for the purpose of instructing them more perfectly in military duty; or habituating them to fatigue; of affording them an opportunity to seeing the world as it is, and learning them to walk well, which I consider as an essential part of elementary education.

8th. To the military exercises should be added fencing and dancing—also, the athletic exercises of running, leaping, and throwing the[696] discus.

9th. All the time not occupied by study and recitation by their regular meals and other necessary duties should be devoted to the Military and other Exercises and duties. No idle time should be allowed.

10th. The pupils should all [illegible] in [illegible], when their diet could be regulated answering to their ages, and other circumstances.

11th. As I consider it of great importance that a youth at an early age should know how to take care of himself, the pupils should live in quarters – a specified number being allowed to each room, where they should be required to keep their rooms and all the apurtenances in order, to make their Beds and also to keep their arms and accoutrements in order.

12th. The grounds for Exercise and parade, should be spacious and enclosed, and no pupil allowed to go out of the Enclosure without permission.

13th. No pocket money to be furnished any pupil except by the Superintendent, or his express order.

14th. A regular attendance on divine Service, on Sundays should be strictly required and enforced – Prayers should be regularly attended Morning and Evening–

16th. Familiar Explanatory lectures on Chemistry Minerology, Botany, Military Science, the Constitution of the United States, political Economy, Commerce, Agriculture, Manufacturers, International law &c &c- should be delivered and comprised in the course of Instruction.

---

[696] Transcript from this point forward sourced from Partridge, Letter by Alden Partridge, 14 October 1826.

17th. In order to render the Seminary complete there should be attached to each a range of [illegible] Shops and also grounds for gardening, and practical agriculture.

18th. All the officers of the Volunteer Corps and Militia in the City to have the privilege of attending the military lectures and practical Military Instruction at a very Moderate Expense. The following are some of the prominent advantages that would result from the adoption of the foregoing plan, viz.

1st. The youth of our Cities Might be placed at an early age, in these Seminaries, where their physical powers would be fully developed and improved, and their Constitutions confirmed, by a regular and Systematic course of Manly healthful and vigorous Exercise to which all would be strictly required to attend.

2nd. Under this system the morals of the pupils would be effectually guarded, and, it is believed, might be easily preserved—

3d. The youths thus Educated would enter upon the duties of life Skillful and accomplished Soldiers — fully qualified to become the safe and efficient defences of their Countrys rights — Each of them would also become an Instructor to the Militia by which means current Military information would be, in a comparatively short time be diffused throughout the great mass of this country bulwark of our liberties.

4th These Seminaries, would also be a most efficient means of improving the militia of the Cities in their vicinity by affording to the officers, the means of being correctly instructed—

I have thus, in an imperfect and hasty manner given, a Synopsis of my views on the subject of a System of Education, which appears to me particularly adapted to our larger Cities, but which might also be extended most beneficially to those of a Second and third Class — an [illegible] to any one which has a population sufficient to support a Seminary—Those youths who are to receive a finished Classical and Scientific Education, Should be transferred from those Seminaries, to Similar Institutions in the interior of the Country, for the Completion of their Course, after they have attained the age of fifteen or Sixteen years. Their morals after this period, could more easily and certainly be preserved at such Institutions, there in the vicinity of a large city, and it would also be better for them to be removed from the [illegible] home, for the completion of their Education, than that it should be done in the immediate vicinity of their families and relatives. The foregoing is submitted with all due deference, to the dispassionate conscientious of our enlightened and liberal public, by one who has long felt a serious and deep interest in the Education of the youth of our Country—

Appendix K: Extract from Daniel Defoe's *Of Academies*[697]

Next to this, which I esteem as the most noble and most useful proposal in this book, I proceed to academies for military studies, and because I design rather to express my meaning than make a large book, I bring them all into one chapter.

I allow the war is the best academy in the world, where men study by necessity and practice by force, and both to some purpose, with duty in the action, and a reward in the end; and it is evident to any man who knows the world, or has made any observations on things, what an improvement the English nation has made during this seven years' war.

But should you ask how clear it first cost, and what a condition England was in for a war at first on this account—how almost all our engineers and great officers were foreigners, it may put us in mind how necessary it is to have our people so practised in the arts of war that they may not be novices when they come to the experiment.

I have heard some who were no great friends to the Government take advantage to reflect upon the king, in the beginning of his wars in Ireland, that he did not care to trust the English, but all his great officers, his generals, and engineers were foreigners. And though the case was so plain as to need no answer, and the persons such as deserved none, yet this must be observed, though it was very strange: that when the present king took possession of this kingdom, and, seeing himself entering upon the bloodiest war this age has known, began to regulate his army, he found but very few among the whole martial part of the nation fit to make use of for general officers, and was forced to employ strangers, and make them Englishmen (as the Counts Schomberg, Ginkel, Solms, Ruvigny, and others); and yet it is to be observed also that all the encouragement imaginable was given to the English gentlemen to qualify themselves, by giving no less than sixteen regiments to gentlemen of good families who had never been in any service and knew but very little how to

---

[697] *An Essay upon Projects* was written in 1693 and published in 1697. Daniel Defoe, *The Earlier Life and the Chief Earlier Works of Daniel Defoe*, ed. Henry Morley (London: George Routledge and Sons, 1889), 133–44.

command them. Of these, several are now in the army, and have the rewards suitable to their merit, being major-generals, brigadiers, and the like.

If, then, a long peace had so reduced us to a degree of ignorance that might have been dangerous to us, had we not a king who is always followed by the greatest masters in the world, who knows what peace and different governors may bring us to again?

The manner of making war differs perhaps as much as anything in the world; and if we look no further back than our civil wars, it is plain a general then would hardly be fit to be a colonel now, saving his capacity of improvement. The defensive art always follows the offensive; and though the latter has extremely got the start of the former in this age, yet the other is mightily improving also.

We saw in England a bloody civil war, where, according to the old temper of the English, fighting was the business. To have an army lying in such a post as not to be able to come at them was a thing never heard of in that war; even the weakest party would always come out and fight (Dunbar fight, for instance); and they that were beaten to-day would fight again to-morrow, and seek one another out with such eagerness, as if they had been in haste to have their brains knocked out. Encampments, intrenchments, batteries, counter-marchings, fortifying of camps, and cannonadings were strange and almost unknown things; and whole campaigns were passed over, and hardly any tents made use of. Battles, surprises, storming of towns, skirmishes, sieges, ambuscades, and beating up quarters was the news of every day. Now it is frequent to have armies of fifty thousand men of a side stand at bay within view of one another, and spend a whole campaign in dodging (or, as it is genteelly called, observing) one another, and then march off into winter quarters. The difference is in the maxims of war, which now differ as much from what they were formerly as long perukes do from piqued beards, or as the habits of the people do now from what they then were. The present maxims of the war are:

"Never fight without a manifest advantage."

"And always encamp so as not to be forced to it."

And if two opposite generals nicely observe both these rules, it is impossible they should ever come to fight.

I grant that this way of making war spends generally more money and less blood than former wars did; but then it spins wars out to a greater length; and I almost question whether, if this had been the way of fighting of old, our civil war had not lasted till this day. Their maxim was:

"Wherever you meet your enemy, fight him."

But the case is quite different now; and I think it is plain in the present war that it is not he who has the longest sword, so much as he who has the longest purse, will hold the war out best. Europe is all engaged in the war, and the men will never be exhausted while either party can find money; but he who finds himself poorest must give out first; and this is evident in the French king, who now inclines to peace, and owns it, while at the same time his armies are numerous and whole. But the sinews fail; he finds his exchequer fail, his kingdom drained, and money hard to come at: not that I believe half the reports we have had of the misery and poverty of the French are true; but it is manifest the King of France finds, whatever his armies may do, his money won't hold out so long as the Confederates, and therefore he uses all the means possible to procure a peace, while he may do it with the most advantage.

There is no question but the French may hold the war out several years longer; but their king is too wise to let things run to extremity. He will rather condescend to peace upon hard terms now than stay longer, if he finds himself in danger to be forced to worse.

This being the only digression I design to be guilty of, I hope I shall be excused it.

The sum of all is this: that, since it is so necessary to be in a condition for war in a time of peace, our people should be inured to it. It is strange that everything should be ready but the soldier: ships are ready, and our trade keeps the seamen always taught, and breeds up more; but soldiers, horsemen, engineers, gunners, and the like must be bred and taught; men are not born with muskets on their shoulders, nor fortifications in their heads; it is not natural to shoot bombs and undermine towns: for which purpose I propose a

Royal Academy for Military Exercises.

The founder the king himself; the charge to be paid by the public, and settled by a revenue from the Crown, to be paid yearly.
I propose this to consist of four parts:

1. A college for breeding up of artists in the useful practice of all military exercises; the scholars to be taken in young, and be maintained, and afterwards under the king's care for preferment, as their merit and His Majesty's favour shall recommend them; from whence His Majesty would at all times be furnished with able engineers, gunners, fire-masters, bombardiers, miners, and the like.

The second college for voluntary students in the same exercises; who should all upon certain limited conditions be entertained, and have all the advantages of the lectures, experiments, and learning of the college, and be also capable of several titles, profits, and settlements in the said college, answerable to the Fellows in the Universities.

The third college for temporary study, into which any person who is a gentleman and an Englishman, entering his name and conforming to the orders of the house, shall be entertained like a gentleman for one whole year gratis, and taught by masters appointed out of the second college.

The fourth college, of schools only, where all persons whatsoever for a small allowance shall be taught and entered in all the particular exercises they desire; and this to be supplied by the proficients of the first college.

I could lay out the dimensions and necessary incidents of all this work, but since the method of such a foundation is easy and regular from the model of other colleges, I shall only state the economy of the house.

The building must be very large, and should rather be stately and magnificent in figure than gay and costly in ornament: and I think such a house as Chelsea College, only about four times as big, would answer it; and yet, I believe, might be finished for as little charge as has been laid out in that palace-like hospital.

The first college should consist of one general, five colonels, twenty captains.

Being such as graduates by preferment, at first named by the founder; and after the first settlement to be chosen out of the first or second colleges; with apartments in the college, and salaries.

	£ per ann.
The general	300
The colonels	100
The captains	60

2,000 scholars, among whom shall be the following degrees:

		allowed £ per ann.
Governors	100	10
Directors	200	5
Exempts	200	5
Proficients	500	
Juniors	1,000	

The general to be named by the founder, out of the colonels; the colonels to be named by the general, out of the captains; the captains out of the governors; the governors from the directors; and the directors from the exempts; and so on.

The juniors to be divided into ten schools; the schools to be thus governed: every school has

100 juniors, in 10 classes.	
Every class to have 2 directors.	
100 classes of juniors is	1,000
Each class 2 directors	200
	1,200

The proficients to be divided into five schools:

Every school to have ten classes of 10 each.	
Every class 2 governors.	
50 classes of proficients is	500
Each class 2 governors is	100
	600

The exempts to be supernumerary, having a small allowance, and maintained in the college till preferment offer.

The second college to consist of voluntary students, to be taken in, after a certain degree of learning, from among the proficients of the first, or from any other schools, after such and such limitations of learning; who study at their own charge, being allowed certain privileges; as—

Chambers rent-free on condition of residence.

Commons gratis, for certain fixed terms.

Preferment, on condition of a term of years' residence.

Use of libraries, instruments, and lectures of the college.

This college should have the following preferments, with salaries

	£ per ann.
A governor	200
A president	100
50 college-majors	50
200 proficients	10
500 voluntary students, without allowance.	

The third and fourth colleges, consisting only of schools for temporary study, may be thus:

The third—being for gentlemen to learn the necessary arts and exercises to qualify them for the service of their country, and entertaining them one whole year at the public charge—may be supposed to have always one thousand persons on its hands, and cannot have less than 100 teachers, whom I would thus order:

Every teacher shall continue at least one year, but by allowance two years at most; shall have £20 per annum extraordinary allowance; shall be bound to give their constant attendance; and shall have always five college-majors of the second college to supervise them, who shall command a month, and then be succeeded by five others, and, so on—£10 per annum extraordinary to be paid them for their attendance.

The gentlemen who practise to be put to no manner of charge, but to be obliged strictly to the following articles:

1. To constant residence, not to lie out of the house without leave of the college-major.

2. To perform all the college exercises, as appointed by the masters, without dispute.

3. To submit to the orders of the house.

To quarrel or give ill-language should be a crime to be punished by way of fine only, the college-major to be judge, and the offender be put into custody till he ask pardon of the person wronged; by which means every gentleman who has been affronted has sufficient satisfaction.

But to strike challenge, draw, or fight, should be more severely punished; the offender to be declared no gentleman, his name posted up at the college-gate, his person expelled the house, and to be pumped as a rake if ever he is taken within the college-walls.

The teachers of this college to be chosen, one half out of the exempts of the first college, and the other out of the proficients of the second.

The fourth college, being only of schools, will be neither chargeable nor troublesome, but may consist of as many as shall offer themselves to be taught, and supplied with teachers from the other schools.

The proposal, being of so large an extent, must have a proportionable settlement for its maintenance; and the benefit being to the whole kingdom, the charge will naturally lie upon the public, and cannot well be less, considering the number of persons to be maintained, than as follows.

FIRST COLLEGE.	
	£ per ann.
The general	300
5 colonels at £100 per ann. each	500
20 captains at 60 „	1,200
100 governors at 10 „	1,000
200 directors at 5 „	1,000
200 exempts at 5 „	1,000
2,000 heads for subsistence, at £20 per head per ann., including provision, and all the officers' salaries in the house, as butlers, cooks, purveyors, nurses, maids, laundresses, stewards, clerks, servants, chaplains, porters, and attendants, which are numerous.	40,000
SECOND COLLEGE.	
A governor	200
A president	100
50 college-majors at £50 per ann. each	2,500
200 proficients at 10	2,000
Commons for 500 students during times of exercises at £5 per ann. Each	2,500
200 proficients' subsistence, reckoning as above	4,000
THIRD COLLEGE.	
The gentlemen here are maintained as gentlemen, and are to have good tables, who shall therefore have an allowance at the rate of £25 per head, all officers to be maintained out of it; which is	25,000
100 teachers, salary and subsistence ditto	4,500
50 college-majors at £10 per ann. is	500
Annual charge	86,300
The building to cost	50,000
Furniture, beds, tables, chairs, linen, &c.	10,000
Books, instruments, and utensils for experiments	2,000
So the immediate charge would be	62,000
The annual charge	86,300
To which add the charges of exercises and experiments	3,700
	90,000

The king's magazines to furnish them with 500 barrels of gunpowder per annum for the public uses of exercises and experiments.

In the first of these colleges should remain the governing part, and all the preferments to be made from thence, to be supplied in course from the other; the general of the first to give orders to the other, and be subject only to the founder.

The government should be all military, with a constitution for the same regulated for that purpose, and a council to hear and determine the differences and trespasses by the college laws.

The public exercises likewise military, and all the schools be disciplined under proper officers, who are so in turn or by order of the general, and continue but for the day.

The several classes to perform several studies, and but one study to a distinct class, and the persons, as they remove from one study to another, to change their classes, but so as that in the general exercises all the scholars may be qualified to act all the several parts as they may be ordered.

The proper studies of this college should be the following:

Geometry.	Bombarding.
Astronomy.	Gunnery.
History.	Fortification.
Navigation.	Encamping.
Decimal arithmetic.	Intrenching.
Trigonometry.	Approaching.
Dialing.	Attacking.
Gauging.	Delineation.
Mining.	Architecture.
Fireworking.	Surveying.

And all arts or sciences appendices to such as these, with exercises for the body, to which all should be obliged, as their genius and capacities led them, as:

1. Swimming; which no soldier, and, indeed, no man whatever, ought to be without.

2. Handling all sorts of firearms.

3. Marching and counter-marching in form.

4. Fencing and the long-staff.

5. Riding and managing, or horsemanship.

6. Running, leaping, and wrestling.

And herewith should also be preserved and carefully taught all the customs, usages, terms of war, and terms of art used in sieges, marches of armies and encampments, that so a gentleman taught in this college should be no novice when he comes into the king's armies, though he has seen no service abroad. I remember the story of an English gentleman, an officer at the siege of Limerick, in Ireland, who, though he was brave enough upon action, yet for the only matter of being ignorant in the terms of art, and knowing not how to talk camp language, was exposed to be laughed at by the whole army for mistaking the opening of the trenches, which he thought had been a mine against the town.

The experiments of these colleges would be as well worth publishing as the acts of the Royal Society. To which purpose the house must be built where they may have ground to cast bombs, to raise regular works, as batteries, bastions, half-moons, redoubts, horn-works, forts, and the like; with the convenience of water to draw round such works, to exercise the engineers in all the necessary experiments of draining and mining under ditches. There must be room to fire great shot at a distance, to cannonade a camp, to throw all sorts of fireworks and machines that are, or shall be, invented; to open trenches, form camps, &c.

Their public exercises will be also very diverting, and more worth while for any gentleman to see than the sights or shows which our people in England are so fond of.

I believe as a constitution might be formed from these generals, this would be the greatest, the gallantest and the most useful foundation in the world. The English gentry would be the best qualified, and consequently best accepted abroad, and most useful at home of any people in the world; and His Majesty should never more be exposed to the necessity of employing foreigners in the posts of trust and service in his armies.

And that the whole kingdom might in some degree be better qualified for service, I think the following project would be very useful:

When our military weapon was the long-bow, at which our English nation in some measure excelled the whole world, the meanest countryman was a good archer; and that which qualified them so much for service in the war was their diversion in times of peace, which also had this good effect—that when an army was to be raised they needed no disciplining: and for the encouragement of the people to an exercise so publicly profitable an Act of Parliament was made to oblige every parish to maintain butts for the youth in the country to shoot at.

Since our way of fighting is now altered, and this destructive engine the musket is the proper arms for the soldier, I could wish the diversion also of the English would change too, that our pleasures and profit might correspond. It is a great hindrance to this nation, especially where standing armies are a grievance, that if ever a war commence, men must have at least a year before they are thought fit to face an enemy, to instruct them how to handle their arms; and new-raised men are called raw soldiers. To help this—at least, in some, measure—I would propose that the public exercises of our youth should by some public encouragement (for penalties won't do it) be drawn off from the foolish boyish sports of cocking and cricketing, and from tippling, to shooting with a firelock (an exercise as pleasant as it is manly and generous) and swimming, which is a thing so many ways profitable, besides its being a great preservative of health, that methinks no man ought to be without it.

1. For shooting, the colleges I have mentioned above, having provided for the instructing the gentry at the king's charge, that the gentry, in return of a favour, should introduce it among the country people, which might easily be done thus:

If every country gentleman, according to his degree, would contribute to set-up a prize to be shot for by the town he lives in or the neighbourhood, about once a year, or twice a year, or oftener, as they think fit; which prize not single only to him who shoots nearest, but according to the custom of shooting.

This would certainly set all the young men in England a-shooting, and make them marksmen; for they would be always practising, and making matches among themselves too, and the advantage would be found in a war; for, no doubt, if all the soldiers in a battalion took a true level at their enemy there would be much more execution done at a distance than there is; whereas it has been known how that a battalion of men has received the fire of another battalion, and not lost above thirty or forty men; and I suppose it will not easily be forgotten how, at the battle of Agrim, a battalion of the English army received the whole fire of an Irish regiment of Dragoons, but never knew to this day whether they had any bullets or no; and I need appeal no further than to any officer that served in the Irish war, what advantages the English armies made of the Irish being such wonderful marksmen.

Appendix L: Military Stoic Reading List

The reading list below is intended to help guide military members on the self-study of Stoic philosophy. This list is not comprehensive but merely a solid listing of resources. Service members will likely be able to identify Stoic themes in other military sources and academic text after reading just a few of the recommendations below.

1.   Archetypal Stories

One of the easiest ways to understand a philosophy is in story form. Below is a list of archetypal stories listed by ease of reading for modern audiences.

a.   *The Martian* by Andy Weir.

The Martian is a modern retelling of Daniel Defoe's *Robinson Crusoe*. In this story, astronaut Mark Watney faces several challenges in surviving on Mars and ultimately reuniting with a spacecraft that can transport him home. While Defoe focuses on the use of *reason*, Weir will use the term *science* to carry an equivalent meaning. Many may know of this story through the 2015 movie of the same name, starring Matt Damon and directed by Ridley Scott.

b.   *Harry Potter* series by J. K. Rowling.

Multiple philosophers have noted the Stoic influence, especially in the early books of the *Harry Potter* series. Edmund M. Kern, professor of history at Lawrence University in Wisconsin, asserted that the *Harry Potter* series "might just comprise the most visible contribution to Stoicism's reemergence as a viable, practical philosophy offering comfort and guidance in these uncertain times."[698] Even though this book series faced some censorship by Christian communities, many Neostoic themes and traits can be seen within the characters of this story.

c.   *The Lord of the Rings* by J. R. R. Tolkien.

Tolkien's *The Lord of the Rings* series displays multiple features of Stoic philosophy, from the dichotomy of control to the focus on optimism in the face of overwhelming odds (what the military community would refer to as the *Stockdale Paradox*). Characters like Aragorn and Gandalf serve as powerful

---

[698] Patricia Cohen, "Think Tank; The Phenomenology of Harry, or the Critique of Pure Potter," *New York Times*, July 19, 2003.

archetypal leaders. Readers should remember that Tolkien was a World War I infantry officer who experienced the Battle of the Somme. While traces of Tolkien's military experiences are incorporated into this story, the *Lord of the Rings* trilogy ultimately demonstrates that military veterans of even extreme campaigns can endure their experiences with the correct philosophy while retaining hope and optimism for the future.

    d.   *The Robinson Crusoe* trilogy by Daniel Defoe.

    Daniel Defoe's *Robinson Crusoe* trilogy is likely the finest example of a Neostoic archetypal story. It is recommended that the first two books of the series (*The Life and Strange Surprizing Adventures of Robinson Crusoe, of York, Mariner: Who Lived Eight and Twenty Years, All Alone in an Un-Inhabited Island on the Coast of America, near the Mouth of the Great River of Oroonoque, Having Been Cast on Shore by Shipwreck, Wherein All the Men Perished but Himself: With an Account How He Was at Last as Strangely Deliver'd by Pyrates* and *The Farther Adventures of Robinson Crusoe*) be read to understand the complete story. The third book of the series (*Serious Reflections during the Life and Surprising Adventures of Robinson Crusoe: With His Vision of the Angelick World*) is written like traditional theology that references the story for purposes of example. When reading this book, it is important to remember that this was one of the most popular adventure books of the 18th and 19th centuries. Generations of Americans, including the Founding Fathers, would have been exposed to this story. This book was frequently used to help children learn how to read, with Captain Alden Partridge of Norwich University using Spanish translations of *Robinson Crusoe* to help instruct his students in a foreign language.[699] While multiple movie adaptations have been made of this book, none capture the full philosophical monologue within the book and therefore are not recommended as a substitute for reading. (Free audiobooks of the first two books of this series can be found on Librivox.org.)

    e.   *The Aeneid* by Publius Vergilius Maro (Virgil).

    The *Aeneid* recounts the legendary founding of Rome as Aeneas and the remaining Trojans find themselves in need of a new homeland after the fall of Troy. This work of poetry was used to instruct students in Latin and teach civic responsibility well into the 20th century. Historically, most American military officers would have been exposed to this work in its original Latin. (A free audiobook is available on Librivox.org.)

---

[699] Partridge, Catalogue of the Officers and Cadets, 1826, p. 22.

2. Primary Sources
   a. Xenophon of Athens.

While Xenophon technically predates Stoic philosophy, many of his works carry proto-Stoic themes and it is his *Apology* (account of the trial of Socrates) that is believed to have inspired Zeno of Citium to found Stoicism. Many of Xenophon's works carry value for the military reader. His *Anabasis* recounts the fighting retreat of a Greek mercenary army out of Persia after the death of Cyrus the Great. This work is notable for its accurate descriptions of battles and the rousing motivational speeches. The exciting tale of the *Anabasis* was used for generations to teach Ancient Greek. Many famous American military officers would have been well-versed in this story. *Cyropaedia* (*Education of Cyrus*) is also a valuable text in terms of leadership theory. Xenophon's Socratic dialogues (*Apology*; *Memorabilia*; *Symposium*; and *Oeconomicus*) are worth the time to read and present Socrates in a slightly different light than the accounts of Plato. Xenophon accounted for the political theories of the time, recording the *Constitution of the Athenians* and the *Polity of the Lacedaemonians* (*Spartans*). The *Hellenica* is an important source of history of the last years of the Peloponnesian War, picking up where Thucydides left off. Of Xenophon's lesser treatises, *On Horsemanship* is highly praised by members of the Stoic community (including by Jock Hutchison, who founded the natural Stoic-based charity HorseBack UK). (Many of Xenophon's works can be found in free audiobook form on Librivox.org.)

   b. Lucius Annaeus Seneca (the Younger).

Lucius Annaeus Seneca was a wealthy Stoic philosopher who became the tutor of Emperor Nero. Seneca's influence was credited for the good governance experienced in the early years of Nero's reign. However, Seneca eventually lost his influence on Nero and Rome descended into chaos. Seneca was forced to take his own life after being accused of participating in a conspiracy to assassinate Nero. Seneca wrote several Stoic essays and letters. Seneca's *Letters from a Stoic* are highly recommended, along with his essay "On the Shortness of Life." (Many of Seneca's works can be found in free audiobook form on Librivox.org.)

   c. Epictetus.

Epictetus was a crippled slave who learned Stoic philosophy under the instruction of Musonius Rufus. Arrian, one of Epictetus' pupils, captured the speeches of Epictetus, compiling them into *The Discources* and *The Enchiridion* (handbook). Both of these works are highly recommended and influential in history. The works of Epictetus pair well with *Thoughts of a Philosophical Fighter Pilot* by James Stockdale, who used Epictetus' philosophy to survive multiple years as a prisoner of war. (The works of Epictetus can be found in free audiobook form on Librivox.org.)

    d.   *The Meditations* by Marcus Aurelius Antoninus.

Marcus Aurelius was literally the most powerful man in the "known world" when he penned *The Meditations*. Succeeding Roman Emperor Hadrian, Marcus Aurelius reigned from 161 to 180 and is recounted as the last of the "Five Good Emperors" of Rome. *The Meditations* are essentially a collection of personal reflections about philosophy and life written by Aurelius while directing the Roman army during the Marcomannic Wars. There is some debate as to whether these writings were ever intended to be shared or if Aurelius would have wanted them destroyed. In either case, history remembers this work as one of the most powerful Stoic texts. While this book can be read independently, those new to Stoic philosophy may find it best to read either *How to Think Like a Roman Emperor* by Donald Robertson or *The Inner Citadel* by Pierre Hadot in tandem with *The Meditations*. (*The Meditations* can be found in free audiobook form on Librivox.org.)

    e.   Justus Lipsius.

Justus Lipsius was a Flemish humanist credited with creating the Neostoic movement. He integrated the works of Seneca and Tacitus into Christian theology, defaulting to Christian ideas in areas of conflict. He is best known for his *De Constantia Libri Duo* (*Two Books on Constancy*). An English translation of this work, *Justus Lipsius on Constancy,* was published by the University of Exeter Press with an introduction and notes by John Sellars. Lipsius also produced several other works on politics and military history, including *Politicorum sive Civilis Doctrinae Libri Sex* (*Six Books of Politics or Political Instruction*); *Monita et Exempla Politica* (*Admonitions and Political Examples*); and *De Militaria Romana*. Lipsius' work had tremendous effects on the political and military reforms of the early modern era.

    f.   *De Jure Belli ac Pacis* (*On the Law of War and Peace: Three books*) by Hugo Grotius.

Hugo Grotius played a critical role in establishing international law, including the conduct of warfare. Built within a Neostoic framework and first published in 1625, *On the Law of War and Peace* evaluates the nature of warfare, just causes for war, and proposed restrictions on the conduct of warfare related to all parties. (A translation by Francis Kelsey is available for free on Google Books.)

    g.   *Vom Kriege* (*On War*) by Carl von Clausewitz.

Compiled and published by Marie von Brühl after the death of Carl von Clausewitz, *On War* would have a major impact on strategic theory from the 19th century to the present. According to Gerhard Oestreich, this work was heavily influenced by Neostoic philosophy and political theory. (Portions of this book can be found in free audiobook form on Librivox.org.)

    h.   *Lecture on Education* by Captain Alden Partridge.

Captain Alden Partridge was the third superintendent of the United States Military Academy and founder of Norwich University. He is likely one of America's greatest educational theorist, with the land-grant colleges largely owing their existence to his efforts. A few of his theories would be used to create the Reserve Officers' Training Corps (ROTC), the largest commissioning source for the U.S. Military. In the early 1820s, he wrote his *Lecture on Education*, in which he lays out the faults of the contemporary college system. While Partridge doesn't specifically reference Stoicism within this text, he draws on many Stoic themes in creating his curriculum, including the careful use of time and Stoic toughening training. (A copy of this lecture is available for free on Google Books.)

    i.   Ralph Waldo Emerson.

Ralph Waldo Emerson was one of the greatest, if not the greatest, of American philosophers. He is credited with leading the transcendentalist movement. Emerson's abolitionist work and support for John Brown helped to spark the American Civil War. His works were deeply affected by Stoic philosophy, with Emerson frequently referencing Antoninus (Marcus Aurelius). Emerson wrote numerous essays and gave frequent lectures. Among his most notable works is the essay "Self-Reliance." The lecture "The Man of Letters," written during the American Civil War, is also highly recommended, as it explicitly calls for instruction for all American scholars in Stoic philosophy and military training. It is believed that Emerson may have been heavily influenced by the educational theories of Captain Alden Partridge, with a recent visit to the United States Military Academy being recorded within this lecture. (Some of Emerson's works are available in free audiobook form on Librivox.org.)

    j.   Thomas Wentworth Higginson.

Thomas Wentworth Higginson was a Unitarian minister and abolitionist. He was a member of the "Secret Six" that funded John Brown's raid on Harper's Ferry. During the American Civil War, he would command the first Union regiment composed of freed slaves. Higginson records his experiences with the First South Carolina Volunteers, including his leadership philosophy and approach to equal opportunity in the work *Army Life in a Black Regiment*. The philosophy he presents in this work could still be used today. It is undoubtable that Higginson was influenced by Stoic philosophy, first by being an understudy of Emerson and immediately after the war translating the *Works of Epictetus*. (*Army Life in a Black Regiment* is available in free audiobook form on

Librivox.org. Wentworth's *Works of Epictetus* (1865) is available on Google Books).

k. *Man's Search for Meaning* by Victor Frankl.

Victor Frankl spent three years imprisoned in Nazi concentration camps. As a trained psychiatrist prior to the war, Frankl witnessed firsthand how the interpretation of one's surroundings could contribute to surviving or perishing in this extreme situation. This led Frankl to develop *logotheraphy*. While Frankl never claimed to be a Stoic, many aspects of *logotheraphy* parallel Stoic principles.

l. *My Hitch in Hell: The Bataan Death March* by Lester Tenney.

Lester Tenney was captured by Imperial Japanese Army forces during World War II. He experienced the brutality of the Bataan Death March and three-and-a-half years of captivity as a prisoner of war. While Tenney was not a Stoic by training or education, many of the lessons he recounts within this book mirror the teachings of ancient Stoicism.

m. *Thoughts of a Philosophical Fighter Pilot* by James Stockdale.

Vice Admiral James Bond "Jim" Stockdale was a naval aviator during the Vietnam War. His A-4 Skyhawk was shot down on September 9, 1965, and he became a prisoner of war held at the infamous "Hanoi Hilton" Hỏa Lò Prison. He was held for seven-and-a-half years before his release. *Thoughts of a Philosophical Fighter Pilot* is a collection of essays and lectures by Stockdale, including those who recall how he employed Stoic philosophy (Epictetus) to help him endure torture and imprisonment. After the war, it was found that prisoners that were exposed to Stockdale had lower rates of lifetime post-traumatic stress disorder (PTSD) than the average American veteran of the conflict because of the positive influence of Stoic philosophy.

3. Secondary Sources

Readers should use caution when engaging with secondary sources. It is unfortunate that not all authors in the current modern Stoicism movement accurately represent the underlying assumptions of ancient Stoic philosophy. In many cases, modern scholars may use elements of ancient Stoicism to justify modern ideologies or view ancient Stoic philosophy through ideologically biased lenses. (This may include substituting 19th-, 20th-, and 21st-century ideas into the philosophy without clearly stating the author's manipulation or making false claims such as Stoicism is incompatible with military service.) Below is a list of recommended secondary sources generally free from ideological bias.

a. *The Inner Citadel* by Pierre Hadot.

The work of the noted historian Pierre Hadot, *The Inner Citadel*, provides a contemporary analysis of Marcus Aurelius' *Meditations*. This text is a useful introduction to Stoic philosophy. For obvious reasons, this would be a good text to read just prior to reading the *Meditations*.

b. *The Stoic Challenge* by William Irvine.

William Irvine, professor emeritus, provides a useful introductory book to Stoicism with *The Stoic Challenge*. Irvine draws quotes from the ancient Stoics while providing modern examples and practical suggestions on how Stoic philosophy might be applied. This is an excellent work for someone first exploring Stoic philosophy.

c. *Neostoicism & the Early Modern State* by Gerhard Oestrich.

Gerhard Oestrich explains the evolution and influence of Neostoic philosophy from the 16th century through the Age of Revolution in the work *Neostoicism & the Early Modern State*. This book is useful in explaining the philosophy and its effects on war, internal law, and the creation of modern nation-states. This text is useful for military readers to come to a better understanding of the philosophical base of the "Western way of war" and how this philosophy affected and continues to affect the conduct of warfare and international relations. Given that military leaders are often not exposed to the larger philosophical schools, this work may assist in further framing and understanding the context of the highly studied 19th-century military thinkers, such as Carl von Clausewitz, Frederick the Great, etc., that are typically reviewed within the military professional education curriculum.

d. Donald Roberston.

Donald Robertson is a cognitive behavioral psychotherapist who has written several important works on Stoicism. *How to Think Like a Roman Emperor* examines the philosophy within and the events surrounding Marcus Aurelius' *The Meditations*. This work is a useful introduction to Stoic philosophy that may be uniquely beneficial to the military audience as it captures the Roman military action that Aurelius was directing while *The Meditations* was being composed. Additionally, Robertson's *The Philosophy of Cognitive Behavioral Therapy: Stoic Philosophy as Rational and Cognitive Psychotherapy* is highly recommended. In this work, Robertson further examines the influence of Stoic philosophy on modern psychology, including the development of cognitive behavioral therapy.

e. *Rome's Last Citizen* by Rob Goodman and Jimmy Soni.

Goodman and Soni's work *Rome's Last Citizen* recalls the events of the fall of the Roman Republic, including the action of Marcus Porcius Cato the Younger, and reveals how this story would shape the founding of the American Republic through Joseph Addison's play *Cato, a Tragedy*. This is a great work that details how the American Founding Fathers were influenced by Stoic philosophy.

f.   "Stoicism for Toughening" by Franklin C. Annis.

Available on Medium.com, this article records the Stoic toughening practices introduced by Captain Alden Partridge during the early years of the United States Military Academy. While many Stoic toughening exercises are still employed within the U.S. military, the purpose and intent of inducing physical and mental discomfort is not always understood by participants. This article records some of the history of Stoic toughening training within the U.S. military and the continued importance of engaging in these activities.

g.   *Cicero Trilogy* by Robert Harris.

Robert Harris recorded the life of Marcus Tullius Cicero with a work of historical fiction. The Trilogy (*Imperium, Lustrum, Dictator*) is told through the perspective of Marcus Tullius Tiro, Cicero's slave and personal scribe. This compelling novel series provides the history of famous Stoics, such as Cato the Younger, during the fall of the Roman Republic. Those beginning their studies into Stoic philosophy may find this trilogy an easier introduction to the fall of the Roman Republic than directly reading primary sources (such as Plutarch).

4.   Stoic Poetry

"Hymn to Zeus" by Cleanthes is the only "true" Stoic poet within the list below. However, Stoic themes are often noticed within the various poems on this list.

a.   "Hymn to Zeus" by Cleanthes
b.   "A Psalm of Life" by Henry Wadsworth Longfellow
c.   "Invictus" by William Ernest Henley
d.   "If" by Rudyard Kipling
e.   "The Proof of Worth" by Edgar Albert Guest
f.   "Desiderata" by Max Ehrmann

Appendix M: Letter from Levi Woodbury to Alden Partridge, 5 April 1824[700]

Capt. A. Patridge [*sic*]
Principal of the Academy
Norwich. VT.

Portsmouth, N.H. April 5th. 1824 –
Dear Sir,

     In my excursion to Norwich last August I had the pleasure to form some practical acquaintance with your system of education. But I was unfortunate in not hearing your lectures on a subject of such unspeakable importance on our free government; and the misfortune is felt by me more sensibly, as I derived much information and happiness from your lectures on other subjects. So far as I understand your system, it has with me a decided preference over any other. In the first place it combines physical with moral & scientific education. This circumstance to those parents, who have in their own persons experienced "the pains and penalties" of body, that most sedentary men are heir to, is an invaluable improvement. That and our in the pursuit of knowledge, which is indispensable to excellence, will never submit to such corporal exercise as is necessary to health, unless the exercise is a part of the system of education & is made to combine pleasure with utility. A neglect of such exercise proves a Pandora's box to students, from which issue dyspepsia, consumptions, headaches, vertigoes and indeed all that host of maladies, which hurry to an untimely grave so great a proportion of the genius and virtue of the rising generation. Your military discipline furnishes a species of exercise admirably adapted to these purposes and invigorating the mind with the body produces that enviable state for the scholar of *sana. Mens in corpore sano*[701] – The second distinguishing feature in your system appeared to me to be the communication to your students, by means of this exercise, of a very considerable degree of military knowledge. This is a species of knowledge not taught, I believe, systematically in any other seminary among us except the

---

[700] Woodbury, Letter from Levi Woodbury to Alden Partridge.
[701] Translates from Latin "a healthy mind in a healthy body."

Academy at West Point; and so far as you make it a mere auxiliary to literary studies and treat it as a polite accomplishment and an essential to form good soldiers & officers in our Militia; your system is altogether original and merits the highest praise. Our government being a government of the people it must by them be defended from both usurpation & invasion. A large standing army agrees neither with its principles of economy, nor the habits of its citizens – Its policy is peace and its great safeguard in times of peril must be the vigorous arm and bayonet of those, whose lives have been devoted to the arts of peace. But force without intelligence is blind – it is a *brutum fulmen*[702] –; and I am aware of no means so well adapted to give efficiency to our Militia as a diffusion, from our Schools & higher seminaries of education, of a due portion of military science – a love of military exercises and an alluring example of these in such young men of talents and wealth as may win others less favoured in these respects to a cheerful performance of every duty, connected with the respectability and usefulness of the Militia system – a third excellence in your academy is the attention bestowed on morals so far as they depend on manliness of character, polish of manners, and the sterner virtues of order, temperance and industry. Without casting any disparagement on the wishes or exertions of other teachers in other institutions it must be admitted, that your course of discipline is in itself more severe and at the same time more attractive; that with you neither wealth nor rank can buy an exemption from obedience & study; that your system tends to inspire pupils with sentiments like those attributed to Alexander of Macedon of its being a "royal virtue to labour"[703]; that you learn them by example as well as precept, the rich and intelligent are born to higher destinies than luxury & indolence – *consumere fruges*;[704] that in aid of the common means of exciting to improvement you bring the strict subordination of military tactics, the pride & vigilance of military office and the strong desire to obtain all those manly accomplishments, which ought to be united in the soldier & scholar. Your young men have before them many examples in the splendid republics of antiquity of the union of moral with literary & military worth; and in their daily devotion to similar pursuits they will be more likely to seek & attain to a kindred excellence, not forgetting that in a government like ours nothing [illegible], but ignorance, effeminacy & vice and that the highest order of nobility is only the highest order of virtue & talents – In this way only can young men become worthy of public confidence & public patronage; to deserve which is a laudable ambition

---

[702] Translates from Latin to "thunderbolt," implying an empty threat.

[703] Alluding to the *De Fortuna Alexandri* by Plutarch.

[704] Translates from Latin to "consume crops."

in any free man, is an ambition indispensable to the safety & glory of any free people and elevates its possesons far above the chance favourites of fortune, intrigue & faction.

At a period of greater leisure, my dear Sir, I may be able to extend these views and in the mean time beg permission to renew assurances of my sincere respect & esteem to Capt. A. Partridge. – Levi Woodbury

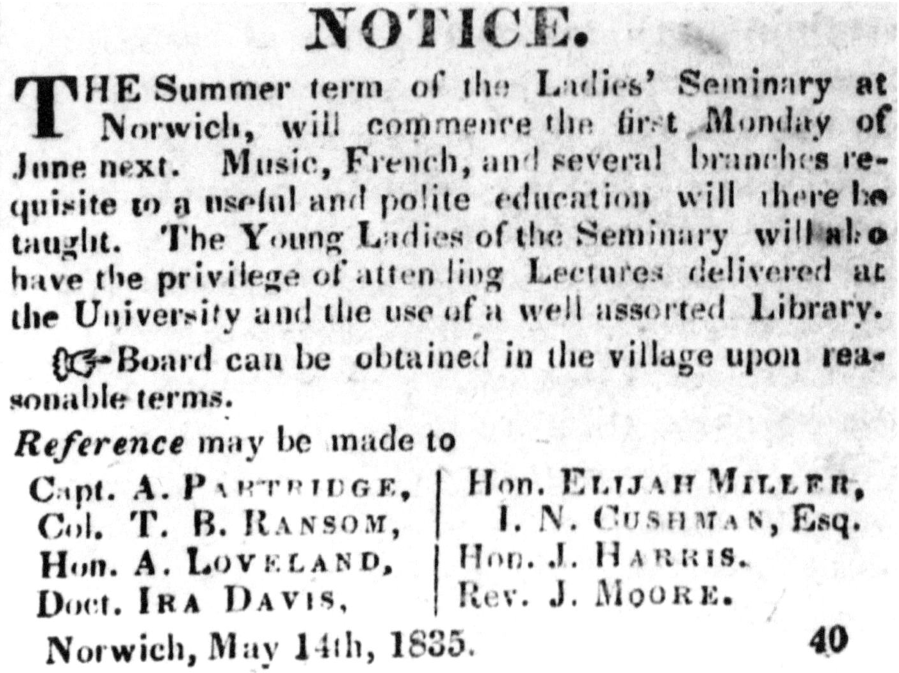

NOTICE.

THE Summer term of the Ladies' Seminary at Norwich, will commence the first Monday of June next. Music, French, and several branches requisite to a useful and polite education will there be taught. The Young Ladies of the Seminary will also have the privilege of attending Lectures delivered at the University and the use of a well assorted Library.

Board can be obtained in the village upon reasonable terms.

Reference may be made to

Capt. A. PARTRIDGE,	Hon. ELIJAH MILLER,
Col. T. B. RANSOM,	I. N. CUSHMAN, Esq.
Hon. A. LOVELAND,	Hon. J. HARRIS.
Doct. IRA DAVIS,	Rev. J. MOORE.

Norwich, May 14th, 1835.   **40**

Figure 19—Notice for the Ladies Seminary at Norwich[705]

---

[705] Notice: The Summer Term of the Ladies' Seminary at Norwich, 1835. Clipping from an unknown publication dated 14 May 1835, Norwich University Archives and Special Collections.

Appendix N: Plan to Reorganize the Militia by Partridge and Burke[706]

Memorial of Alden Partridge and Edmund Burke in behalf of the State Military Convention of Vermont praying the adoption of a plan proposed by them for the reorganization of the militia of the United States.

February 9, 1839. Referred to the Committee on the Militia and ordered to be printed and that 5,000 additional copies be furnished for the use of the Senate.

To the honorable Congress of the United States:

The memorial of Alden Partridge and Edmund Burke

Respectfully showeth:

That your memorialists were appointed a committee by the State military assembled at Montpelier, State of Vermont, on the 12th day of October, 1838, to address your honorable body on the subject of providing more efficient system for the organization and discipline of the militia of United States.

The whole history of the world, from the earliest period down to the time, proves that an efficient system of national defence has been considered by all independent nations, of the first importance. The wise and patriotic framers of the constitution of the United States duly appreciated the importance of this subject. In the preamble to that sacred document, "provision for the common defence" is expressly declared to be one of the great and leading objects to be accomplished by its adoption. In the fifteenth article of the eighth section it is declared that Congress have power to provide for organizing, arming, and disciplining the militia; and in the second article of the amendments to the constitution it further declared, "that a well regulated militia being necessary to the of a free State, the right of the people to keep and bear arms not be infringed." By the first of these provisions Congress is invested with full power to render the militia, in every respect, effective for military service; while, by the second, the absolute necessity of this great constitutional power, for the security of a free State, is unequivocally asserted. The only question, then, to be considered is, by what means can Congress carry

---

[706] Partridge and Burke, Memorial of Alden Partridge and Edmund Burke, in behalf of the State Military Convention of Vermont.

out into practice, the power conferred by the constitution for rendering the militia effective for military service? In answer to this question your memorialists would beg leave, respectfully to refer your honorable body to the accompanying proceedings of the military convention assembled at Norwich, State of Vermont, on the 4th of July, 1838, and of its adjourned sessions at the same place on the 15th of the succeeding August, and at Montpelier on the 12th of the following October. At the session of the 15th of August, the accompanying plan for the discipline of the militia, and address to the people of the United States on the same subject, were reported by a committee previously appointed for that purpose, and unanimously adopted by the convention. To this plan and address your memorialists would respectfully refer your honorable body for a full exposition of their views on this important subject. Your memorialists believe that if Congress would pass a law embracing the provisions contained in the proposed plan and address, the great body of the militia of the United States would, in a short time, under a vigorous and correct administration of the law, be brought to such a state of discipline as would enable them to contend successfully with any troops in the world. The cultivation and dissemination of military science would, also, on the proposed plan be extended to every portion of the United States, instead of being confined to a single isolated institution, supported at public expense, the advantages of which but a comparatively very small portion of our citizens can enjoy, though all are taxed to support it. Our citizen-soldiers possess a physical and moral energy which, combined with enough of mechanical discipline to enable them to act together, would render them irresistible when opposed by mere *mercenary* soldiers.

The whole expense of carrying the proposed plan into successful, practical operation, it is believed, would not exceed six and a half millions of dollars, only about one-half the estimated expense of our military establishment for the year 1838. This amount, your memorialists believe, would bear but a small ratio to the great advantages that would result from its adoption. When, however, it is considered that this money would be distributed among the patriotic yeomanry of our country as a reasonable compensation for their time and expenses, while preparing themselves for the correct and efficient discharge of the important duties imposed upon them by the constitution, all objections on account of the expense, even were it much greater, must immediately vanish.

The adoption of the proposed plan would also render a standing army, beyond a small peace establishment just sufficient to protect our necessary fortifications, arsenals, &c., totally unnecessary, and place the military defence of the country where the framers of the constitution evidently intended it to be placed, in the hands of the people themselves, the only safe guardians of our national independence and of our civil and religious liberties

All of which is respectfully submitted

Alden Partridge
Edmund Burke

January 31, 1839

MILITARY CONVENTION

To the officers, non-commissioned officers, musicians, and privates of the militia of Vermont, New Hampshire, New York, and adjacent States; and, also, to all others who are friendly to an efficient and well-organized militia:

GENTLEMEN: You are hereby requested to meet in *military convention*, either personally or by delegation, at Norwich, in the State of Vermont, on the 4th of July next, at 10 o clock, A.M., to take into consideration the present militia system of the United States, and to devise and propose such alterations and amendments thereto as shall appear best calculated to render the system efficient, equitable, and just in its operation; and to prepare the militia, who are the *constitutional* guardians of our civil liberties, both against foreign invasion and domestic insurrection and rebellion, for the correct and efficient discharge of the important duties imposed upon them by the constitution of the United States and of the several States. It is believed that the General Government not only possesses the *constitutional power*, but that on it devolves the duty and responsibility of providing an efficient organization for the militia, and, also of furnishing an adequate system of instruction for the officers, non-commissioned officers, &c.; and that the expenses of such instructive institution, including a reasonable compensation to the officers, non-commissioned officers, &c., for their time and expenses, while attending said instructive institution, should be defrayed from the *Treasury of the United States.* It is also believed that the officers, non-commissioned officers, musicians, and privates, when called out to discharge military duty, in conformity with the requisitions of the existing laws, whether at the regular trainings, musters, or otherwise, should be allowed a compensation for their time and expenses. Is it not as reasonable, is it not as just, that the militia should receive a compensation for their public services, as it is that those who are engaged in the discharge of duties purely civil should be compensated? What are our liberties worth, if we are incapable of defending them against all aggressions? And how can this be so certainly and so safely accomplished, as by means of a well-organized and well-disciplined body of citizen soldiers? Let the history of every nation answer this question. Is it not, then, much better and much more in accordance with principles of our free institutions, that a portion of the public revenue (the people's money) should be expended in encouraging and perfecting this great *constitutional defence* of our liberties and independence, than that it should be absorbed in building up a *standing army*, which will finally become in the United States, as it has in every other country, a most powerful engine in breaking down the liberties of the people? Let an intelligent and free people answer. The proposed convention is intended to be preparatory to a *national military convention*, to be holden at the city of New York, or some other convenient place, at such time as this convention shall designate. It is earnestly requested that all who feel an interest in the improvement of the militia will exert themselves in advancing the objects of the proposed convention, and that they will also attend the same, so far as their circumstances will permit. As I have devoted a large portion of my life to the dissemination of military science amongst the people, and a considerable portion of time to the *practical* instructions of the militia, I trust I shall not be considered intrusive in making the foregoing request.

A. PARTRIDGE

NORWICH, June 20, 1838.

P.S.—Editors of newspapers friendly to the improvement of the militia, are respectfully requested to publish the foregoing notice. A.P.

MILITARY CONVENTION

A military convention, of which previous notice was given assembled at Norwich on the 4th July, instant at which were present about 300 citizens of New Hampshire and Vermont, friendly to the improvement of the militia.

The convention was called to order in the University Hall, by Captain A. Partridge; when on motion, James Udall, Esq., of Hartford, was chosen president, and Colonel Nathaniel Miller, of Bridgewater, vice president; and Major Wm. E. Lewis, of Norwich, and Josiah Sweet, jr., of Claremont, New Hampshire, secretaries.

On motion, a committee of three, to wit: Captain A. Partridge, Doctor I. Davis, and Colonel N. Miller, were appointed a committee to report resolutions to this convention.

After a short recess the committee reported the following resolutions, which were unanimously adopted:

1. *Resolved,* That when this convention adjourns, it adjourn to meet at this place on Wednesday, the 15th day of August next, at 2 o'clock, P.M.

2. *Resolved,* That a committee of five be appointed to prepare an address to the people of the United States on the subject *of the militia.*

3. *Resolved*, That the aforesaid committee prepare to report to this convention, at its adjourned session, a general plan for the improvement of the militia of the United States.

Whereas the whole history of the world, from the earliest periods down to the present time, proves that wars have existed and been frequent amongst all nations, whether in the savage, barbarous, or civilized state; and whereas, the same causes which have hitherto produced wars and bloodshed, viz: a disposition on the part of tyrannical Governments to trespass on the rights of others still exist, and are in active operation, it must be evident to every reflecting, candid, and patriotic mind, that an efficient and appropriate military defence is absolutely necessary for every nation, which is not prepared to surrender its independence and liberties to lawless aggression; and whereas, it appears by the preamble to the present constitution of the United States, that one great and leading object in forming and adopting said constitution was "to provide for the *common defence*;" and whereas such *"common defence"* is, by the aforesaid constitution, vested in the great body of the *people*, capable of bearing arms; and whereas, it is expressly declared in the 2d article of the amendments to the constitution aforesaid, that "a well regulated militia being necessary to the security of a free State, the right of the people to keep and bear arms shall not be infringed;" therefore,

Resolved, That an efficient and well-regulated militia is absolutely necessary for the preservation of our national independence, and of our political, civil, and religious liberty; and that, consequently, all attempts to discourage the militia from the prompt and efficient discharge of the high and important duties devolving upon them, as *citizen soldiers* and to depress their martial energies, dampen their patriotic efforts to discharge their duties, are anti American, anti-constitutional, hostile to civil and religious liberty, and well calculated to prostrate our national independence at the footstool of tyrannical and lawless aggressions, or to build up a mercenary standing army, to eat out the people's substance and trample down their rights.

Resolved, That a correct and efficient system of instruction, for the officers and non-commissioned officers, is absolutely necessary to the improvement of the militia in discipline, and to the rendering them efficient for military service.

Resolved, That the officers and non-commissioned officers of the militia ought to receive a reasonable compensation for their time and expenses attending the course of *military instruction*.

Resolved, That whenever the militia, or any portion of them, are called out in a legal manner to attend to the regular trainings, muster, or any military duty, they ought to be allowed a reasonable compensation for time and expense.

Resolved, That due encouragement ought to be given to the formation of companies of cavalry, artillery, light infantry, and riflemen; inasmuch as such companies, being well organized, disciplined, and equipped, would always be ready for service on any sudden emergency.

Resolved, That that portion of the rules and articles of war which deprives the officers of the militia of their relative rank, when serving in conjunction with the regular troops of the United States, by placing them under the command of the officers of said regular troops of the same grade, although the commissions of such regular officers may be of *later* date than the commissions of the said militia officers, is anti-constitutional in principle, and has a direct tendency to depress the pride and laudable ambition of the citizen soldier, by degrading him in comparison with the mercenary, and ought to be repealed.

And whereas, it is absolutely necessary to a proper organization of the militia that a prompt and correct system of returns should be established; and that cannot be established otherwise than by an efficient organization of the Adjutant General's Department in every State; therefore,

Resolved, That the Adjutant General in each State ought to be allowed an established salary, and such number of assistants as the magnitude and importance of his duties may render necessary.

Resolved, That, as efficient organization of the Quartermaster and Paymaster General's Department is absolutely necessary to the efficient organization of the militia, the same provision ought to be made for the Quartermaster and Paymaster Generals of each State, as is purposed to be made for the Adjutant General, in the preceding resolution.

Resolved, That inasmuch as every American citizen has a deep and abiding interest in the military defence of the country, (the security of his liberty, his person, and his property, being involved therein,) each and every one ought to contribute his due proportion, either by actual service or pecuniary aid, to said defence; and that, consequently, all exemptions from rendering such service, or contributing such pecuniary aid, are anti-constitutional, anti-republican, and ought to be abolished.

And whereas, it is expressly declared in the 15th article of the 8th section of the constitution of the United States, that Congress shall have power to provide for organizing, arming, and disciplining the militia; therefore,

Resolved, That Congress is not only vested with all the necessary powers by the foregoing provisions of the constitution to provide an efficient system for the organization and discipline of the militia of the United States, but that to neglect to exercise those powers is betraying the trust the people have reposed in them, as the constituted guardians of the public safety.

Resolved, therefore, That it is the duty of Congress to provide, by law, such an organization for the militia and system of instruction for the officers and non-commissioned officers, as will insure a correct and efficient discipline, and thereby prepare them for the discharge of the important duties imposed upon them by the constitution.

And whereas, the public revenue of the United States was evidently intended, by the framers of the constitution, to enable Congress to carry out into practical effect the provisions of that sacred instrument; and whereas, none of these provisions are of greater importance than that which empowers Congress to provide for "the common defence;" and whereas, by the 2d article of the amendments to the constitution it is expressly declared, "that *a well regulated militia is necessary to the security of a free State;*" therefore,

Resolved, That all the expenses necessary to provide such an organization for the militia, and to furnish such instruction for the officers and non-commissioned officers as will insure a correct and efficient discipline, ought to be paid from the public Treasury of the United States.

And whereas, it is a fundamental principle in every free Government that stations of honor, trust, and emolument shall be equally open to all; and whereas, by the regulations of the Military Academy at West Point, (sanctioned by the War Department,) all the offices of honor, trust, and emolument in the military service of the United States are monopolized by the *proteges* of that institution, who are *educated at the public expense,* to the *utter exclusion* of those who are equally well qualified, and who are *educated at their own expense;* therefore,

Resolved, That the aforesaid regulations are anti-constitutional, an invasion of the equal rights of American citizens, and ought to be rescinded.

Resolved, That, in the opinion of the members of this convention, the Military Academy at West Point is well calculated to establish a *military aristocracy* in the United States and to constitute the *nucleus* of a permanent *standing army,* and that, consequently, it ought to be abolished, and the amount which it costs the people to support it (about *two hundred thousand dollars per annum*) ought to be appropriated to the furnishing of instruction, and improving the discipline of the militia of the United States.

Resolved, That, in the opinion of the members of this convention, it is much more in accordance with the letter and spirit of our constitution, that a portion of the public revenue (the people's money) should be appropriated to defraying the expenses for instructing and improving the militia (the only national military force known to the constitution) in discipline, than that it should be absorbed in a military aristocracy, and building up a standing army, which, it is believed all our constitutious declare to hostile to civil liberty.

Resolved, That it be recommended to the militia, and also to all those who are in favor of an efficient militia system throughout the United States, to hold conventions for the purpose of devising the best means of rendering the militia efficient for service, and of placing them on the high and honorable ground originally contemplated by the constitution of the United States; also, for the purpose of making the necessary arrangements for holding a *national military convention*, either at New York or some other place, at a future convenient time.

Resolved, That it be earnestly recommended to the militia of Vermont, New Hampshire, and adjacent States, to appoint delegates (at least one from each regiment) to attend the adjourned meeting of this convention, on the 3d Wednesday (15th day) of August next, at Norwich, Vermont.

Resolved, That the editors of the Globe, the National Intelligencer, and all the other papers published at the City of Washington, also, all the editors of newspapers in Vermont, and New Hampshire, and likewise all editors, in every part of the United States, who are friendly to an efficient organization and discipline of the militia, be respectfully requested to publish the proceedings of this convention.

The following gentlemen were then appointed to draught an address to the people of the United States, to wit: General Jacob Washburn, of Proctorsville, Captain Alden Partridge, of Norwich, Colonel N. Miller, of Bridgewater, General G. Blackmer, of Bennington, and Colonel S.B. Hazeltine, of Bakersfield.

Dr. Davis introduced the following resolution; which was adopted:

Resolved, That it is the constitutional duty of Congress to provide for arming, and disciplining, and instructing the militia, and should they continue to neglect that sacred duty, we pledge ourselves to use every honorable exertion to convince them of their error, through the *ballot box*.

The convention was ably addressed by Captain Partridge, N. Robinson, Esq., and others:

When on motion the convention adjourned.

JAMES UDALL, *President*

Wm. E. LEWIS,
J. SWEET, Jr.,   *Secretaries*

MILITARY CONVENTION

The adjourned meeting of the military convention at Norwich, was held on the 16th August. The meeting was called to order by Colonel N. Miller, vice president; when on motion, General Jacob Washburn was appointed president pro tempore, J. Udall, Esq. being absent. A large number of gentlemen were present from Vermont, New Hampshire, New York, Connecticut, and Massachusetts.

Captain Partridge, from the committee appointed for that purpose, reported an address on the subject of the militia, to the people of the United States; which was read and adopted. The committee also reported a plan for the organization and improvement of the militia of the United States; which was also read and adopted.

The following resolutions were then introduced by Dr. Davis, read, and unanimously adopted:

Resolved, That when this convention adjourns, it adjourn to meet at Montpelier, on Friday, the 12th of October, at 2 o'clock in the afternoon.

Resolved, That it be recommended to the militia, and friends of an efficient militia system, in Vermont and adjacent States, to send delegates to the adjourned meeting at Montpelier.

It was then on motion, voted that the doings of this convention, including the address, be published in all the newspapers in the United States friendly to the improvement of the militia, and that are willing to publish the same.

On motion, voted to adjourn.

J. WASHBURN, *President pro tem.*

Wm. E. LEWIS,
J. SWEET, Jr.,   *Secretaries*

To the military convention, adjourned to meet at Norwich on the 15th of August, 1838:

GENTLEMEN: The committee appointed at your session on the 4th of July last, to propose a plan for the improvement of the militia of the United States, and report the same at your adjourned session, beg leave respectfully to report: That they had the important subject committed to them under consideration, and respectfully submit the following plan, which appears to them to be simple in its execution, and well calculated to accomplish the object contemplated:

In the 15th article of the 8th section of the constitution of the United States, it is declared that Congress shall have power to provide for *organizing, arming,* and *disciplining* the militia; and in the second article of amendments to the constitution it is further declared, that a "well regulated militia being necessary for the security of a free State, the right of the people to keep and bear arms shall not be infringed." Your committee feel well assured that the clause above quoted from the 15th article and 8th section of the constitution, invests Congress with full power to make all laws and adopt all measures necessary for organizing and disciplining the militia of the United States, so as to render them efficient for military services, and thereby enable them to discharge, in the best possible manner, the important duties imposed upon them, as the guardians of our national independence, and of our civil and religious liberties; while the second article of the amendments is an unequivocal and decided expression of the importance which the wise and patriotic founders of our republic attributed to a well regulated militia. For the purpose, then of carrying into effect the powers thus vested in Congress, your committee submit the following propositions, which they believe ought to be embodied into general law, viz:

1st. Let the militia be composed of all citizens capable of bearing arms between the ages of twenty and fifty years.

2d. Let the enrolled militia be divided into three classes; the first, or junior class, to be composed of those between twenty-one and thirty-one years of age; the second class, of those between thirty-one and forty-one years of age; and the third or senior class, of those between forty-one and fifty years of age.

3d. Let the junior class be required to *encamp* six successive days each year, either by battalions or brigades, for the purpose of military instruction; and let each individual be paid from the Treasury of the United States one dollar and fifty cents per day during each encampment.

4th. Let the class be officered during the time of encampment, either by officers selected from this particular class, or by such as may be detailed in rotation from the officers of the whole body of the militia as Congress shall by law prescribe.

5th. When any member of the first class shall have attained the age thirty-one, he would pass into the second class, and thence, at the age of forty-one, into the third class; the members of which classes should be kept enrolled, but not required to do any further military duty in time of peace, except to turn out once a year to be mustered, unless the *Legislature* of the States should otherwise prescribe by law.

6th. All details for active service, to be made from the junior class, until all the members of that class shall, in rotation, have pursued one term of duty; after which detail should be made from the second class, and ultimately from the third class, should emergencies require it. In time of great public danger, however, details should be made from all the classes, simultaneously, should the public safety render it necessary.

7th. Let the Adjutant and Inspector General the Quartermaster General, and the Paymaster General of each State be allowed an annual salary of twelve hundred dollars each.

8th. Let officers of the whole body of the militia be called out into camp of rendezvous, either by battalion or brigade, three successive days in the year, for the purpose of military instruction, and be allowed two dollars per day each during the time.

9th. Let the Military Academy at West Point be *abolished*, and *State institutions* substituted instead thereof, for the cultivation of, and general examination into, military science, agreeably to the plan contained in the accompanying address to the people of the United States.

10th. Let the professors of the proposed State institutions be required to attend, annually, all the camps of instruction in their respective States, in the capacity of military instructors.

11th. As every citizen has a deep and abiding interest in the military defence of the country, the security of liberty, person, and property, therefore every one ought to contribute his due proportion of service for that purpose. If, however, anyone be prevented from rendering such service by personal disability, or other imperative necessity, he ought to pay an equivalent in place thereof, in money. By adopting the foregoing plan, your committee feel confident that the following, amongst other advantages, would result, viz:

1st. The officers of the militia would, in the course of ten years, become well acquainted with the practical details of discipline and military duty, which would inspire confidence in the soldiers, and enable them to command them successfully in active service.

2d. The great body of the militia would, in like manner, become so well versed in mechanical discipline, that, combined with their superior physical and moral energies, it would enable them to combat successfully, like the republican legions of Rome, among enemies to which they might be opposed.

3d. An uniformity of discipline would be established among the militia of all the States, as the military instructors would be required to instruct agreeably to one uniform system.

4th. Military science would be uniformly and equally disseminated throughout the United States.

5th. A proper military spirit would be cherished among the people without which civil liberty will not long exist in any country. The history of every republic proves the truth of this assertion.

6th. All necessity for a standing army would be done away, as the militia would be perfectly competent to the effective military defence of the country.

7th. It would constitute the safest guarantee of peace. What nation would attempt to invade our rights, when they were defended by a well-disciplined force of six hundred thousand freemen which could be brought into the field almost at a moment's warning, and this backed by a reserve of at least thirteen hundred thousand more? We believe none. All of which is respectfully submitted.

JACOB WASHBURN,
ALDEN PARTRIDGE,
N. MILLER,
S.B. HAZELTINE.

To the people of the United States:

FELLOW CITIZENS: The undersigned having been appointed by the military convention assembled at Norwich, Vermont, on the 4th of July, 1838, a committee to address you on the subject of the organization, discipline, &c., of the militia of the United States, enter on the discharge of the duties assigned them with a full conviction of its great importance to the welfare of the country. The history of the whole world, from the earliest period down to the present time, proves that wars have existed in every age, and amongst all nations, whether in the savage, the barbarous, or the civilized state. The cause which has produced those wars and bloodshed, viz: a disposition on the part of the *unjust* and *tyrannical* Governments and lawless aggressors to trample on the rights and liberties of mankind, still exists and is in active operation; and so long as the same cause does exist and continue in active operation, it is absurd to suppose that the same results will not flow from it. This consideration clearly points out and most forcibly enjoins upon every nation which is not prepared to surrender its rights and liberties to lawless and tyrannical aggression, the duty of being well prepared to protect them, by a competent and appropriate system of military defence. Indeed, one great and leading object in framing the present constitution of the United States was to provide for the common defence. What would all our boasted liberties be worth if we are incapable of defending them? The military defence of countries is of two descriptions; the one is, when a portion of the population is equipped with all the habiliments of war, organized into the regular standing army, and maintained as such in time of peace as well as in time of war; the other is when the great body of the people capable of bearing arms, and organized into militia, constitute the military force of the country. As to standing armies, it is believed there is but one opinion amongst the people of the United States on the subject. They consider them hostile to civil liberty, and the history of many ages proves them to have been so. They are the necessary appendages of monarchy, and constitute the right arm of tyranny. Take away the standing armies of Europe, and what monarch could sustain himself six months on his throne? Our revolutionary fathers well understood their character, and witnessed their power in combatting against liberty, and in shedding innocent blood. Do we wish further proof the recent tragic scenes at St. Charles, St. Eustache, and in other parts of Canada, will furnish it. It is evident, then, that the attention of a free people should not, even for one moment, be directed to a standing army, as the guardian of their liberties and independence. For the protection of these they should rely only on themselves. The great and good men who framed the constitution of the United States were well aware of the absolute necessity of a well regulated and efficient national militia, for the preservation of our national independence and of public liberty. They, therefore, wisely invested Congress with full power to adopt all the necessary and proper measures for establishing and rendering efficient, for the military defence of the country such national force.

In the 15th article of the 8th section of the constitution, it is declared that Congress shall have power to provide for organizing, arming, and disciplining the militia; and, as a proof of the importance which was attached to the militia, it is further declared in the same article of amendment to the constitution, that "a well regulated militia being necessary to the security of a free State, the right of the people to keep and bear arms shall not be infringed." The necessity of a well regulated militia to the security our free institutions is established by the high authority of the framers the Government, and sustained by the voice of the people themselves, when they adopted the present constitution, and Congress was invested with full power to make this institution what it was evidently designed to be, the great and safe constitutional military force of the nation, to repel all foreign invasion, to aid in the execution of the laws, and to suppress insurrections. But in order that the national force should be efficient, it should be organized and disciplined. An efficient and appropriate organization must contain the following provisions, viz:

1st. A proper arrangement into companies, battalions, brigades, and divisions, with the appropriate number of officers, non-commissioned officers, musicians, &c., to each of the foregoing divisions.

2d. There should be a due proportion of infantry, artillery, cavalry, light infantry, and riflemen.

3d. Provisions should be made for an efficient organization of the staff, embracing the departments of the adjutant and inspector general, the quartermaster general, the commissary general, the paymaster general, and the medical department, in each State.

4th. It is believed that it would be a great improvement to divide the militia into three classes, for the purpose of military instruction and also for active service, in the following manner: The first class to embrace those from twenty-one to thirty-one years of age, the second to embrace those from thirty-one to forty-one years of age, and the third class, which should constitute a corps of reserve, those from forty-one to forty-five years of age. The details under each of the foregoing articles should be so arranged as to correspond with what has been found most efficient and useful in active service, in the present improved state of the military art.

We will now proceed to the subject of disciplining the militia and the first thing here to be done is, to ascertain the true meaning of the word *discipline*, as it is used in the constitution. Does it simply mean to instruct them in the manual exercise, to face, to march, to wheel or turn, and to execute a series of mechanical movements, (many of them unnecessary and calculated merely for show) called manoeuvres, or evolutions? An may have all this, and still be comparatively inefficient when opposed by a scientific and skilful enemy. It is evident that the framers of the constitution intended the militia to constitute the grand military defence of the nation and that they intended to invest Congress with full power to adopt all the necessary measures to prepare them for the correct and efficient discharge of the important duties devolving upon them; and that, consequently, under the head of disciplining them they intended to embrace all those branches of instruction, both theoretical and practical, that are necessary to qualify both officers and soldiers for the appropriate discharge the duties of their respective stations, in the efficient military corps to which they may belong. We then consider a proper system of military instruction, both for officers and soldiers, as absolutely necessary to disciplining, or rendering them effective for military service; and that, consequently, any system of legislation on the subject that does not provide for this, will utterly fail of accomplishing its object. This will lead to an examination into the character of the proposed instruction, and the best and most economical means of communicating it. Military, like every other kind instruction, naturally divides into two grand divisions, viz: theoretical and practical; the former comprising such investigations as are necessary to a correct understanding of the principles of the different branches of military science; the second, the practical application of those principles to every kind of active military service. This second division, then, evidently comprises all the duties that the great body of the militia the rank and file, would be called upon to discharge; the first division being more particularly necessary for the higher grades of officers, engineers, &c. And we will now proceed to point out what appears to be the best manner of communicating the necessary instruction in it.

3d. Let there be appointed in each State military instructors, in such manner and in such numbers as will be hereafter proposed, whose duty it shall be to attend the several camps, and afford all the necessary instruction to the battalion or brigade there assembled. These instructors should have received a regular scientific military education, and be competent to instruct not only in all the established manœuvres, embracing the evolutions of the line, and other field, garrison, and camp duties, but also to explain the use of the whole in active service and the principles on which they are founded.

4th. The course of instruction would consequently embrace every thing that is necessary to prepare troops for active service. Now one course of instruction would better prepare those subjected to it for active military service, than they are, or ever will be, under the present system, in the whole period from eighteen to forty-six years of age, and by continuing under such instruction for ten years, they would become well drilled soldiers. They would then pass into the second class, and be relieved from military duty, except in time of war, or other urgent necessity. As the duties of the officers are of a higher order than those of the rank and file they should be called out either into camp or rendezvous, by brigades, at least three days in each year, in addition to the time above mentioned, where they should be instructed more particularly in the duties of their respective stations. This would prepare them for the correct discharge of their duties in the camps of general instruction.

We next proceed to an examination of the first, or theoretical division of instruction, which embraces the investigation of principles, and the cultivation and improvement of the military art, and which, probably, comprises within its wide range a greater variety of branches, and a greater amount of science, than any other department of human knowledge. It is based on the mathematics, and embraces within its scope almost the whole circle of the sciences. The military art has undergone a great change, and been much improved within the last fifty years. The French revolution has had the principal agency in effecting this. Every nation, therefore, which expects or wishes to meet its enemies on equal terms, must cultivate this, as they would cultivate any other branch of useful knowledge, viz: by making it a branch of study and instruction at seminaries of learning established for that purpose. The question now arises, in what way, or by what means, can Congress provide most effectually, most economically, most certainly, and most in accordance with the principles of our republican institutions, for the cultivation and general dissemination of military science throughout the country? for such provision we are confident is absolutely necessary to enable us to keep pace with other nations in the improvement of the great art of national defence, and without which our militia system can never be perfected. We are aware many will answer that such provision has already been made by means of the Military Academy at West Point. We will proceed to examine this subject. The institution at West Point is wholly supported at the expense of the people of the United States, who have already paid about half a million of dollars for erecting buildings, procuring apparatus, &c., &c. In addition to this, the annual expense of the institution is about $300,000. Provision is made by law for educating 250 cadets at this seminary, and to complete the course of instruction requires four years. The number of graduates therefore, for each year, (supposing all who entered to complete the full course,) would be *sixty-two*. The people, therefore, at this rate are taxed about $4,851 for every officer educated at West Point. But as a considerable portion of the cadets never complete the course of instruction, the cost of each graduate is proportionably increased, and $5,000 would probably be a low estimate. But, notwithstanding this heavy tax on the people for its support, the institution does not, in the most remote degree, aid in the discipline of the militia, or in the dissemination of military science. Those cadets who graduate are immediately attached to the army, to the limited sphere of which all their operations are confined. The organization of the seminary is also in direct violation of the fundal mental principles of our republican institutions, viz: That stations of honor, trust, and emolument, shall be equally open to all. The law establishing the seminary on its present plan prohibits the appointment of any cadet who is under fourteen or above twenty-one years of age; and the regulations of the War Department require that all the vacuums for officers that occur in the army shall be filled by the graduates from West Point. The consequence is, that every young man in the United States who has attained the age of twenty one years, unless he happens to be one of the selected 250, whom executive favor has placed in this public charity school to be educated at the public expense, is utterly excluded from holding any office of honor, trust, or emolument, (above that of a non-commissioned officer,) in the military service of the country.

Suppose Congress should pass a law excluding from such offices of honor, trust, and emolument, all American citizens, except such select number of 250 as the President might think proper to choose, between the ages of fourteen and twenty-one years; would not the whole people pronounce it unconstitutional, and as a most high-handed and flagrant invasion of their rights? Yet the War Department has done the same thing by regulation, and it has been acquiesced in for nearly twenty years. By this means the Military Academy is converted into a grand monopoly, (supported by the people's money,) and constitutes a *nucleus* around which is rapidly concentrating a military aristocracy. But is the course of education at this institution such as to prepare young men to endure cold and heat, hunger, thirst, fatigue, and all the hardships incidental to war? Let facts answer. A few years ago a bill was introduced into Congress making an appropriation of money (it is believed $60,000 was the sum) to erect a building for the purpose of protecting these hardy soldiers from the weather during their military exercises. The bill passed; and, during the discussion, a member gravely moved an additional appropriation to furnish them with umbrellas also. Is this the way, fellow citizens, to *form soldiers* for active and efficient services? We hear you respond: No! What, then is the use of it? We answer to form military pedants and military dandies. What, then, you may ask, should be done with the institution? We answer, abolish it—totally abolish it—and substitute something much better in its place. If you ask what better substitute we would propose, we answer, State institutions. Let these State institutions be adopted for the cultivation and general dissemination of military science throughout our country.

We will now proceed to state, in detail, the manner in which we feel confident this important object can be most certainly, most economically, and most beneficially accomplished. Let Congress pass a law containing the following provisions, viz:

1st. That whenever the Legislature of any State will organize seminaries in said State, (not to exceed the number hereinafter proposed,) in which military science and instruction shall be made an appendage to the usual civil education of the students; and shall provide competent professors for that department; such professors shall be paid from the public Treasury of the United States, and no charge be made to the students for such military instruction.

2d. That the number of such seminaries, to which each State shall be entitled shall be according to the following ratio of population, viz: each State the population of which does not exceed 250,000, shall be entitled to one seminary; each State, the population of which exceeds 250,000, but does not exceed 600,000, shall be entitled to two seminaries; each State the population of which exceeds 600,000, but does not exceed 1,000,000, shall be entitled to three seminaries; each State, the population of which exceeds 1,000,000, but does not exceed 1,500,000 shall be entitled to four seminaries, and each State whose population exceeds 1,500,000 shall be entitled to five seminaries.

3d. That there should be allowed to each seminary, one professor of tactics, and one professor of military and civil engineering, each professor to be allowed a salary of eighteen hundred dollars per annum. These professors, in addition to the duties at their respective seminaries, should be required to attend, in the capacity of instructors, the camps of instruction, both for the officers and for the junior class of the militia, as has been before mentioned.

4th. The course of instruction at those seminaries should embrace the practical drill of the infantry and rifleman; permanent and field fortifications, the construction of marine batteries, mining and counter mining, the principles of gunnery, civil and military architecture, the construction of bridges, roads, canals, and locks; castrametation, field and garrison duty, embracing the turning off, mounting, and relieving guards, sentinels, &c.

5th. Each seminary shall be furnished with at least two field pieces (six pounders) completely equipped, and the requisite number of muskets and accoutrements, knapsacks, &c. At these institutions would be formed our scientific engineers and artillerists, and skilful tacticians and able generals. The students, after graduating, would be dispersed over the whole United States, and disseminate the knowledge they had thus acquired, and on any sudden emergency, there would be found at every exposed point, in the ranks of the people themselves, able engineers and skilful leaders to give an efficient and successful direction to the martial energies of the militia. There would be, according to the foregoing plan, about seventy-five of the seminaries in the United States, which would require one hundred and fifty professors; the amount of their salaries, at eighteen hundred dollars per annum, would be two hundred and seventy thousand dollars. Suppose now that each of these seminaries should number one hundred students, there would be constantly, under the proposed system of education, in the United States, seven thousand five hundred young men; and in case the whole course of instruction, civil and military, should require the usual period of four years, the number of graduates each year would be one thousand eight hundred and seventy-five, These young men, by mingling with the people, would disseminate their military information to every point of the United States, and render the most efficient aid in disciplining the militia.

Let us now compare the results of this plan with those of the Military Academy at West Point.

And first, the cost of obtaining sixty graduates at that institution is about three hundred thousand dollars; the cost of obtaining one thousand eight hundred and seventy-five, equally well qualified, on the proposed plan, would be about two hundred and seventy thousand dollars. The military knowledge acquired at West Point is confined to the army, and consequently does not in the least aid in the discipline of the militia. That acquired on the proposed plan, would be generally disseminated, and aid most essentially in the discipline of the militia. The Military Academy at West Point has a direct tendency to build up a standing army, and establish a military aristocracy in the country. The proposed plan is entirely free from this objection, and is in perfect accordance with the principles of our republican institutions, which constitute every citizen a soldier, where his services as such are required. Its tendency is to retain the military defence of the country in the hands of the people, where the constitution has placed it, by rendering a standing army unnecessary, and thereby keeping down a military aristocracy, which is the necessary and inevitable consequence of a standing army in every country. It does not encroach in the least degree on the rights of the States, for if the Legislatures of any of the States do not wish such institutions within their limits, they will not avail themselves of the provisions of the law by organizing them. But as the militia of each State constitutes its military defence, as well as a portion of the military defence of the nation, every State has a deep interest in the discipline of its militia, and would consequently have an interest in co operating with the General Government in all constitutional measures which be best calculated to render such discipline most perfect. The system education at the proposed seminaries (combining the civil and military) would also prepare those who should have the advantages of it, much better than a purely civil one, for the discharge of important public duties. Is not the President of the United States commander-in-chief of the whole force of the country? Ought he not, then, even if he should never exercise any active military command, to be well acquainted with all military subjects to enable him to judge and decide correctly respecting the military defence of the nation? The governors of the States are the constitutional military commanders of the military force of their respective States; ought they not, then, to possess similar qualifications? Are not the members of Congress called upon annually to vote away large amounts of their constituents' money for military purposes? And can they judge correctly of the necessity or expediency of such appropriations, if they are entirely ignorant of military subjects? We think not. We do not hold that ignorance is any qualification for the discharge of important public duties.

We will next proceed to make an estimate of the expense necessary to carry into effect the whole of the preceding plan. We will assume the whole number of militia in the junior class (from 21 to 31 years of age) to be six hundred thousand, and propose to allow each individual one dollar and fifty cents per day; this, for six days, the proposed time of instruction, would amount to nine dollars per annum for each, and five million four thousand dollars for the whole. We next assume the whole number officers (estimating the whole body of the militia) to be one hundred thousand, and allowing each two dollars per day for the three days it is proposed they should be under instruction each year, the amount be six dollars each, and six hundred thousand dollars per annum for the whole. The number of adjutant and inspectors general, quartermasters general, and paymasters general would be seventy-eight, and allowing a salary of twelve hundred dollars per annum, the whole amount be ninety-three thousand six hundred dollars. The expense of the professors at the proposed seminaries has been estimated at two hundred and seventy thousand dollars per annum. Add together the foregoing items and the result will be as follows, viz:

For the instruction of the junior class
$5,400,000.00
For the instruction of officers
$600,000.00
For the salaries of the adjutant, quartermaster, and pay-master generals
$93,600.00
For the salaries of the professors at the seminaries
$270,000.00
For contingent expenses
$136,400.00
Whole amount
$6,500,000.00

We will next present an estimate of the annual expense of a standing army of twenty-four thousand rank and file, (including musicians,) as follows, viz:

24,000 men, at an expense of $210 per man
$5,040,000.00
6 major generals, at $4,000 each
$24,000.00
12 brigadier generals, at $2,500 each
$30,000.00
24 colonels, at $1,800 each
$43,200.00
24 lieutenant colonels, at $1,600 each
$39,400.00
24 majors, at $1,400 each
$34,200.00

1,250 company officers, average expenses $1,000
$1,250,000.00
For the expense of the staff and contingencies
$129,200.00
Whole amount
$6,600,000 00

This estimate is believed to be below, rather than above the real expense.

Thus it appears that a standing army of 25,350 men, including officers, would involve the country in a greater annual expense than the proposed plan for organizing and disciplining the militia in such a manner as to render them effective for any kind of military service. Let us now, fellow-citizens, candidly compare the foregoing plans (involving nearly the same amount of expense) for the military defence of the United States, and make our decisions on the true merits of the case. By the former, you have a well organized and effective military force of *six hundred thousand citizen soldiers*, distributed over every inhabited portion of the United States, prepared to turn out in defence of their rights at a moment's warning; by the latter, you would have *twenty-five thousand mercenaries*, distributed along a sea-coast and inland frontier of more than six thousand miles, and consequently, so scattered as to be incapable of resisting a powerful invading force at any point. By the former you would have *military science* cultivated and disseminated amongst the people, in every section of the country and skilful officers residing at and in the vicinity of all places exposed to an attack, and capable of giving an efficient direction to the physical and moral energies of the militia; by the latter, nothing of this would take place. But it may be argued that it is not intended to keep up an army in time of peace, that shall be adequate to the defence of the country in time of war; then, we ask, why not reduce it to the minimum peace establishment of three thousand men, (commanded by a brigadier general,) as was done in Mr. Jefferson's administration? This would be sufficient to station a public guard at every necessary military post, for the protection of public property, &c. Why keep up, at a great public expense, a military force greater than this, but which, nevertheless, is totally inadequate for defence in case of war? But it may be said that a larger peace establishment would constitute a *nucleus* around which an army could soon be formed in the event of war. The whole history of the world, from the time of the Romans to the present period, proves that troops which are kept, for any considerable time, on a peace establishment, become enervated and broken down by indolence and dissipation, and, consequently, in a great measure unfit for the active duties of the field. The Romans alone were able, during the earlier periods of the empire, to sustain a peace establishment which was fit for the emergencies of war. This was accomplished by encamping their legions on the remote frontiers of the empire, far removed from cities and towns, and subjecting them to a discipline more strict, and to duties more severe, than had been required of them in time of war. But can this system finally sink under the enervating influence of peace, and those renowned legions eventually become degenerate and incapable of resisting the hardy, but undisciplined barbarians of the north? The celebrated republican legions which established the reputation of the Roman arms and laid the foundation of the glory and future greatness of their country, were *citizen soldiers*, draughted for a single campaign; at the expiration of which they returned home, and were replaced by new recruits. We feel confident, fellow citizens, that the plan we propose to you for the discipline of the militia, if adopted and carried out into practice, would fully accomplish the object contemplated. It will furnish a military defence composed of *citizen soldiers*, that will be adequate to all emergencies, both of peace and of war. Such a force can never become dangerous to the liberties of the country; for it is composed of the people themselves, (who can have no motive for enslaving themselves,) and will render entirely unnecessary a standing army, which the whole history of the world proves to be hostile to civil liberty. The expense we do not consider as constituting any objection to its adoption. The ordinary revenue in time of peace will be amply sufficient to meet this and the other necessary expenses of the Government. The money would also be distributed among the industrious and patriotic yeomanry of the country, instead of going to support idlers and speculators. We conclude, fellow-citizens, by urging you to a serious and candid examination of this subject, for we feel that none of greater importance can claim your attention. It rests upon you to decide whether you will intrust the protection of your liberties and independence to a mercenary army, which may ultimately—led on by some future Cæsar—wrest them from you; or whether you will confide them to a well disciplined militia. We think you will not hesitate to choose the latter.

JACOB WASHBURN
ALDEN PARTRIDGE
N. MILLER
S.B. HAZELTINE

MILITARY CONVENTION

The adjourned State military convention assembled at the court-house in Montpelier, on the 12th day of October, 1838, at 2 o'clock P.M.

The convention was called to order by the president; when, on motion General J. Washburn, General John Kellogg, Colonel J.G. Dudley, Major Earl, were appointed vice presidents, and Major J.W. Curtiss was chosen secretary.

On motion of Doctor Davis, a committee of four was appointed to prepare and present resolutions to the convention, and to suggest the business necessary to be transacted. The committee consisted of General J. Washburn, Colonel Thomas, Major I. Davis, Captain A. Partridge, and Captain G.A. Grant: when the convention adjourned to meet at 10 o'clock A.M. tomorrow.

OCTOBER 13, 10 O'CLOCK A.M.

Convention met pursuant to adjournment.

The committee on resolutions reported the following which were adopted Whereas the constitution of the United States contemplates a full and complete organization and disciplining of the whole body of the militia, by the General Government; and whereas, also, Congress has hitherto neglected, in a great measure, that important duty, by reason of which our only constitutional defence is becoming dispirited and inefficient; therefore,

Resolved, That it is the imperative duty of Congress to enact such laws as shall afford the necessary facilities for a full and perfect organization of the whole body of the militia of the United States.

Resolved, That a committee of three be appointed to present to Congress the plea and address heretofore adopted by this convention, together with an appropriate memorial on the subject of the militia.

Resolved, That this convention recommend that a military convention of the New England States be holden in the city of Boston, on the twenty-second day of February next, and that it be recommended to the militia of those States immediately to elect delegates to attend the same.

Resolved, That this convention appoint a committee of three, to present an address to the militia of New England, and also to the militia of the United States.

Resolved, That, in the opinion of this convention, it is expedient to hold a national military convention in New York, in June next. (Subject to the control of the convention at Boston).

Resolved, That when this convention adjourns, it will adjourn to meet at Norwich, on the 4th day of July next, at 10 o'clock, A.M., and that this convention elect some person to deliver an address on the occasion.

Resolved, That the secretary of this convention be directed to furnish his excellency the Governor with a copy of the proceedings of this convention, with a request that he immediately lay the same before both houses of the Legislature Resolved That the secretary of this convention be directed to furnish each of our Senators and Representatives in Congress with a copy of the proceedings of this convention and that they be requested to use all necessary exertions to promote the objects of this convention.

[A committee was here appointed to take into consideration the present militia law of this State and recommend such alterations as might seem necessary. The committee made a somewhat voluminous report; and as the Legislature has considered and acted upon some of its recommendations, it is not deemed necessary to insert it here]

Resolved, That the provision of the militia law which exempts from taxation the polls of those who are equipped and who do military duty, and, also, which deducts ten dollars from the list of each parent, master, or guardian, for each minor by them equipped, is unequal in its operation, and should be amended, so that none should be exempt from taxation, but instead thereof shall receive an equivalent for their services in money.

In accordance with a resolution to appoint a committee to memorialize Congress, the following gentlemen were appointed, viz:

Captain A. Partridge, Norwich, Vermont; Hon. Asa Aikens, Windsor, Vermont; Major Edmund Burke, Newport, New Hampshire.

The following gentlemen were appointed on a committee to address the militia, viz:

Colonel J.P. Miller, Montpelier, Vermont; General James Wilson, jr., Keene, New Hampshire; Hon. S.S. Brown, St. Albans, Vermont. Captain A. Partridge was appointed to deliver the address at Norwich, on the 4th day of July next, and General Isaac Fletcher was appointed a substitute.

When, on motion the convention adjourned.

JAMES UDALL, *President.*

J.W. CURTISS, *Secretary pro tem.*

Appendix O: Military Defence of the United States[707]

MEMORIAL

Of a committee of the military convention at Norwich, Vt. Praying the revision and alteration of the military defences of the United States, to the honorable Congress of the United States, which,

RESPECTFULLY SHOWETH:- That your memorialists were appointed a committee by the military convention assembled at Norwich, State of Vermont, in August 1939, to present to your honorable body a general plan for the military defence of the United States. Your memorialist enters on the discharge of the duty assigned them, with a full conviction of the great importance of the subject to the safety and general welfare of the country. In examining the preamble of the Constitution of the U. States, your memorialist finds, that one of the important objects proposed to be accomplished by the adoption of this instrument, was "to provide for the common defence;" and if any one of the objects enumerated in this preamble can be considered more important than any other, this doubtless would be entitled to the pre-eminence. It is, indeed, the safe guarantee of all the rest. What would our proud independence, and the advantages we derive from our republican institutions be worth, if we were incapable of defending them? The military defence of all countries, so far as such defense depends on physical force, naturally divides into two divisions, viz: a standing army, and a militia, or citizen soldiery. Your memorialist will proceed to examine each of those separately; and

I.   A STANDING ARMY.- Standing armies have, in all ages and in all countries, been considered and found to be hostile to civil liberty. Their very organization and profession separate them, both in feeling and in interest, from the great body of the people. Their immediate dependence on the Executive of a country, and subjection to his mandates, makes them the ready instruments of his will to oppress the people and establish tyranny. Besides, if the people are brought to repose their confidence in a standing army for the defense of their independence they will inevitably lose their own martial spirit,

---

[707] Partridge et al., "Memorial of a Committee of the Military Convention at Norwich, Vt."

another name for the spirit of liberty, gradually sink down to a state of supine indifference and inability to defend themselves, ere long, fall an unresisting and helpless prey to the power in which they had reposed their confidence for protection. The history of every republic which has preceded ours, proves the truth of these assertions. Your memorialists, therefore, consider any system which vests the national defence in a standing army, as anti-constitutional, anti-republican, and ultimately destructive to our free institutions. But it may be asked, shall we then have no regular force in them of peace? Your memorialists answer yes, and will proceed, respectfully, to exhibit their views as to the objects, magnitude, and organization of such force; and

1. *Its object.-* Your memorialists would pointedly reject the idea that such force should be sustained in the United States with any intention that it should be relied upon for national defence in time of war: this should be considered the peculiar and exclusive prerogative of the militia. The only object then, for which a peace establishment should be sustained in this country, ought to be, that it constitute, simply, a police-guard, to protect, in time of peace, the necessary fortifications on the seaboard and inland frontier, as well as other public property that requires such protection from lawless depredation. Whenever it is allowed to go beyond this limit, the people will begin to rely upon it for the protection of their independence; the militia will be neglected, and gradually sink into insignificance, and the peace establishment soon swell to such magnitude as to become a standing army, in the appropriate sense of the word, involving an immense expense for its support, with the necessary accompaniments of taxes and evils growing out of similar establishments in Europe, and ultimately, with the prostration of our republican institutions.

2. *Its magnitude.-* The magnitude of our peace establishment should be determined by the duties it is necessarily required to discharge. During the first term of Mr. Jefferson's administration, it was reduced to three thousand men, commanded by a brigadier-general. Our seacoast was then, with the exception of the coast of Florida, as extensive as it is at the present time; and the inland frontier, after the acquisition of Louisiana, the same.- Now, your memorialists believe, that the whole, including all the public property, was just as well protected then as it has been since, and the expense of the military establishment was not one-fifth what it now is. But it is often alleged, as a justification for the increase of our peace establishment, that the population of the country has much increased since that time, and that, consequently, should have a larger regular force. Your memorialists cannot discover the correctness of such an inference from such data. If our population has increased has not the Militia also? And is such increase of the power of the people to protect

their inland and maritime frontier, a substantial reason for the increase of our regular force, unless we mean to rely on a standing army for protection? Your memorialist cannot conceive that it is; they would come to a conclusion directly the reverse; they can view all such suggestions in no other light than as specious pretexts for gradually withdrawing the protection of our independence from the citizen-soldiery, where the framers of the Constitution placed it, and vesting it in a standing army. From a due consideration of all these circumstances, your memorialist would respectfully, but urgently, recommend, that the peace establishment be reduced to a number not exceeding six thousand men, (officers included) being fully convinced that all the duties necessarily devolving on such establishment, can be just as well discharged for the public benefit as by the present number.

      3.   *Organization.* Your memorialist would recommend, that the proposed force should be divided into ten battalions, organized as infantry of the line, but instructed also, in the duties of artillerists; that the United States be divided into ten military departments, in which the troops should be distributed in such proportion as the President of the United States may direct; that each department be commanded by a colonel; that there be located at the city of Washington an inspector general, with the rank, pay and emoluments of a colonel, though whom, all orders from the war Department should be transmitted to the commandants of departments, and to whom all reports should be made. The pay and emoluments of officers and soldiers to be the same as were allowed during the first term of Mr. Jefferson's administration. No grade of officer above that of colonel to be allowed. Brevet rank to be abolished. The corps of engineers to be reduced in every respect to its first organization of 1802. The military academy at West Point (excepting the corps of engineers) to be entirely abolished; and in lieu thereof, a general system of military instruction to be established in the several states agreeable to the plan proposed by the military convention assembled at Norwich, Vermont, in August 1838 (a copy of which accompanies this memorial,) by which means, military science would be disseminated throughout every part of the United States.- By the organization her proposed, your memorialists are confident that all the appropriate duties of a peace establishment, would be just as well discharged as they are by the present military establishment, and that the expense of supporting it would be less than one-half of what is now required.

      II.   THE MILITIA.- The militia, or citizen-soldiery of the United States, comprises all the free white able-bodied citizens of the country (excepting such as are legally exempted from being enrolled), between the ages of eighteen and forty-five, amounting, probably, to about 1,500,000 men. This force ought to be regarded, exclusively, as constituting the grand constitutional military defence of the nation. Composed of the citizens themselves, with feelings and interests identified with the support of our free institutions, it can never prove dangerous to civil liberty; while its number, if properly organized and disciplined, would set foreign invasion at defiance, and constitute the surest guarantee of the peace. The patriotic framers of the constitution of the United States were fully impressed with the absolute necessity of such a force for the security of the country; and it is consequently declared, in the second article of the amendments to the Constitution, that "A well regulated militia being necessary to the security of a free state, the right of the people to keep and bear arms shall not be infringed." They also conferred on Congress all the necessary powers to render this force efficient for active service, by declaring, in the 15th article of the 3th section of the Constitution, that "Congress shall have power to provide for organizing arming, and disciplining the militia," &c. The only question then is, in what manner can the power thus conferred be exercised, so as to accomplish, in the best possible manner, the important object contemplated? In answer to this inquiry, your memorialists would be leave respectfully to submit to your honorable body the plan for the organization and discipline of the militia adopted by the military convention assembled at Norwich, Vermont, on the 15th of August 1838, with the address to the people of the United States explanatory of the principles of the aforesaid plan, and which was also adopted by the same convention, both of which documents accompany this memorial. This plan appears to your memorialist to be economical, simple in detail, easily reduced to practice, and in every respect well calculated to render the militia effective for any kind of service. They would, therefore, respectfully but urgently, recommend that it be adopted in all its provisions. Your memorialists are, however, well aware that it has been urged against the militia, by professional military men, that they are unfit for military service, except to act as an appendage to regular troops, or when called out on sudden emergencies, and for short tours of duty. Your memorialists are very far from coinciding with such opinions, and will proceed to prove, by facts and arguments which, it is believed, cannot be invalidated, that they are capable of making the best soldiers in the world without losing their character of militia; and

1st. That they are more expert in the use of fire-arms, and better marksmen, than the disciplined regulars of Europe, even without the aid of regular mechanical discipline, it is believed, is a fact that cannot be disputed. In all the battles of the Revolutionary, as well as of the late war, when the militia, alone were opposed to British regulars, the proportional loss of the regulars was much the greatest. Witness the battles of Lexington, Bunker's Hill, Bennington, and King's Mountain, in the Revolutionary war; and the battles of

New Orleans, the Thames, North Point, and Plattsburg, in the late war. This superiority is a natural consequence of their being accustomed to the use of fire arms from early youth in hunting, turkey-shoots, firing at targets, &c.

2d. They possess more physical activity and energy than regular troops. This is a consequence of the active and laborious duties which the great body of our militia are called daily to discharge. The firmness and vigor of constitution, and strength and elasticity of muscle, thus acquired, render them far superior to regular troops (who are broken down, and their energies paralyzed by the enervating influence of idleness and the inactivity incidental to a garrison or camp in time of peace), for the performance of the vigorous and active duties of the field in time of war.

3d. They possess the moral of the soldier (which is the great degree stimulus to deeds of valor) a much greater degree than mercenary troops. What comparison can exist between the motives that influence the citizen-soldier, battling for liberty, for independence--yea, for everything he holds dear, and those which influence the mercenary, who fights alone for pay in obedience to the stern mandates of a master? The superiority of our citizen-soldiers over a veteran, gallant, and disciplined foe, even without the aid of mechanical discipline was fully exhibited both during the Revolutionary and the late war. The battles of Lexington, Bunker's Hill, of Bennington, of King's mountain, of New Orleans, of the Thames, and of Plattsburg, together with the defence of Fort Moultrie, and of Stonington, rank among the most brilliant achievements in the annuals of military history.

But if our citizen-soldiers have done so much without the aid of mechanical discipline, what could they not accomplish, what could they not do, with it? We answer, they could accomplish every thing, except impossibilities. No mercenary troops could withstand their prowess.—The great advantage which mechanical discipline confers is, that it ensures an accuracy of movement, and enables thousands to concentrate their force, and act as one man. In maneuver, and close combat with the bayonet, such discipline gives a decided superiority to troops that possess it over the loose ranks and uncertain movements of troops which are destitute of it. The plan for the organization and discipline of our militia, which your memorialists have recommended to your honorable body for adoption, it is confidently believed, would confer all the mechanical discipline which is necessary to render our citizen soldiers superior to the regular troops of any nation in Europe. But if further proofs are required to show the efficiency of citizen-soldiers, your memorialists would respectfully refer your honorable body to the French Revolution, which opens a wide field from the contemplation of the statesman and the soldier. Your memorialists would select the principal events of the year 1793, as best calculated to illustrate their views. The situation of France, in the spring and summer of that year, was as follows: on the northern, eastern, and southern frontiers she was environed by the veteran troops of England, Austria, Sardinia, and Spain; in la Vendee a destructive and violent civil war was raging, which required a large force to check its progress; in the south, three principal cities, Marseilles, Lyons, and Toulon, had revolted against the authority of the convention: and one of them (Toulon), the great naval arsenal of France, was surrendered into the hands of the allied powers; while her armies, disorganized and disheartened by the defection of Dumouriez,[708] and the treason and incapacity of other officers, made, comparatively but a feeble resistance to the overwhelming forces of the enemy, which had passed the frontier and laid siege to Dunkirk, Conde, Valenciennes, and Mayenne, the last three of which were compelled to surrender. During the fearful events which literally "tried men's souls," the project was devised and matured for raising the nation *en masse*, for the purpose of expelling the victorious enemy from French territory. On the 23d of August, 1793, Barrere, the reporter of the committee of public welfare,[709] submitted to the convention his celebrated report on this subject, accompanied by the plan of a decree, which was to carry out into practical effect the principles of the report. The following extracts from this masterly report, and from the accompanying decree (which were adopted, with some amendments, by the convention,) will fully illustrate the spirit and principles which actuated the French at this momentous crisis. The reporter observes: "For this week past, we have been deliberating on the best mode of organizing the great movement unanimously demanded by all the commonalities of the republic, and to give it that regularity without which it would only become an additional weapon in the hands of our domestic and foreign enemies. We began by seizing the true and proper sense and import of the words: 'rising in one body.' We first asked ourselves does this mean a contingent? Does it mean a certain number of men, to be furnished by every division of the republic, in proportion to its population? Is it a recruiting levy? Is it a departure in one body—that is to say, comprising the totality of all the fighting men of the republic? A contingent? The contingent of the republic means all French citizens. A recruiting levy? Aristocracy attends it. Citizens, all Frenchmen are soldiers. We ought, therefore, to rise in one body. But I declare a rising in one body, in the sense which they [meaning the aristocrats and counter-revolutionists] endeavor to attach to it, the rising in a body of ten millions of armed men, organized, ranged, and marching in order of battle. Such a levy is

---

[708] Charles-François du Périer Dumouriez (1739–1823 A.D.)
[709] Bertrand Barère de Vieuzac (1755–1841 A.D.)

impracticable, to be made regular. All Frenchmen are, doubtless in a state of service: but, in this sense, some of them own their industry to their country, others, their fortunes; others, their arms; others, their cares; and all, the blood which circulates in their veins" The following extracts from the decree which was adopted by the national convention, consequent on this report, will show the manner in which its principles were carried into effect: "From the present moment, till that when all the enemies shall have been driven from the territory of the republic, all Frenchmen shall be in permanent readiness for the service of the armies; the young men shall march to the contest: the married men forge arms and transport the provisions; the women shall make tents and clothes, and wait in hospitals; the children shall make lint of old linen; the old men shall cause themselves to be carried to the public squares to excite the courage of the warriors, and preach hatred against the enemies of the public; the battalions, which shall be organized in every district, shall be ranged under a banner with this inscription: 'The French nation risen against tyrants."

This celebrated decree was no sooner passed, than the most energetic measures were adopted to carry it into execution. Three hundred thousand men, full of the same patriotic ardor that fired the Americans at Bunker's Hill, were soon enrolled, and marched without delay to the different military depots, where they were placed under accomplished military instructors to be taught the mechanical part of discipline, and in a few weeks were prepared to march to the frontiers, and combat successfully the veteran and disciplined troops of Europe. The effects of the patriotic valor of these citizen soldiers were soon felt by the allied powers as well as by the rebels of la Vendee, of Lyons, and of Toulon. On the 7th and 8th of September, the allied army, under the command of Marshal Feytag[710] and the Duke of York,[711] which was besieging Dunkirk, was entirely defeated and routed, and the siege raised, with the loss of all the duke's heavy artillery. On the 9th October, the city of Lyons was compelled to surrender. On the 15th and 16th of October, the Austrians were defeated with great loss at Maubeuge, by the French, under the command of General Jourdan,[712] after a desperate conflict of two days. On the 21st of November, the Austrians, under the command of veteran Wurmser,[713] were defeated with great loss at Deux Ponts, by the Fench, under the command of General Hoche.[714] On the 8th of December, the French under the command of

---

[710] Heinrich Wilhelm von Freytag (1720–1798 A.D.)

[711] Prince Frederick, Duke of York and Albany (1764–1803 A.D.)

[712] Jean-Baptiste Jourdan, 1st Count Jourdan (1762–1833 A.D.)

[713] Dagobert Sigismund, Count von Wurmser (1724–1797 A.D.)

[714] Louis Lazare Hoche (1768–1797 A.D.)

General Pichegru,[715] stormed (with fixed bayonets) all the redoubts of the allies which covered Haguenou; and on the 22d of the same month, they were drive from Bischoilers, Duschein, and Haguenou, with great slaughter, notwithstanding the almost continued lines by which those points were joined. The entrenchment on the heights of Reishoffen, Jauderhauffen, and Freywellers, Rad-Neuth, are said to have been not less formidable than those of Jemappe, three rows of redoubts, which the allies considered as impregnable. They were, however, stormed by the army of the Moselle, under General Hoche, who had joined Pichegru, and carried sword in hand. From the 27th of December, there was a continued series of desperate conflicts, when, on the 27th, the French took possession of the celebrated lines of Weissenbourg in triumph. On the 28th, the siege of Landau was raised, and the Austrians retreated to the Rhine, and the Prussians to cover Mayenee.

On the 19th December, Toulon was evacuated by the allies, after a siege of about three months, and taken possession of by the republicans, under General Dugamier.[716] Such were some of the brilliant achievements performed by the citizen soldiers of France, in the short space of three months. And here your memorialists would ask, did the regular troops of France, under the dynasty of the Bourbons, ever accomplish more in the same space of time, even when commanded by a Turenne,[717] a Conde,[718] a Luxembourge,[719] a Saxe,[720] or a Villars?[721] We believe not: no, not in a whole campaign. And are the militia of the United States less intelligent, than were the great body of the people of France in 1793? It would be a libel on their reputation to say they are. Your memorialists fell, then, that they have shown clearly that the militia of the United States, under a proper system of organization and discipline) such as they have recommended to your honorable body), are fully competent without any aid from a standing army to the defense of the country under all emergencies, and can be safely relied upon for that purpose.

III.   FORTIFICATIONS.—Your memorialist will next proceed to examine the subject of fortifications, as a branch of the military defence of the United States. They presume it is not even contemplated to introduce into this country the extensive and complicated system which prevails on the continent

---

[715] Jean-Charles Pichegru (1761–1804 A.D.)

[716] Jacques François Coquille (1738–1794 A.D.)

[717] Henri de La Tour d'Auvergne, vicomte de Turenne (1611–1675 A.D.)

[718] Louis II de Bourbon, Prince of Condé (1621–1686 A.D.)

[719] François-Henri de Montmorency-Bouteville, Duke of Piney-Luxembourg (1628–1695 A.D.)

[720] Maurice, Count of Saxony (1696–1750 A.D.)

[721] Claude Louis Hector de Villars, Prince de Martigues, Marquis (1715–1718 A.D.)

of Europe, where nearly all the principal cities are surrounded by fortifications; and will, consequently, confine their remarks to such works only as are necessary to aid in the defence of our maritime and inland frontier. Your memorialists would respectfully present to your honorable body the extract from a lecture, delivered by one of their number (Captain Partridge) at Windsor, Vermont, in June, 1821, as expressing their views on this important subject. (See page 23). Engineers differ in opinion as the description of works best calculated for maritime defence.—Your memorialists, however, are inclined to decide in favor of the *Montalembert* castles, mounting from two to four tiers of guns. Some of the prominent advantages which those castles possess over open batteries, are the following, viz:

1st. Being casemated and bomb proof, the perishable parts, such as the gun-carriages, platforms, &c. are entirely secured from the weather, and consequently, much less liable to decay, and the troops much better protected from the shots and shells of an enemy than in open batteries.

2. The exterior wall being circular, and presenting the convex surface towards the enemy, is much stronger than a rectilinear wall of the same thickness, for the reason that the effect of shot impinging on a convex surface of such wall is to wedge the materials, more compactly, by driving them from the circumference towards the centre; and, also, from the reason that a portion of the force with which all shot, that are not projected in the direction of the radii of the curve would impinge on the wall, would be destroyed in consequence of striking obliquely on the same.

3. In a work on the size of Castle Williams in the harbor of New York, five guns from each tier can be brought to bear on a ship of war, which should lie at the usual battering distance; and from all the tiers, fifteen, or twenty, according to the number of tiers. Before such a battery, if well manned and the guns well serviced, no ship could lie. Your memorialist are aware, however, it has been objected to these castles that the smoke will accumulate in the casemates to such a degree, as to render it impossible for the men to stand the guns. The experiments made at Castle Williams, soon after it was completed, proved the fallacy of this objection; and indeed, as these casemates are boen[722] in the rear, there appears to be no reasonable ground for apprehension on this account. In addition to the foregoing, your memorialists would recommend the construction of Martella towers at the more exposed points on the seaboard. These towers cost but little comparatively; are of very solid and permanent construction, and capable of withstanding the battering from heavy artillery from a long time.—Their proper position is in the vicinity of landing-places for

---

[722] Possibly meaning gusty or windy.

the purpose of preventing the disembarking and landing of troops. Their height may be from twenty to thirty-five or forty feet, and guns which are mounted on the top have consequently, an extensive command. The strength and efficiency of these towers were fully proved by the defence which one of them (mounting two eighteen-pounders, and manned by thirty-three men) made against the British forces, under the command of Lord Hood,[723] when they attacked Bastia in the island of Corsica. (See Lord Hood's official account, dated February 22, 1794). This tower, situated on the point of Martella, was attacked by two ships (the Fortitude and Juno) both of which were beaten off much damaged, after a cannonade of two and a half hours. It was subsequently attacked from the land side, but withstood a continued and severe cannonade, from batteries only 150 yards distant, for two days, without the loss of a man killed, and only two wounded.—When England was threatened with an invasion by Napoleon,[724] the Government ordered the construction of a number of these towers on the more exposed parts of the coast, and in the vicinity of the landing-places, to prevent, or at least, to render very hazardous, the disembarking and landing of troops. For our island frontier, your memorialists believe no fortifications to be necessary, except stockades or earthen works of simple construction, which are sufficient to withstand the attacks of savages.

MARINE DEFENCES.—Under marine defences, your memorialists would embrace gunboats, steamships, and fireships, all of which may be used in harbor and seacoast defence.

1.   *Gunboats.* Gunboats are in many respects, well adapted to aid in the defense of harbors and other inland waters. Lying low on the water and presenting but a small object for ships of war to fire at, while the shots from their guns if well directed, can scarcely miss striking a ship between wind and water, their fire is very dangerous to larger vessels. As they draw but little water, they can take their station in shoal water, where ships cannot approach them, and throw their shot in safety from such distances as may be thought best. They may, therefore, be considered in many situations, an important appendage to fortifications in the defence of harbors, bays, &c. The Swedes and Danes, it is believed, have been so well convinced of their efficiency, as to have uniformly kept flotillas of them on the Baltic.

2.   *Steamships.* Steamships of war are of modern invention, but will doubtless be found the most efficient kind of marine defence for harbors, bays, &c., that has yet been tried. Constructed in the strongest and most solid

---

[723] Samuel Hood, 1st Viscount Hood (1724–1816 A.D.)

[724] Napoleon Bonaparte (Emperor Napoleon I) (1769–1821 A.D.)

manner, with their wheels and machinery well secured from the enemy's shot, mounted with the heaviest artillery, and moved by the irresistible power of steam, they possess most decided advantages over ordinary ships of war, especially when such ships are becalmed or at anchor. In such situations, they would be enabled to take their station under the stern or bow of their antagonist, which would be entirely disabled by a few well directed shots. Time will determine whether steamships are adapted for service on the ocean, but of their superiority or harbor and seacoast defence, it is believed there can be no doubt.

3. *Fireships*. Fireships have been used from the earliest period of naval warfare; sometimes successfully, but more generally the reverse. When sent against an enemy's ship, after being abandoned by their crews, they cease to be under control; and, consequently, are very liable to run aground, to miss their object, to be towed off, or to explode before coming in contact with their enemy; in all of which contingencies they fail in producing the intended effect. It is evident these defects cannot be remedied unless they can be kept under control until they have produced the contemplated effect. And here your memorialists would beg leave to call attention to your honorable body to the plan of a fireship proposed by Uriah Brown, Esq., late of the city of Washington. One of your memorialists (Captain Partridge) has the opportunity, a few years ago, carefully to examine a model of a proposed ship, constructed by Mr. Brown, and a strongly impressed in its favor. The proposed ship would be seventy-five feet in length, but comparatively narrow. It would be low in the water, and the sides and deck composed of solid oak timber, two feet in thickness. All the parts above the water to be covered with thick sheet-iron. The deck, and sides down to the water-line, to be formed of inclined planes, making such angles at their common section, as that no shot fired from a ship of war can strike on any part of the proposed ship, at a greater angle than 15 degrees; the consequence would be, that all shot striking the ship would ricochet without penetrating in the least, the sides or deck. That shot striking on a smooth and hard surface, such as in presented by the proposed ship, at an angle not greater than 15 degrees, would ricochet, is evident from the fact that in ricochet-firing in sieges, the shot, impugning on the surface of the ground at an angle of 13 degrees, do not bury themselves, but ricochet very well. The sides and deck, then, of the proposed ship may be safely considered ball-proof. The ship is to be propelled by steam, and the machinery, wheels, &c., would be well secured as in any steamship; and it would also be perfected under control, the crew being at the same time all below deck. Mr. Brown has invented a liquid composition, which can be projected by mechanical power through tubes passing through orifices in the deck, to the distance of from 250 to 300

yards. This liquid, when it leaves the tubes and comes in contact with the air, takes fir, and on whatever combustible substance it falls, it immediately sets it on fire, burning so furiously that water will not extinguish it. The certain effect of fluid, in producing combustion, was proved by several experiments made at Baltimore, during the late war, and subsequently, by others at Washington, one of which was witnessed by one of your memorialists. Mr. Brown has also devised a very ingenious method of directing this ship towards its antagonist, and measuring its distance therefrom with a good degree of accuracy (although below deck), by means of reflected light, thrown on a graduated surface. To prevent the ship from being boarded he has devised a method for throwing the liquid flame over the deck, so that in case an enemy should succeed in getting possession of the deck, so that in case the enemy should succeed in getting possession of the deck, their destruction would be inevitable. But even should the deck be possessed by an enemy, the possession of the ship would remain with the crew, who would be enabled to carry off the boarders as prisoners. Suppose, now, that a ship constructed and equipped as the one proposed, was to bear down on a ship of war at anchor or becalmed and coming within striking distance, should project a mass of this liquid flame on the rigging, masts, decks, &c., all of which are highly combustible; your memorialist believe the destruction of a ship so situated would be inevitable. The balls fired from the ship attacked, glancing from the sides and deck of the fire-ship, would produce no effect, and surrender or destruction must be the certain result.

From a due consideration of the foregoing reasons, your memorialist are convinced that the proposed plan would remedy all the principal defects which have hitherto attached to fireships, and prove this the most efficient engine for defence for all our harbors and inland waters.—Indeed, what enemy would presume to approach an antagonist which would ensure his surrender or immediate destruction? The late General Bernard,[725] after due examination, gave his opinion decidedly in favor of the proposed ship; but actual experiment is the only certain means of testing the success of this, as well as of most other projects. Your memorialist are informed that Mr. Brown asked an appropriation of only ten thousand dollars to enable him to prove, by experiment, to the satisfaction of any committee the Government might appoint, that the proposed ship would answer all the purposes he contemplated. Would it not be advisable, then, to authorize such an appropriation as would be required to make all the experiments necessary to test, satisfactorily, the practical utility of such ships for seacoast and harbor defence? Your memorialists believe it would, and, accordingly respectfully

---

[725] Baron Simon Bernard (1779–1839 A.D.)

recommend that such appropriation be made. Should the result of the experiment be favorable, it would, undoubtedly, be not only far the cheapest, but most efficient means of defence—as one or two of these ships would be amply sufficient to protect any of our harbors, and other inland waters, against any inland force whatever. The composition intended by Mr. Brown would seem to be as potent as that which produced the celebrated liquid fire of the Greeks (invented by Callinicus,[726] an engineer and chemist of Heliopolis, in Syria); by which the almost innumerable fleet of the Saracens was twice destroyed, before Constantinople, in the early part of the eight century, and the siege raised. According to Gibbon,[727] this Greek Fire was projected through copper tubes planted on the prows of the fireships, and involved in flames every combustible substance on which it fell.

Your memorialists will now proceed to make some remarks on the recent report of the Secretary of War, and regret that a sense of duty compels them to record their dissent to its leading feature, from the following reasons, Viz:

1st. Because it places the army entirely in advance of the citizen-soldiery, as constituting the main defence of the country: this, your memorialists contend, is in direct hostility to the principles of our republican institutions.

2d. Because the plan proposed for the organization and discipline of the militia would, if carried out, reduce the effective military to two hundred thousand men, less than one-seventh of the whole number now subject to military duty.

To vest the military defence of the country in so comparatively small a portion of the constitutional force, would, in the opinion of your memorialists, be more dangerous to liberty, than to limit the right of voting at elections to an equally small proportion of the freemen of the United States. Your memorialists believe that all measures which are calculated to draw power, whether civil or military, from the many, and concentrate it in the few, are adverse to civil liberty, and not, for a moment, to be tolerated. Although the plan proposed by the Secretary, for the discipline of the select two hundred thousand militia, appears specious, yet, if your memorialists do not much mistake, it would, ere long, assume the character of a standing army in disguise, similar to the yeoman corps in England; and that, under its influence, the great body of the militia would be neglected, and sink into disrepute, its martial energies paralyzed, and the institution finally destroyed.

---

[726] Kallinikos from Heliopolis of Syria (ca. 650 A.D.)
[727] Historian Edward Emily Gibbon (1737–1794 A.D.)

3. Because it proposes fully to sustain the Military Academy at West Point, not on the account of any advantage it is to the militia or to the people, but on account of its importance to the character of the army. Now, in the opinion of your memorialists, there is not on the face of the earth a more aristocratic, anti-republican, and monopolizing, establishment, than the institution at West Point. The whole people are taxed to support it, while scarcely one in six hundred and forty thousand can derive any advantage from it. Its whole organization is in direct violation of the principles of our republican institutions.

4. Because it proposes to prosecute the present expensive, and comparatively useless system of fortifications, on which so large an amount of the people's money is annually expended, instead of being appropriated to aid the discipline of our citizen-soldiery, who constitute the real defence of the country, and ultimately to involve the nation in an enormous annual expense, to keep these works in repair. Mounds of earth, or of masonry, thrown up in the form of fortifications, are nothing but the result of labor, directed by science, and can always be constructed on the spur of the occasion, by the troops that are to defend them, provided they have competent engineers to direct. The plan for general dissemination of military science among the people, proposed by the military convention at Norwich Vermont, on the 15th of August, 1838, and recommended by your memorialists to your honorable body for adoption, would furnish an ample supply of scientific engineers in every section of the United States who would be perfectly competent to direct the energies of our citizen-soldiers in the construction of fortifications in all cases of emergency. Your memorialists have now discharged, according to the best of their ability, the important duty imposed upon them by a highly respectable convention of their fellow citizens, and would most respectfully and urgently recommend the result of their labors to the favorable consideration of your honorable body.

All of which is respectfully submitted

A. PARTRIDGE,
T. B. RANSOM,
JOHN WRIGHT,
N. B. CUTTING,
C. L. NEWTON
S. B. HAZELTINE.

Bibliography

"Academic Program." United States Military Academy West Point. https://www.westpoint.edu/academics/academic-program.

Adams, John. From John Adams to Benjamin Rush, 28 July 1789. National Archives. https://founders.archives.gov/documents/Adams/99-02-02-0705.

Adams, Samuel. Epilogue: Securing the Republic: Samuel Adams to John Scollay, 30 Dec. 1780. http://press-pubs.uchicago.edu/founders/documents/v1ch18s14.html.

Adams, Samuel. *An Oration Delivered at the State-House, in Philadelphia, to a Very Numerous Audience on Thursday the 1st of August, 1776; by Samuel Adams.* Philadelphia, PA: E. Johnson, 1776.

Addison, Joseph. *Cato, a Tragedy.* Edinburgh: John Wood, 1713.

Allen, Joseph Dana. *A Journal of an Excursion, Made by the Corps of Cadets of the A.L.S. & M. Academy, Norwich, Vt. under Command of Capt. A. Partridge, June, 1824.* Windsor, VT: Simeon Ide, 1824.

"Alternative College Entrance Exam." Classic Learning Test, June 21, 2023. https://www.cltexam.com/.

Ambrose, Stephen E. *Duty, Honor, Country: A History of West Point.* Baltimore, MD: Johns Hopkins University Press, 1999.

American Literary, Scientific and Military Academy. *A Journal of an Excursion Made by the Corps of Cadets, of the American Literary, Scientific and Military Academy, under Capt. Alden Partridge. June, 1822.* Concord, NH: Hill and Moore, 1822. https://archives.norwich.edu/digital/collection/p16663coll2/id/16934/rec/3.

Americanus. "District Milita." *National Intelligencer*, April 21, 1814. George Washington University Special Collections Research Center.

Anderson, Marcus Aurelius. "Donald Robertson on Marcus Aurelius the Stoic Emperor, Fatherhood, Errors in ChatGPT, Stoic Meta Lessons, and Why the World Needs Philosophy Now More than Ever." *Acta Non Verba*, February 14, 2024. https://www.actanonverbapodcast.com/donald-robertson-on-marcus-aurelius-the-stoic-emperor-fatherhood-errors-in-chat-gpt-stoic-meta-lessons-and-why-the-world-needs-philosophy-now-more-than-ever/.

Anderson, Terry L. & Fred S. Mc Chesney, "Raid or Trade? An Economic Model of Indian-White Relations." *Journal of Law and Economics* 37, no. 1 (April 1994): 39–74. https://doi.org/10.1086/467306.

Annas, Julia. "Ethics in Stoic Philosophy." *Phronesis* 52, no. 1 (2007): 58–87. https://doi.org/10.1163/156852807x177968.

------. "Is Plato a Stoic?" *Méthexis* 10, no. 1 (March 30, 1997): 23–38. https://doi.org/10.1163/24680974-90000268.

Annis, Franklin C. "Future Army Leaders: Expert Specialists or Master Generalists?" From the Green Notebook, June 5, 2018. https://fromthegreennotebook.com/2018/06/05/future-army-leaders-expert-specialists-or-master-generalists/.

------. "Krulak Revisited: The Three-Block War, Strategic Corporals, and the Future Battlefield." Modern War Institute at West Point, February 3, 2020. https://mwi.westpoint.edu/krulak-revisited-three-block-war-strategic-corporals-future-battlefield/.

------. "Stoicism for Toughening." Medium, December 27, 2022. https://medium.com/stoicism-philosophy-as-a-way-of-life/stoicism-for-toughening-8a5c66c2e5bc.

Antoninus, Marcus Aurelius. *The Meditations of the Emperor Marcus Aurelius Antoninus*. Translated by George W. Chrystal. London: Schulze & Co., 1902. Gutenberg Project. https://www.gutenberg.org/files/55317/55317-h/55317-h.htm.

Arney, David C., Martin Vozzo, William Ebel, George Mitroka, Allen Geishecker, and Louis Yuengert. "Alden Partridge." Math Biographies of Department Heads. https://www.westpoint.edu/sites/default/files/pdfs/Math/Partridge.pdf.

"Articles of Confederation (1777)." National Archives. https://www.archives.gov/milestone-documents/articles-of-confederation.

"The Art of Living: Professor John Sellars on Stoicism as a Medicine for the Mind." Daily Stoic, n.d. https://dailystoic.com/john-sellars/.

Aquinas, Thomas. "Whether Sins Are Fitting Divided into Sins of Thought, Word, and Deed." Translated by Laurence Shapecote. Summa Theologica, 2020. https://aquinas.cc/la/en/~ST.I-II.Q72.A7.SC.

Aurelius, Marcus. *Meditations: a New Translation, with an Introduction, by Gregory Hays*. Translated by Gregory Hays. New York, NY: Modern Library, 2002.

Bailyn, Bernard, ed. *The Debate on the Constitution: Federalist and Antifederalist Speeches, Articles, and Letters During the Struggle Over Ratification*. Vol. I. New York, NY: Literary Classics of the United States, 1993.

------. *The Debate on the Constitution: Federalist and Antifederalist Speeches, Articles, and Letters During the Struggle Over Ratification*. Vol. II. New York, NY: Literary Classics of the United States, 1993.

Barnard, Henry, ed. *The American Journal of Education*. Vol. XIII. Hartford, CT: Henry Barnard, 1863.

------. *Military Schools and Courses of Instruction in the Science and Art of War*. Rev. ed. New York, NY: E. Steiger, 1872.

Barr, Michael. "Shido and Stoicism: An Exploration of Commonalities in Principles and Practices." Shido and Stoicism, September 19, 2024. https://waveman21.blogspot.com/2024/09/normal-0-false-false-false-en-us-x-none.html.

Bowra, C. M. "Aeneas and the Stoic Ideal." *Greece and Rome* 3, no. 7 (1933): 8–21. https://doi.org/10.1017/s0017383500002175.

Brammer, Robert. "General James Wilkinson, the Spanish Spy Who Was a Senior Officer in the U.S. Army during Four Presidential Administrations: In Custodia Legis." Library of Congress, April 21, 2020. https://blogs.loc.gov/law/2020/04/general-james-wilkinson-the-spanish-spy-who-commanded-the-u-s-army-during-four-presidential-administrations/.

Brennan, Tad. *The Stoic Life: Emotions, Duties, and Fate*. Oxford: Clarendon Press, 2006.

"A Brief History of Stoicism." Stoic Journey, November 9, 2019. https://stoicjourney.org/2016/07/28/a-brief-history-of-stoicism/.

"A Brief History of West Point." United States Military Academy West Point. https://www.westpoint.edu/about/history-of-west-point/brief-history-of-west-point.

Brown, Eric. "Socrates in the Stoa." In *A Companion to Socrates*, edited by Sara Ahbel-Rappe and Rachana Kamtekar, 275–84. Malden, MA: Blackwell Publishing Ltd., 2006. https://doi.org/10.1002/9780470996218.ch17.

Buck, David Arzo Ashley. Letter from D. A. A. Buck to Alden Partridge, 15 January 1824, Norwich University Archives and Special Collections. https://archives.norwich.edu/digital/collection/p16663coll4/id/6510/rec/4.

Bush, George Gary. *History of Education in Vermont*. Washington, D.C.: Government Printing Office, 1900.

Butler, William F. *Charles George Gordon*. New York, NY: MacMillan and Co., 1889.

Carl, Marissa. "West Point's Legacy Families." *West Point Magazine*. Fall 2012, https://alumni.westpointaog.org/file/history/Legacy-Article-WPM-FA12.pdf.

Child, Hamilton. *Gazetteer of Orange County, Vt., 1762–1888*. Vol. I. Syracuse, NY: Syracuse Journal Company, 1888.

Cicero, Marcus Tullius. *Cicero de Officiis*. Translated by Walter Miller. London: Heinemann, 1913.

------. *Cicero's Select Orations, Translated into English; with the Original Latin, from the Best Editions, in the Opposite Page; and Notes Historical, Critical, and Explanatory*. Translated by William Duncan. A New Edition, Corrected. New Haven, CT: Sidney's Press, 1811.

Clausewitz, Carl von. *On War*. Vol. I. Translated by Frederic Natusch Maude. London: Kegan Paul, Trench, Trübner & Company, 1908.

Clerici, Alberto. "*In publicis malis*. Justus Lipsius and the 'Double Face' of Neostoicism in the European Wars of Religion." In *Crisis and Renewal in the History of European Political Thought*, edited by Cesare Cuttica, László Kontler, and Clara Maier, 259–79. Boston, MA: Brill, 2021. https://doi.org/10.1163/9789004466876_014.

Cohen, Patricia. "Think Tank; The Phenomenology of Harry, or the Critique of Pure Potter." *New York Times*, July 19, 2003. https://www.nytimes.com/2003/07/19/books/think-tank-the-phenomenology-of-harry-or-the-critique-of-pure-potter.html.

Cohler, Anne M., Harold S. Stone, and Basia C. Miller, trans. *Montesquieu: The Spirit of the Laws*. Cambridge: Cambridge University Press, 1989.

Collins, Jim, and James C. Collins. *Good to Great: Why Some Companies Make the Leap... and Others Don't.* London: Random House Business, 2001.

Cooke, Oliver Dudley. Letter from O. D. Cooke & Co. to Isaac Partridge, Approximately 1825-1829., 1826. https://archives.norwich.edu/digital/collection/p16663coll4/id/23012/rec/1.

Crockett, David. "Resolution to Abolish West Point," February 25, 1830. United States House of Representatives: History, Art & Archives. https://history.house.gov/Records-and-Research/Listing/lfp_030/.

Cui, Siwen. "The Marian Military Reform and Its Effects on the Roman Republic." *Proceedings of the 2021 International Conference on Public Relations and Social Sciences (ICPRSS 2021)* (2021): 992–98. https://doi.org/10.2991/assehr.k.211020.294.

Cullum, George W. *Biographical Register of the Officers and Graduates of the U.S. Military Academy, from 1802 to 1867*. Rev ed. 3 vols. New York, NY: J. Miller, 1879.

Cunliffe, Marcus. *Soldiers & Civilians: The Martial Spirit in America 1775–1865*. Boston, MA: Little, Brown, and Company, 1968.

Davis, Richard Beale. *Intellectual Life in Jefferson's Virginia 1790–1830*. Knoxville, TN: University of Tennessee Press, 1972.

Defoe, Daniel. *The Earlier Life and the Chief Earlier Works of Daniel Defoe*. Edited by Henry Morley. London: George Routledge and Sons, 1889.

------. *The Life and Strange Surprizing Adventures of Robinson Crusoe, of York, Mariner: Who Lived Eight and Twenty Years, All Alone in an Un-Inhabited Island on the Coast of America, near the Mouth of the Great River of Oroonoque, Having Been Cast on Shore by Shipwreck, Wherein All the Men Perished but Himself: With an Account How He Was at Last as Strangely Deliver'd by Pyrates*. London: Printed for W. Taylor at the Ship in Pater-Noster-Row, 1719.

------. *Serious Reflections during the Life and Surprising Adventures of Robinson Crusoe with His Vision of the Angelick World*. Ship and Black-Swan in Pater-nofter-Row: Printed for W. Taylor, 1720.

Degen, Robert. *The Evolution of Physical Education at the United States Military Academy*. West Point, NY: United States Military Academy, 1967.

Department of the Army. *Field Manual (FM) 21-18: Foot Marches*. Washington, D.C.: Government Printing Office, 1950.

------. *USAIS Pamphlet 350-6: Expert Infantryman Badge*. Fort Moore, GA: Department of the Army United States Army Infantry School, 2023. https://www.moore.army.mil/infantry/EIB/content/PDF/20240214%20USAIS%20350-6%2030%20Aug%202023%20signed.pdf.

Dupuy, R. Ernest. "Mutiny at West Point." *American Heritage* 7, no. 1 (December 1955). https://www.americanheritage.com/mutiny-west-point.

------. *The Story of West Point*. Washington, D.C.: The Infantry Journal, 1943.

East, Whitfield B. *A Historical Review and Analysis of Army Physical Readiness Training and Assessment*. Fort Leavenworth, KS: Combat Studies Institute Press, U.S. Army Combined Arms Center, 2013.

Eddy, Edward Danforth, Jr. *Colleges for Our Land and Time: The Land-Grant Idea in American Education*. New York, NY: Harper & Brothers, 1957.

"Editorial." *Norwich Guidon* IV, no. 31 (June 4, 1926): 2.

Edwards, Mark W. "The Expression of Stoic Ideas in the 'Aeneid.'" *Phoenix* 14, no. 3 (1960): 151. https://doi.org/10.2307/1086300.

Ellis, Charles. "The Student Diary of Charles Ellis March 10 to June 25, 1835." Edited by Ronald B. Head. http://juel.iath.virginia.edu/exist/cocoon/juel/juel_one?doc=/db/JUEL/ellis_diary/Ellis_typed_version.xml&key=P44390.

Ellis, William A. *Norwich University, 1819–1911: Her History, Her Graduates, Her Roll of Honor.* 3 vols. Montpelier, VT: Capital City Press, 1911.

Elting, John Robert, ed. *Military Uniforms in America: Years of Growth, 1796–1851.* Vol. II. San Rafael, CA: Presidio Press, 1977.

Emerson, Ralph Waldo. *Essays.* Boston, MA: James Munroe and Company, 1840.

------. *Lectures and Biographical Sketches by Ralph Waldo Emerson.* New York, NY: Houghton, Mifflin, and Co., 1895. https://archive.org/details/lecturesbiographemer/page/n5/mode/2up.

Emerson, William K. *Encyclopedia of United States Army Insignia and Uniforms.* Norman, OK: University of Oklahoma Press, 1996.

Enfield, William. *The History of Philosophy from the Earliest Times to the Beginning of the Present Century: Drawn up from Brucker's* Historia Critica Philosophiae. Vol. I. Dublin: Printed for P. Wogan, P. Byne, A. Grueber, W. M'Kenzie, J. Moore, J. Jones, R. M'Allister, W. Jones, J. Rice, R. White, and G. Draper, 1792.

Epictetus. *Discourses, Books 3–4. Fragments. The Encheiridion.* Translated by W. A. Oldfather. Cambridge, MA: Harvard University Press, 1928.

------. *Discourses, Fragments, Handbook.* Edited by Christopher Gill. Translated by Robin Hard. Oxford: Oxford University Press, 2014.

------. *The Enchiridion,* 2nd ed. Translated by Thomas W. Higginson. New York, NY: Liberal Arts Press, 1955. https://www.gutenberg.org/files/45109/45109-h/45109-h.htm.

------, and P. E. Matheson. *Epictetus: The Discourses and Manual.* Oxford: Clarendon Press, 1916. https://archive.org/details/MN40058ucmf_2/.

Evans, Michael. "Stoicism and the Profession of Arms." Quadrant Online, April 25, 2018. https://doi.org/https://quadrant.org.au/magazine/2010/01-02/stoic-philosophy-and-the-profession-of-arms/.

Fallows, Samuel. "Report of Bishop Fallows." In *Journal of the Proceedings of the Twenty-First General Council of the Reformed Episcopal Church,* 79–86. Philadelphia: Reformed Episcopal Publication Society, 1897. http://www.recus.org/documents/GCJournals/GCREC15.pdf.

Flint, K. R. B. "A Norwich Man's Creed." *Norwich Guidon* VI, no. 1 (September 16, 1927): 2. Norwich University Archives and Special Collections. https://archives.norwich.edu/digital/collection/p16663coll2/id/9662/rec/1.

------., C. V. Woodbury, and C. N. Barber, eds. *Norwich University: A Record of the Celebration of the One Hundredth Anniversary of Its Founding.* Northfield, VT: Norwich University, 1930.

Ford, Christopher A. "Preaching Propriety to Princess: Grotius, Lipsius, and Neo-Stoic International Law." *Case Western Reserve Journal of International Law* 28, no. 2 (1996): 313–66.

Forman, Sidney. "The United States Military Philosophical Society, 1802–1813: Scientia in Bello Pax." *The William and Mary Quarterly* 2, no. 3 (1945): 273–85. https://doi.org/10.2307/1921452

Fourcy, Ambroise. *Histoire de l'École Polytechnique*. Paris: L'Auteur, 1828.

Frankl, Viktor E., and Gordon W. Allport. *Man's Search for Meaning*, 3rd ed. Translated by Ilse Lasch. New York, NY: Simon & Schuster, 1984.

Franklin, Benjamin. *The Compleated Autobiography by Benjamin Franklin*. Edited by Mark Skousen. Washington, D.C.: Regnery Publishing, 2005.

Fretwell, Peter, and Taylor Kiland. "Review of Leadership Lessons from Hanoi Hilton: Vice Admiral James Stockdale's Principles Can Inspire Any Organization's Leaders." *Proceedings* 135, no. 11 (November 2009). https://www.usni.org/magazines/proceedings/2009/november/leadership-lessons-hanoi-hilton.

Fry, James B. *The History and Legal Effect of Brevets in the Armies of Great Britain and the United States: From Their Origin in 1692 to the Present Time*. New York, NY: D. Van Nostrand, 1877.

Gainey, Sean A. "Army ROTC History." United States Army Cadet Command. https://armyrotc.army.mil/history/.

"Gardening for the Common Good." Smithsonian Libraries. https://library.si.edu/exhibition/cultivating-americas-gardens/gardening-for-the-common-good.

Gignilliat, Leigh Robinson. *Arms and the Boy: Military Training in Schools and Colleges, Its Value in Peace and its Importance in War, with Many Practical Suggestions for the Course of Training, and with Brief Descriptions of the Most Successful Systems Now in Operation*. Brooklyn, NY: Bobbs-Merrill Co., 1916.

Goddard, Merritt Elton, and Henry Villiers Partridge. *A History of Norwich, Vermont*. Hanover, NH: Dartmouth Press, 1905.

------. "Reminiscences of Captain Alden Partridge," ca. 1884. Dartmouth Library Archives.

Goodman, Rob, and Jimmy Soni. *Rome's Last Citizen: The Life and Legacy of Cato, Mortal Enemy of Caesar*. New York, NY: Saint Martin's Griffin, 2014.

Greenleaf, Walter J. *Federal Laws and Rulings Affecting Land-Grant Colleges and Universities*. Washington, D.C.: United States Department of the Interior, Office of Education, 1930.

"Guidon Reprint Creed as Written for Cadets by Prof. K. R. B. Flint." *The Norwich Guidon* XXVI, no. 20 (March 5, 1940): 1. Norwich University

Archives and Special Collections. https://archives.norwich.edu/digital/collection/p16663coll2/id/26361/rec/1.

Guidotti, John A. "Controversy: The Legacy of Alden Partridge." 1991. Eisenhower Leader Development Program Papers. United States Military Academy Library. http://usmalibrary.contentdm.oclc.org/cdm/singleitem/collection/p16919coll1/id/32/rec/12.

Hadot, Pierre. *The Inner Citadel: The Meditations of Marcus Aurelius*. Translated by Michael Chase. Cambridge, MA: Harvard University Press, 2001.

Hancock Jr., John. Journal, 1824. Norwich University Archives and Special Collections. https://archives.norwich.edu/digital/collection/p16663coll4/id/31526/rec/1.

Hardy, Rob. "Cato." George Washington's Mount Vernon. https://www.mountvernon.org/library/digitalhistory/digital-encyclopedia/article/cato/.

Harmon, Ernest N. *Norwich University, Its Founder and His Ideals*. New York, NY: Newcomen Society in North America, 1951.

Hart, B. H. Liddell. *Greater than Napoleon: Scipio Africanus*. Edinburgh: William Blackwood & Sons Ltd., 1930.

Holbrook, John. *Military Tactics: Adapted to the Different Corps in the United States, According to the Latest Improvements*. Middletown, CT: E. & H. Clark, 1826.

Holden, Edward S., and W. L. Ostrander. *The Centennial of the United States Military Academy at West Point, New York, 1802–1902*. Vol. I, *Addresses and Histories*. Washington, D.C.: Government Printing Office, 1904.

Hook, Brian S. "Oedipus and Thyestes among the Philosophers: Incest and Cannibalism in Plato, Diogenes, and Zeno." *Classical Philology* 100, no. 1 (January 2005): 17–40. https://doi.org/10.1086/431428.

Ierodiakonou, Katerina, ed. *Topics in Stoic Philosophy*. Oxford: Clarendon Press, 2007.

"The Institution of the Society of the Cincinnati." The Society of the Cincinnati. https://www.societyofthecincinnati.org/institution-the-society-of-the-cincinnati/.

Ioachim, Gabriela, et al. "Evaluating the Before Operational Stress Program: Comparing In-Person and Virtual Delivery." *Frontiers in Psychology* 15 (July 25, 2024). https://doi.org/10.3389/fpsyg.2024.1382614.

Irvine, William B. *The Stoic Challenge: A Philosopher's Guide to Becoming Tougher, Calmer, and More Resilient*. New York, NY: W.W. Norton & Company, Inc., 2019.

Jefferson, Thomas. December 2, 1806: Sixth Annual Message. Miller Center, August 23, 2023. https://millercenter.org/the-presidency/presidential-speeches/december-2-1806-sixth-annual-message.

------. Doctrines of Jesus Compared with Others, 21 April 1803. National Archives. https://founders.archives.gov/documents/Jefferson/01-40-02-0178-0002#:~:text=In%20a%20comparative%20view%20of,I%20learned%20among%20it's%20professors.

------. From Thomas Jefferson to Peter Carr, 19 August 1785. National Archives. https://founders.archives.gov/documents/Jefferson/01-08-02-0319.

------. Notes of Cabinet Meeting on the President's Address to Congress, 23 November 1793. National Archives. https://founders.archives.gov/documents/Jefferson/01-27-02-0399.

------. Thomas Jefferson to John Adams, 22 August 1813. National Archives. https://founders.archives.gov/documents/Jefferson/03-06-02-0351.

------. To John Adams from Thomas Jefferson, 12 October 1813. National Archives. https://founders.archives.gov/documents/Adams/99-02-02-6238.

Johncock, Will. "Bringing People Closer: Cicero, Hierocles, and Cosmopolitanism." *Epoché Magazine*, May 30, 2022. https://epochemagazine.org/28/bringing-people-closer-cicero-hierocles-and-cosmopolitanism/.

Kern, Andrew. "Why the Founding Fathers Feared a Standing Army." Foundation for Economic Education, February 6, 2024. https://fee.org/articles/why-the-founding-fathers-feared-a-standing-army/.

Konstantakos, Leonidas. "On Cleomenes and Sphaerus: How Stoic Was the Spartan King?" *Anais de Filosofia Clássica* 7, no. 14 (2013): 74–81. https://revistas.ufrj.br/index.php/FilosofiaClassica/article/view/1367/2256

Krulak, Charles C. "The Crucible: Building Warriors for the 21st Century." In *The Legacy of Belleau Wood: 100 Years of Making Marines and Winning Battles, an Anthology*, edited by Paul Westermeyer and Breanne Robertson, 313–18. Quantico, VA: History Division, United States Marine Corps, 2018.

Laërtius, Diogenes. *The Lives and Opinions of Eminent Philosophers*. Translated by C. D. Yonge. London: Henry G. Bohn, 1853.

Lang, Daniel W. "Amos Brown and the Educational Meaning of the American Agricultural College Act." *History of Education* 31, no. 2 (2002): 139–65. https://doi.org/10.1080/00467600110109258.

Larkin, Daphne. "Norwich Museum Presents Historian on Partridge's Rugged Outdoor Lessons." VTDigger, February 15, 2018.

Larson, Carlton F. W. "Titles of Nobility, Hereditary Privilege, and the Unconstitutionality of Legacy Preferences in Public School Admissions." *Washington University Law Review* 84, no. 6 (January 2006): 1375–1440.

Lawson, Kenneth E. "Religion and the U.S. Army Chaplaincy and the Florida Seminole War 1817–1858." 2004. The College of William & Mary Special Collections Research Center.

Lee, Arthur. Arthur Lee to Franklin and Silas Deane, 15 June 1777. National Archives. https://founders.archives.gov/documents/Franklin/01-24-02-0131.

Leira, Halvard. "Justus Lipsius, Political Humanism and the Disciplining of 17th Century Statecraft." *Review of International Studies* 34, no. 4 (2008): 669–92. https://doi.org/10.1017/S026021050800822X.

Leonard, Fred Eugene. *Guide to the History of Physical Education*. Philadelphia, PA: Lea & Febiger, 1923. https://archive.org/details/guidetohistoryof00leon/page/n7/mode/2up.

Linklater, Andro. *An Artist in Treason: The Extraordinary Double Life of General James Wilkinson*. New York, NY: Walker, 2010.

List of Cadets and Where They are Billeted, [1824?]. Norwich University Archives and Special Collections. https://archives.norwich.edu/digital/collection/p16663coll4/id/17782/rec/1.

List of Student Actors and the Roles They Are Playing in Dramatic Productions, 1825–1828. Norwich University Archives and Special Collections. https://archives.norwich.edu/digital/collection/p16663coll4/id/23368/rec/2.

Lochée, Lewis. *An Essay on Military Education*. 2nd ed. London: T. Cadell, 1776. https://www.google.com/books/edition/An_Essay_on_Military_Education/gy5cAAAAQAAJ?hl=en&gbpv=1.

------. *A System of Military Mathematics*. 2 vols. London: T. Cadell, 1776.

Long, A. A. *Stoic Studies*. Berkeley, CA: University of California Press, 2001.

------. Stoicism Ancient and Modern. lecture presented at the Stoicon 2018, September 29, 2018. https://www.youtube.com/watch?v=_xuQ4i46K_M.

Lord, Gary T. Alden Partridge's Proposal for a National System of Education: A Model for the Morrill Land-Grant Act. *History of Higher Education*

Annual: 1998: The Land-Grant ACT and American Higher Education: Contexts and Consequences, vol. 18, edited by Roger L. Geiger, 11–24. University Park, PA: Pennsylvania State University, 1998.

Maass, John R. *Defending a New Nation, 1783–1811*. Washington, D.C.: Center of Military History, United States Army, 2013.

MacDonald, Hugh P. *The Power of Emerson's Wisdom*. New York, NY: Pageant, 1954.

Manuscripts and Archives Division, The New York Public Library. "ferson, Thomas. Monticello. To Capt. Alden Partridge. Envelope."*e New York Public Library Digital Collections.* https://digitalcollections.nypl.org/items/1a7cdef0-3e10-0133-8029-00505686d14e

Marenbon, John. Anicius Manlius Severinus Boethius. Stanford Encyclopedia of Philosophy, September 21, 2021. https://plato.stanford.edu/Entries/boethius/.

McArthur, John C. Civil Institutions and Military Instruction. *Journal of the United States Infantry Association* VI, no. 1 (July 1909): 860–68.

McElheran, Megan, Franklin C. Annis, Hanna A. Duffy, and Tessa Chomistek. Strengthening the Military Stoic Tradition: Enhancing Resilience in Military Service Members and Public Safety Personnel through Functional Disconnection and Reconnection. *Frontiers in Psychology* 15 (September 23, 2024). https://doi.org/10.3389/fpsyg.2024.1379244.

Michell, H. "The Iron Money of Sparta." *Phoenix* 1 (1947): 42. https://doi.org/10.2307/1086107.

"Militarism." Oxford Reference. https://www.oxfordreference.com/display/10.1093/acref/9780199891580.001.0001/acref-9780199891580-e-5102.

Miller, Jonson. *Engineering Manhood: Race and the Antebellum Virginia Military Institute*. Amherst, MA: Lever Press, 2020.

Milton, John. "Of Education." https://milton.host.dartmouth.edu/reading_room/of_education/text.shtml.

The Miriam and Ira D. Wallach Division of Art, Prints and Photographs: Print Collection, The New York Public Library. "Capt. Alden Partridge." *The New York Public Library Digital Collections.* https://digitalcollections.nypl.org/items/510d47da-fcb9-a3d9-e040-e00a18064a99.

"The Modern Stoicism Team." Modern Stoicism, September 17, 2019. https://modernstoicism.com/contributors/.

Monroe, James. Letter from James Monroe to Alden Partridge, 22 January 1815, Norwich University Archives and Special Collections. https://archives.norwich.edu/digital/collection/p16663coll4/id/2818/rec/1.

Morford, Mark. *Stoics and Neostoics: Rubens and the Circle of Lipsius*. Princeton, NJ: Princeton University Press, 2017.

Morn, Frank. *Academic Politics and the History of Criminal Justice Education*. Westport, CT: Greenwood Press, 1995.

"Morrill Act (1862)." Milestone Documents. National Archives. https://www.archives.gov/milestone-documents/morrill-act.

Morris, John F. "Crucibles of Virtue and Vice: The Acculturation of Transatlantic Army Officers, 1815–1945." PhD diss., Columbia University, 2020.

Morse, Edward Clarke. *Blood of an Englishman*. Abilene, TX: Jones of Texas, 1943.

Moten, Matthew. *The Delafield Commission and the American Military Profession*. College Station, TX: Texas A&M University Press, 2000.

New York Military Magazine: Devoted to the Interests of the Militia Throughout the Union. New York, NY: Labree and Stockton, 1841.

Nietzsche, Friedrich Wilhelm. *Schopenhauer as Educator*. Translated by James W. Hillesheim and Malcolm R. Simpson. South Bend, IN: Regnery/Gateway, Inc., 1980.

Noel, Rebecca. "'No Wonder They Are Sick, and Die of Study': European Fears for the Scholarly Body and Health in New England Schools before Horace Mann." *Paedagogica Historica* 54, no. 1–2 (2017): 134–53. https://doi.org/10.1080/00309230.2017.1355327.

Norton, Herman Albert. *Struggling for Recognition: The United States Army Chaplaincy, 1791–1865*. Washington, D.C.: Dept. of the Army, Office of the Chief of Chaplains, 1977.

Norwich Bicentennial. *200 Things about Norwich*. Northfield, VT: Norwich University, 2019. https://alumni.norwich.edu/file/200_things_book/NU_200Things_Chapters_1.pdf.

Norwich University. Catalogue of the Corporation, Officers, and Cadets of Norwich University, for the Academical Year 1848–9. Norwich University Archives and Special Collections. https://archives.norwich.edu/digital/collection/p16663coll2/id/17277/rec/9.

------. Catalogue of the Corporation, Officers and Cadets of Norwich University, for the Academical Year, 1852–'53. Norwich University

Archives and Special Collections.
https://archives.norwich.edu/digital/collection/p16663coll2/id/17457
/rec/5.

------. "History of NU."
https://www.norwich.edu/index.php?option=com_content.

Norwich University Alumni & Family. "NU Club of the Upper Valley Legacy
March Breakfast," 2019.
https://alumni.norwich.edu/UpperValleyLegacyMarch.

Notice: The Summer Term of the Ladies' Seminary at Norwich, 1835. Norwich
University Archives and Special Collections.
https://archives.norwich.edu/digital/collection/p16663coll8/id/124.

Oestreich, Gerhard. *Neostoicism and the Early Modern State*. Edited by Brigitta
Oestreich and Helmut Georg Koenigsberger. Cambridge: Cambridge
University Press, 2008.

Opsomer, Jan. "Is Plutarch Really Hostile to the Stoics?" In *From Stoicism to
Platonism: The Development of Philosophy, 100 BCE–100 CE*, edited by
Troels Engberg-Pedersen, 296–321. Cambridge: Cambridge University
Press, 2021.

Painter, Jacqueline S., ed. *The Trial of Captain Alden Partridge, Corps of Engineers:
Proceedings of a General Court-Martial Convened at West Point in the State of
New York, on Monday, 20th October 1817, Major General Winfield Scott,
President*. Northfield, VT: Friends of the Norwich University Library,
1987.

Pappas, George S. *The Cadet Chapel, United States Military Academy*. West Point,
NY: Association of Graduates, United States Military Academy, 1987.

Parker, Daniel. General Order Appointing a Court of Inquiry, 19 February
1816. Norwich University Archives and Special Collections.
https://archives.norwich.edu/digital/collection/p16663coll4/id/14894
/rec/4.

Partridge, Alden. "Captain Partridge's Lecture on Education." In *The Art of
Epistolary Composition, or Models of Letters, Billets, Bills of Exchange… to
Which Are Added, a Collection of Fables… for Pupils Learning the French
Language; a Series of Letters between a Cadet and His Father, Describing the
System Pursued at the American, Literary, Scientific and Military Academy at
Middletown, Conn. … and a Discourse on Education, by Capt. Alden
Partridge…*, ed. François Peyre-Ferry, 263–80. Middletown, CT: E. & H.
Clark, 1826.
https://books.google.com/books?id=rVQCAAAAYAAJ&source=gbs_
navlinks_s.

------. Catalogue of the Officers and Cadets of the American Literary, Scientific and Military Academy. Norwich, Vermont, August, 1821. Norwich University Archives and Special Collections. https://archives.norwich.edu/digital/collection/p16663coll2/id/13839/rec/15.

------. Catalogue of the Officers and Cadets of the American Literary, Scientific and Military Academy: To Which Is Subjoined the Prospectus and Internal Regulations of the Institution, &c. &c. Norwich, Vermont, November, 1821. Norwich University Archives and Special Collections. https://archives.norwich.edu/digital/collection/p16663coll2/id/13856/rec/2.

------. A Catalogue of the Officers and Cadets of Norwich University, for the Academic Years 1838–39: 39–40: 40–41. Norwich University Archives and Special Collections. https://archives.norwich.edu/digital/collection/p16663coll2/id/14139/rec/2.

------. Catalogue of the Officers and Cadets: Together with the Prospectus and Internal Regulations of the American Literary, Scientific and Military Academy at Middletown, Connecticut. 1826. Norwich University Archives and Special Collections. https://archives.norwich.edu/digital/collection/p16663coll2/id/13927/rec/5.

------. Catalogue of the Officers and Cadets: Together with the Prospectus and Internal Regulations of the American Literary, Scientific and Military Academy at Middletown, Connecticut. 1827. Norwich University Archives and Special Collections. https://archives.norwich.edu/digital/collection/p16663coll2/id/13972/rec/9.

------. A Catalogue of the Officers and Students of Norwich University for the Academic Year, 1835–6. Norwich University Archives and Special Collections. https://archives.norwich.edu/digital/collection/p16663coll2/id/14075/rec/1.

------. "Communication." *National Intelligencer.* 1826. Norwich University Archives and Special Collections.

------. Enclosure: Summary of Meteorological Observations by Alden Partridge, [ca. 9 December 1815]. National Archives. https://founders.archives.gov/?q=Ancestor%3ATSJN-03-09-02-0152&s=1511311111&r=2.

------. Essay on Luxury by Alden Partridge, approximately 1804. Norwich University Archives and Special Collections. https://archives.norwich.edu/digital/collection/p16663coll4/id/23653/rec/1.

------. Experiments on Artillery and Infantry Firing at West Point, 8 November 1810. Norwich University Archives and Special Collections. https://archives.norwich.edu/digital/collection/p16663coll4/id/9490/rec/3.

------. Experiments on Artillery and Infantry Firing at West Point, November 1814. Norwich University Archives and Special Collections. https://archives.norwich.edu/digital/collection/p16663coll4/id/23734/rec/6.

------. Guidelines for Establishing Multiple Military Academies, Approximately 1815–1817. Norwich University Archives and Special Collections. https://archives.norwich.edu/digital/collection/p16663coll4/id/9515/rec/4.

------. Internal Regulations for the Military Academy at West Point, State of New York, Approximately 1814–1817. Norwich University Archives and Special Collections. https://archives.norwich.edu/digital/collection/p16663coll4/id/9490/rec/3.

------. An Investigation of Sir Isaac Newton's Binomial Theorem by Alden Partridge, Approximately 1810–1815. Norwich University Archives and Special Collections. https://archives.norwich.edu/digital/collection/p16663coll4/id/10065/rec/1.

------. *Journal of a Tour, of a Detachment of Cadets, from the A.L.S.&M. Academy, Middletown, to the City of Washington, in December, 1826.* Middletown, CT: W. D. Starr, 1827.

------. Letter from Alden Partridge to Daniel Tompkins, 8 August 1816, Norwich University Archives and Special Collections. https://archives.norwich.edu/digital/collection/p16663coll4/id/3266/rec/3.

------. Letter from Alden Partridge to Gales & Seaton of the National Intelligencer Newspaper, 3 February 1829, Norwich University Archives and Special Collections. https://archives.norwich.edu/digital/collection/p16663coll4/id/16002/rec/2.

-----. Letter from Alden Partridge to Samuel Partridge, 2 October 1807, Norwich University Archives and Special Collections.

https://archives.norwich.edu/digital/collection/p16663coll4/id/1279/r
ec/4

-----. Letter from Alden Partridge to Samuel Partridge, 12 April 1813, Norwich
University Archives and Special Collections.
https://archives.norwich.edu/digital/collection/p16663coll4/id/1527/r
ec/2.

------. Lecture on Education, 1825. The British Library.

------. Lecture on National Defense, 18 July 1821, Norwich University Archives
and Special Collections.
https://archives.norwich.edu/digital/collection/p16663coll4/id/7729/r
ec/1.

------. Letter by Alden Partridge, 14 October 1826. Norwich University
Archives and Special Collections.
https://archives.norwich.edu/digital/collection/p16663coll4/id/12104
/rec/6.

------. Letter from Alden Partridge to John Wright, 16 February 1819. Norwich
University Archives and Special Collections.
https://archives.norwich.edu/digital/collection/p16663coll4/id/3454/r
ec/1.

------. Letter to Jonathan Williams, 19 January 1812. Norwich University
Archives and Special Collections.
https://archives.norwich.edu/digital/collection/p16663coll4/id/1379/r
ec/1.

------. Letter to the Public, 1820, Norwich University Archives and Special
Collections.
https://archives.norwich.edu/digital/collection/p16663coll4/id/15512
/rec/3.

------, T. B. Ransom, John Wright, C. L. Newton, and S. B. Hazeltine.
"Memorial of a Committee of the Military Convention at Norwich, Vt.
Praying the Revision and Alternation of the Military Defense of the
United States, to the Honorable Congress of the United States." *Citizen
Soldier*, July 22, 1840.

------, and Edmund Burke. Memorial of Alden Partridge and Edmund Burke,
in behalf of the State Military Convention of Vermont, praying the
adoption of a plan proposed by them for the reorganization of the
militia of the United States, February 9, 1939. U.S. Congressional Serial
Set. Sen. Doc. No. 197, 25th Congress, 3rd Session (1839). Law Library
of Congress.

------. Memorial of Alden Partridge, praying Congress to adopt measures with a
view to the establishment of a general system of education for the

benefit of the youth of this nation, January 21, 1841. U.S. Congressional Serial Set. H.R. Doc. No. 69, 26th Congress, 2nd Session (1841). Law Library of Congress.
https://www.govinfo.gov/content/pkg/SERIALSET-00383_00_00-050-0069-0000/pdf/SERIALSET-00383_00_00-050-0069-0000.pdf.

------. Memorial of Alden Partridge, relating to the Military Academy at West Point, and praying that young men educated at other military schools may have an equal chance for admission to the army as those young men have who are educated at West Point, January 21, 1841. U.S. Congressional Serial Set. H.R. Doc. No. 68, 26th Congress, 2nd Session (1841). Law Library of Congress.
https://www.govinfo.gov/content/pkg/SERIALSET-00383_00_00-049-0068-0000/pdf/SERIALSET-00383_00_00-049-0068-0000.pdf.

------. *The Military Academy at West Point Unmasked: or, Corruption and Military Despotism Exposed.* Washington, D.C.: Sold at the Bookstore of J. Elliot, 1830.

------. "The Next President," by Alden Partridge, 21 May 1841, Norwich University Archives and Special Collections.
https://archives.norwich.edu/digital/collection/p16663coll4/id/20509/rec/3

------. Nomination of Capt. Partridge for Representative in Congress, and Expression of His Views on Public Matters, Approximately 1838, Norwich University Archives and Special Collections.
https://archives.norwich.edu/digital/collection/p16663coll4/id/23625/rec/5

------. Norwich, Vt. Sept. 17, 1829., The Norwich Female Seminary Is This Day Opened to Receive Young Ladies under the Care of Miss Mary B. Ware, as Principal ...., Norwich University Archives and Special Collections.
https://archives.norwich.edu/digital/collection/p16663coll8/id/30/rec/1.

------. "Notice." *Vermont Republican & Courier*, May 29, 1835.

------. Notice to Parents and Guardians of Students at the Academy, Approximately 1820–1825, Norwich University Archives and Special Collections.
https://archives.norwich.edu/digital/collection/p16663coll4/id/23362/rec/12.

------. Observations on the Reorganization of the Military Academy at West Point, New York, 12 January 1821. Norwich University Archives and Special Collections.

https://archives.norwich.edu/digital/collection/p16663coll4/id/9734/rec/1.

------. Prospectus and Internal Regulations of the American Literary, Scientifick and Military Academy; to be Opened at Middletown, in the State of Connecticut in the Month of August, 1825. Norwich University Archives and Special Collections. https://archives.norwich.edu/digital/collection/p16663coll2/id/13894/rec/1

------. Prospectus of the American Literary, Scientific and Military Academy in Norwich, Vermont, 1820. Norwich University Archives and Special Collections. https://archives.norwich.edu/digital/collection/p16663coll4/id/15492/rec/4

------. Prospectus of the Norwich University and Norwich Seminary for Young Ladies, at Norwich, State of Vermont. 1835. Norwich University Archives and Special Collections. https://archives.norwich.edu/digital/collection/p16663coll2/id/14091/rec/2.

------. Records Relating to the Death of Cadet Thomas T. Hubbart on 25 October 1822. Norwich University Archives and Special Collections. https://archives.norwich.edu/digital/collection/p16663coll4/id/16254/rec/1

------. Regulations for Governing the United States Military Academy in New York, Approximately 1810–1814. Norwich University Archives and Special Collections. https://archives.norwich.edu/digital/collection/p16663coll4/id/9499/rec/1.

------. Regulations for the Military Academy at West Point, 3 January 1815. Norwich University Archives and Special Collections. https://archives.norwich.edu/digital/collection/p16663coll4/id/14822/rec/1

------. Regulations for the Military Academy at West Point during Vacation, 20 December 1813. Norwich University Archives and Special Collections. https://archives.norwich.edu/digital/collection/p16663coll4/id/14787/rec/4.

------. Talking Points by Alden Partridge for a Speech or Publication, Approximately 1820-1829. Norwich University Archives and Special Collections. https://archives.norwich.edu/digital/collection/p16663coll4/id/23419/rec/9.

------. "Walking." *Journal of Health* II, no. 8 (December 22, 1830): 121–22.
https://books.google.com/books?id=PvEhAQAAMAAJ&dq=%22eac
h+year+shoulder+your+knapsack+and+with+your+barometer%22&s
ource=gbs_navlinks_s.

"Pershing at UNL." Nebraska U: A Collaborative History from the Archives of
the University of Nebraska–Lincoln.
https://unlhistory.unl.edu/exhibits/show/generalpershing/pershingteac
hing.

Pigliucci, Massimo. "Stoicism and the Military: An Unvirtuous Coupling."
Modern Stoicism, April 22, 2023.
https://modernstoicism.com/stoicism-and-the-military-an-unvirtuous-
coupling-by-massimo-pigliucci/.

Pincus, Lionel and Princess Firyal Map Division, The New York Public
Library. "Plan des forts, batteries et poste de West-Point, 1780." *The
New York Public Library Digital Collections*.
https://digitalcollections.nypl.org/items/510d47da-ef75-a3d9-e040-
e00a18064a99

Plato. *The Republic of Plato: Books I–V*. Translated by T. Herbert Warren. New
York, NY: MacMillian and Co., 1892.
https://archive.org/details/republicofplato00plat_0/page/n7/mode/2u
p.

Plumley, Charles A. Ms. Traditions of Norwich. Norwich University, n.d.,
Norwich University Archives and Special Collections.

Plutarch. "The Life of Cato the Younger." In *Parallel Lives*, vol. VIII, translated
by Bernadotte Perrin, 237–411. Cambridge, MA: Harvard University
Press, 1919.

------. "The Life of Lycurgus." In *Parallel Lives*, vol. I, translated by Bernadotte
Perrin, 205–303. Cambridge, MA: Harvard University Press, 1914.

------. *Plutarch's Lives*, vol. IV. Edited by Arthur Hugh Clough. Translated by
John Dryden. Boston, MA: Little, Brown and Co., 1859.

Pratt, Richard A. "Essence of Military Education: Contributions of Captain
Alden Partridge to the United States Military Academy, 1806–1817."
1997. Eisenhower Leader Development Program Papers. United States
Military Academy Library.
http://usmalibrary.contentdm.oclc.org/cdm/singleitem/collection/p16
919coll1/id/16/rec/5.

Priestley, Joseph. *Socrates and Jesus Compared*. Philadelphia, PA: P. Byrne, 1803.

Reid, Heather L. "Plato's Gymnastic Dialogues." In *Athletics, Gymnastics, and
Agōn in Plato*, edited by Heather L. Reid, Mark Ralkowski, and Colleen
P. Zoller, 15–30. Fonte Aretusa: Parnassos Press, 2020.

Reeves, Ira L. *Military Education in the United States*. Burlington, VT: Free Press Printing Co., 1914.

Richards, Penny L., and George H. S. Singer. "To Draw Out the Effort of His Mind." *Journal of Special Education* 31, no. 4 (1998): 443–66. https://doi.org/10.1177/002246699803100403.

Richardson, Robert D. *Emerson: The Mind on Fire*. Berkeley, CA: University of California Press, 2015.

Rimell, Victoria. "The Best a Man Can Get: Grooming Scipio in Seneca Epistle 86." *Classical Philology* 108, no. 1 (2013): 1–20. https://doi.org/10.1086/669787.

"The Rise of the University of Nebraska: And the Celebration of the Silver Anniversary." Nebraska U: A Collaborative History from the Archives of the University of Nebraska–Lincoln. https://unlhistory.unl.edu/exhibits/show/silver-anniversary/silver-anniversary.

Robertson, Donald J. "The Art of Stoic Walking." *Journal of the Stoic Gym* 1, no. 8 (August 2019). https://thestoicgym.com/the-stoic-magazine/article/196.

------. "The Difference between Stoicism and Stoicism." Stoicism—Philosophy as a Way of Life, February 19, 2021. Medium. https://medium.com/stoicism-philosophy-as-a-way-of-life/the-difference-between-stoicism-and-stoicism-907ee9e35dc5.

------. "The Greatest Documentary on Stoicism?" Stoicism: Philosophy as a Way of Life, March 14, 2023. Substack. https://donaldrobertson.substack.com/p/the-greatest-documentary-on-stoicism.

------. *How to Think Like a Roman Emperor: The Stoic Philosophy of Marcus Aurelius*. New York, NY: St. Martin's Press, 2019.

------. "How to Walk Like a Stoic." Stoicism—Philosophy as a Way of Life, March 1, 2020. Medium. https://medium.com/stoicism-philosophy-as-a-way-of-life/how-to-walk-like-a-stoic-e1a41c8d5af0.

------. *The Philosophy of Cognitive-Behavioural Therapy (CBT): Stoic Philosophy as Rational and Cognitive Psychotherapy*. Abingdon, Oxon: Routledge, 2020.

------. "Stoicism as a Spartan Philosophy of Life." Stoicism—Philosophy as a Way of Life, October 31, 2019. Medium. https://medium.com/stoicism-philosophy-as-a-way-of-life/stoicism-as-a-spartan-philosophy-of-life-646d2c87d8d9.

------. "The Stoicism of Thomas Jefferson." Stoicism—Philosophy as a Way of Life, September 18, 2021. Medium. https://medium.com/stoicism-

philosophy-as-a-way-of-life/the-stoicism-of-thomas-jefferson-
e9266ebcf558.

Robertson, John K. "Some Preliminary Insights into the Formative Years at
the United States Military Academy." Presented to the Eisenhower
Fellows, West Point, NY, September 12, 1991.

Romaneski, Jonathan M. "Importing Napoleon: Engineering the American
Military Nation, 1814–1821." PhD diss., The Ohio State University,
2017.

Ross, Earle D. *Democracy's College: The Land-Grant Movement in the Formative Stage*.
Ames, IA: Iowa State College Press. 1942.

Rufus, Gaius Musonius. *Musonius Rufus, "The Roman Socrates"*. Translated by
Cora E. Lutz. New Haven, CT: Yale University Press, 1947.

Sadler, Gregory. "Symposium: What Is Modern Stoicism?" Modern Stoicism,
July 30, 2017. https://modernstoicism.com/symposium-what-is-
modern-stoicism/.

Sanftleben, Kurt Allen. "A Different Drum: The Forgotten Tradition of the
Military Academy in American Education." EdD diss., The College of
William & Mary, 1993. *W&M ScholarWorks*.
https://dx.doi.org/doi:10.25774/w4-8h3r-mn48.

Sayers, Dorothy L. *The Lost Tools of Learning*. London: Methuen, 1948.

Sellars, John. "Neo-Stoicism." Internet Encyclopedia of Philosophy.
https://iep.utm.edu/neostoic/.

------. *Stoicism*. London: Routledge, Taylor & Francis Group, 2014.

Seneca, Lucius Annaeus. *Dialogues and Essays*. Translated by John N. Davie.
Oxford: Oxford University Press, 2007.

------. *Letters from a Stoic*. Translated by Robin Campbell. Harmondsworth:
Penguin Books, 1969.

------. *Moral Epistles*. Translated by Richard M. Gummere. Vol. I. Cambridge,
MA: Harvard University Press, 1917.

------. "On the Shortness of Life," translated by Gareth D. Williams. *In
Hardship and Happiness*, translated by Elaine Fantham, Harry M. Hine,
James Ker, and Gareth D. Williams, 105–39. Chicago, IL: University of
Chicago Press, 2014.

------. *Seneca's Morals of a Happy Life, Benefits, Anger and Clemency*. Translated by
Roger L'Estrange. New ed. Chicago, IL: Belford. Clarke & Co., 1881.
http://www.gutenberg.org/files/56075/56075-h/56075-
h.htm#SENECA_OF_BENEFITS.

Smith, John C. "Documents Accompanying the Report of the Secretary of War
to the Committee of the Senate Respecting the Conflicting Jurisdiction
of the General and State Governments over the Militia—Continued.

Sharon, Connecticut, July 2, 1814." *National Intelligencer*, May 2, 1815. George Washington University Special Collections Research Center.

Spartianus, Aelius. "The Life of Hadrian: Part I." In *Historia Augusta*, vol. I, translated by David Magie, X. Cambridge, MA: Harvard University Press, 1921.

Spencer Collection, The New York Public Library. "West Point." *The New York Public Library Digital Collections*. https://digitalcollections.nypl.org/items/510d47d9-7e48-a3d9-e040-e00a18064a99

Spiegel, Ted. "The Cadet Uniform Factory." West Point Association of Graduates. https://www.westpointaog.org/west-point-magazine-uniform-factory.

Stelnicki, Andrea M., Laleh Jamshidi, Amber J. Fletcher, and R. Nicholas Carleton. "Evaluation of Before Operational Stress: A Program to Support Mental Health and Proactive Psychological Protection in Public Safety Personnel." *Frontiers in Psychology* 12 (2021). https://doi.org/10.3389/fpsyg.2021.511755.

Stevens, Neil E. "America's First Agricultural School." *Scientific Monthly* 13, no. 6 (December 1921): 531–40.

Stewart, Richard W. *The United States Army and the Forging of a Nation, 1775–1917*, vol. I. Washington, D.C.: Center of Military History, United States Army, 2005.

"Stoicism Noun." Oxford Advanced American Dictionary. https://www.oxfordlearnersdictionaries.com/us/definition/american_english/stoicism#.

Strobridge, Truman R. "West Point, Thayer & Partridge." *Military Review*, October 1989, 78–86. https://www.google.com/books/edition/Review_of_Current_Military_Literature/adesdJDSlXgC?hl=en&;gbpv=1.

Sullivan, Gordon R. "Remarks by General (Ret.) Gordon R. Sullivan." West Point Association of Graduates, October 1, 2003. https://alumni.westpointaog.org/page.aspx?pid=508.

Swift, Joseph Gardner. *Memoirs of Gen. Joseph Gardner Swift: First Graduate of the United States Military Academy, West Point, Chief Engineer U.S.A. from 1812 to 1818, 1800–1865*. Worcester, MA: F. S. Blanchard & Co., 1890. https://archive.org/details/memoirsofgenjoseph00swif/mode/2up

Taylor, Sheridan, and Andrew Carlquist. The Voice of Experience: Retired Master Corporal, Sheridan Taylor. Other. In *Before Operational Stress – Military*. Calgary: Wayfound Mental Health Group, 2023. https://education.beforeoperationalstress.com/courses/take/bos-military/lessons/50547004-the-voice-of-experience-retired-master-corporal-sheridan-taylor.

Temple, Julien, dir. *The Ecstasy of Wilko Johnson*. BBC Imagine, 2015. 1 hr., 32 min.

Tenney, Lester I. *My Hitch in Hell: The Bataan Death March*. Lincoln, NE: Potomac Books, 2018.

Tilden, Arnold. *The Legislation of the Civil-War Period Considered as a Basis of the Agricultural Revolution in the United States*. Los Angeles, CA: University of Southern California, 1937.

To Captain Alden Partridge of the Military Institute, Norwich, Vermont, from Volume 3 of the U.S. Military Magazine, 1842. Hand-colored lithograph on wove paper. 11 1/8 × 9 in. Hood Museum, Dartmouth College, Hanover, NH. https://hoodmuseum.dartmouth.edu/objects/pr.950.5.1.

Tocqueville, Alexis. *Democracy in America*. Translated by Henry Reeve. 4th ed. Vol. I. Philadelphia, PA: J. & H. G. Langley, 1841. https://archive.org/details/democracyiname01tocq/mode/2up.

Todd, Frederick P., and Frederick T. Chapman. *Cadet Gray: A Pictorial History of Life at West Point as Seen through Its Uniforms*. New York, NY: Sterling Publishing Co., 1955.

True, Alfred Charles. *A History of Agricultural Education in the United States 1785–1925*. Washington, D.C.: Government Printing Office, 1929.

------. "Military Instruction." In *Proceedings of the Eighteenth Annual Convention of the Association of American Agricultural Colleges and Experiment Stations Held at Des Moines, Iowa, November 1–3, 1904*, edited by A. C. True, W. H. Beal, and H. C. White, 91–106. Washington, D.C.: Government Printing Office, 1904.

Trulock, Alice Rains. *In the Hands of Providence: Joshua L. Chamberlain and the American Civil War*. Chapel Hill, NC: The University of North Carolina Press, 2013.

Turner, Judith. *"Man's Search for Meaning* Teacher's Guide." PenguinRandomhouse.com. n.d. https://www.penguinrandomhouse.com/books/206272/mans-search-for-meaning-by-viktor-e-frankl/9780807000007/teachers-guide/.

United States Census Office. *Compendium of the Enumeration of the Inhabitants and Statistics of the United States: As Obtained at the Department of State, from the Returns of the Sixth Census, by Counties and Principal Towns*. Washington, D.C.: Blair & Rives, 1841.

U.S. Congress. *U.S. Statutes at Large*, vol. 2 (1799–1813), *6th through 12th Congress*. Boston: Charles C. Little and James Brown, 1845. Law Library of Congress.

https://www.loc.gov/resource/llsalvol.llsal_002/?sp=1&r=-0.355,0.279,1.893,1.074,0

U.S. Congress. *U.S. Statutes at Large*, vol. 6, *Private Laws and Resolutions (1789–1845)*, edited by Richard Peters. Boston: Charles C. Little and James Brown, 1846. Law Library of Congress. https://www.loc.gov/item/llsl-v6/.

U.S. Congress. *U.S. Statutes at Large*, vol. 40, *65th Congress*. Washington, D.C.: Government Printing Office, 1919. Law Library of Congress. https://www.loc.gov/item/llsl-v40/

Vedder, Richard K. "The Failure of Federal Higher Education: 'The Morrill Land-Grant Act: Fact and Mythology.'" YouTube, April 9, 2019. https://www.youtube.com/watch?v=YknUchmxSpw.

------. "The Morrill Act." *In Unprofitable Schooling: Examining Causes of, and Fixes for, America's Broken Ivory Tower*, edited by Todd J. Zywicki and Neal P. McCluskey, 31–64. Washington, D.C.: Cato Institute, 2019.

------. *Restoring the Promise: Higher Education in America*. Oakland, CA: Independent Institute, 2019.

Wade, Kathryn Lindsay Anderson. "The Intent and Fulfillment of the Morrill Act of 1862: A Review of the History of Auburn University and the University of Georgia." Master's thesis, Auburn University, 2005. https://etd.auburn.edu/xmlui/bitstream/handle/10415/677/WADE_KATHRYN_23.pdf?sequence=1.

Washington, Booker T. "Atlanta Compromise." Speech presented at the Cotton States and International Exposition, September 18, 1895.

Washington, George. Washington's Sentiments on a Peace Establishment, 1 May 1783. National Archives Administration. https://founders.archives.gov/documents/Washington/99-01-02-11202.

Waterman, Laura, Guy Waterman, and Bill McKibben. *The Green Guide to Low-Impact Hiking and Camping*. New York, NY: Countryman Press, 2016.

Waugh, Elizabeth Dey Jenkinson. *West Point: The Story of the United States Military Academy Which Rising From the Revolutionary Forces Has Taught American Soldiers the Art of Victory*. New York, NY: Macmillan Co., 1944.

Webb, Lester A. *Captain Alden Partridge and the United States Military Academy, 1806–1833*. Northport, AL: American Southern, 1965.

------. "The Origin of Military Schools in the United States Founded in the Nineteenth Century." PhD diss., University of North Carolina at Chapel Hill, 1958.

Weber, William. *Neither Victor nor Vanquished: America in the War of 1812*. Washington, D.C.: Potomac Books, 2013.

Wettemann, Robert P., Jr. *Privilege vs. Equality: Civil-Military Relations in the Jacksonian Era, 1815–1845*. Santa Barbara, CA: Praeger Security International, 2009.

Whewell, William. *The Platonic Dialogues for English Readers*. Cambridge, MA: MacMillan & Co., 1859.

Winder, R. H. "General Order." *National Intelligencer*, November 29, 1817. George Washington University Special Collections Research Center.

Woodbury, Levi. Letter from Levi Woodbury to Alden Partridge, 5 April 1824. Norwich University Archives and Special Collections. https://archives.norwich.edu/digital/collection/p16663coll4/id/6770/rec/1.

Wrone, David R. "The People's Choice College: The Movement behind the Morrill Land Grant Act of 1862." Illinois Periodicals Online. https://www.lib.niu.edu/1998/iht519842.html.

Xenophon. *The First Part of Xenophon's Memorabilia of Socrates: With a Literal Interlinear Translation*. London: John Taylor, 1831. https://archive.org/details/firstpartxenoph00xenogoog/page/n4/mode/2up?q=body.

------. *The Works of Xenophon*. Vol. II. Translated by Henry Graham Dakyns. London: Macmillan & Co., 1892. https://books.google.com/books?id=3_REAQAAIAAJ&pg=PR73#v=onepage&q&f=false.

Figure 20—The Partridge House[728]

---

[728] Located at 383 Main Street, Norwich, Vermont, the Partridge House was built in 1820 for Milton Partridge (Alden Partridge's cousin) by architect Joseph Emerson. Milton sold the house to Alden Partridge shortly after construction. Alden Partridge would die in this house as a result of an abdominal aneurysm. The house is now run as a bed and breakfast. For information about staying at the Partridge House, visit https://partridgehousevermont.com/. (Artist Gala S.)

Index

B

C

D

H

I

J

433